D1452428

Philosophical Studies Series

Volume 143

Philosophical Studies Series aims to provide a forum for the best current research in contemporary philosophy broadly conceived, its methodologies, and applications. Since Wilfrid Sellars and Keith Lehrer founded the series in 1974, the book series has welcomed a wide variety of different approaches, and every effort is made to maintain this pluralism, not for its own sake, but in order to represent the many fruitful and illuminating ways of addressing philosophical questions and investigating related applications and disciplines.

The book series is interested in classical topics of all branches of philosophy including, but not limited to:

- Ethics
- Epistemology
- Logic
- Philosophy of language
- Philosophy of logic
- Philosophy of mind
- Philosophy of religion
- Philosophy of science

Special attention is paid to studies that focus on:

- the interplay of empirical and philosophical viewpoints
- the implications and consequences of conceptual phenomena for research as well as for society
- philosophies of specific sciences, such as philosophy of biology, philosophy of chemistry, philosophy of computer science, philosophy of information, philosophy of neuroscience, philosophy of physics, or philosophy of technology; and
- contributions to the formal (logical, set-theoretical, mathematical, information-theoretical, decision-theoretical, etc.) methodology of sciences.

Likewise, the applications of conceptual and methodological investigations to applied sciences as well as social and technological phenomena are strongly encouraged.

Philosophical Studies Series welcomes historically informed research, but privileges philosophical theories and the discussion of contemporary issues rather than purely scholarly investigations into the history of ideas or authors. Besides monographs, *Philosophical Studies Series* publishes thematically unified anthologies, selected papers from relevant conferences, and edited volumes with a well-defined topical focus inside the aim and scope of the book series. The contributions in the volumes are expected to be focused and structurally organized in accordance with the central theme(s), and are tied together by an editorial introduction. Volumes are completed by extensive bibliographies.

The series discourages the submission of manuscripts that contain reprints of previous published material and/or manuscripts that are below 160 pages/88,000 words.

For inquiries and submission of proposals authors can contact the editor-in-chief Mariarosaria Taddeo via: mariarosaria.taddeo@oii.ox.ac.uk

More information about this series at https://link.springer.com/bookseries/6459

Björn Lundgren • Nancy Abigail Nuñez Hernández
Editors

Philosophy of Computing

Themes from IACAP 2019

Editors
Björn Lundgren
The Department of Philosophy and Religious Studies
Utrecht University
Utrecht, The Netherlands

Institute for Futures Studies
Stockholm, Sweden

Department of Philosophy
Stockholm University
Stockholm, Sweden

Nancy Abigail Nuñez Hernández
Institute of Philosophy
Czech Academy of Sciences
Prague, Czech Republic

ISSN 0921-8599 ISSN 2542-8349 (electronic)
Philosophical Studies Series
ISBN 978-3-030-75266-8 ISBN 978-3-030-75267-5 (eBook)
https://doi.org/10.1007/978-3-030-75267-5

This Springer imprint is published by the registered company Springer Nature Switzerland AG
The registered company address is: Gewerbestrasse 11, 6330 Cham, Switzerland

This book is dedicated to the memory of Gunilla Lundgren (June 20, 1954–October 6, 2020).

Foreword

In the late 1970s, when I was just starting to think about the ethical issues related to computers, I was struck by how little was being written on the topic and how shallow much of it was. I remember thinking that the discourse could benefit greatly from philosophical concepts and deeper philosophical analysis. At the same time, a small number of philosophers turned their attention to the work of Herbert Simon and Norbert Wiener, seeing in the work of these two big thinkers a way to understand the significance of modern computers. These philosophers thought that computation had enormous implications for, and might even resolve, many fundamental philosophical issues. I was especially taken with Simon's statement that: "Perhaps the most important question of all about the computer is what it has done and will do to man's view of himself and his place in the universe" (Simon, H.A. 1977. "What computers mean for man and society." *Science* 195(4283): 1186-1191, p. 1190). Also, during this time, a small number of philosophers began to see in computers the opportunity to teach logic in a better way than it had ever been taught before; they began developing software for this purpose.

While modern computers attracted the interest of this small group of philosophers, in those early days and for the next several decades, very few scholars in the humanities and social sciences recognized the significance of modern computers and the prescience of Simon's statement. That small group of philosophers was the exception. In the mid-1980s, with an abundance of foresight, they hatched the idea of annual workshops and conferences focused on the connections between philosophy and computing. These meetings brought together computer scientists with philosophers from a variety of fields within philosophy. Over the years, the conferences grew and took various forms; they began as small local events and grew to international events. Eventually, the activities and the organization that made them happen morphed into what is now known as the International Association for Computing and Philosophy (IACAP).

The papers gathered together in this volume are from the 2019 annual conference of IACAP. They testify to the rich and robust territory created by bringing philosophy and computers together. The papers illustrate how traditional philosophical questions have been expanded and extended as a result of the combination. Although it is tempting to seek a broad yet pithy account of the connection between philosophy and computers, a simple or singular account is not possible. Philosophy and computing have a myriad of connections.

The *mere idea* of computers, as suggested in Simon's statement, has had an influence. It has, for example, presented novel ontological questions. Philosophers are now asking whether it would ever be possible (or even obligatory) to recognize robots as persons with rights comparable to those of humans. The question arises because the future trajectory of robot technology seems to point in the direction of robots that, although silicon-based or a form of artificial life, will think and act in ways that could be considered equivalent to human thinking and behavior. There is even research underway to produce behavior in robots that would manifest as something like human emotions. These possibilities have led philosophers to look at traditional philosophical notions differently, that is, to rethink such notions as personhood, equivalence, and ontological status. Grappling with the ontological and ultimately the moral status of robots is evident in this volume in a chapter that asks, should robots be allowed to punish us?

Computers are not, of course, just an idea, they are machines that operate in the physical world. Indeed, since the invention of modern computers, their use has come to permeate many, if not most, domains of human activity. Because of the use of computers in science, the fundamental question of epistemology – how do we know? – has been taken in new directions. Several of the chapters in this volume take up issues relating to computationally produced forms of knowledge in addressing computer simulation, computer-assisted proofs, and simplicial complexes.

Because computers operate in the physical world, and especially because they change how we relate to one another and how social institutions are constituted, an endless array of ethical questions and concerns have been raised. Although ethics has to do with how people behave, computers affect when, where, and how we behave and interact with one another. The use of computers affects individual behavior as well as the nature of human organizations and institutions. Hence, the issues, especially the situations and cases, taken up by moral philosophy must – to be relevant to the experience of those alive today – address a computerized world. The chapter on predictive fairness in this volume is a good example of this. There is a long and deep tradition of philosophical thinking about fairness and the capacity of computers to combine and process huge quantities of data and to use the results to make fine-grained predictions about human behavior has required the extension of notions of fairness to the use of these new and powerful predictive computations.

The chapters in this volume extend the long tradition of philosophical thinking by bringing computers and computing into the conversation. The volume carries forward conversations that began centuries ago and are now infused with the idea of powerful computers and the experience of living in a world in which they operate.

University of Virginia, Charlottesville, VA, USA Deborah G. Johnson
October 2020

Preface

Alan Turing's "On computable numbers, with an application to the Entscheidungsproblem" (1936) and "Computing Machinery and Intelligence" (1950) are among the foundational works on computer science and artificial intelligence. Any reader of these papers can see that Turing confronted deep philosophical worries. Since then, computer science and artificial intelligence have developed at a blistering pace with a tendency to highly technical specialization, growing them apart from philosophy. However, the legacy of Turing should remind us of the significance of bringing together philosophy, computer science, and artificial intelligence. The chapters contained in this volume represent efforts in that fertile direction that seeks to foster collaborations between philosophers, computer scientists, and experts on artificial intelligence.

This book is the official proceedings of the 2019 meeting of the *International Association for Computing and Philosophy* (IACAP). Since 1986, IACAP has organized either international or regional conferences yearly. So far, 2020 is the only exception.[1] The 2019 conference was arranged in Mexico City at the *National Autonomous University of Mexico* by Nancy Abigail Nuñez Hernández, Sergio Rajsbaum, and Luis Estrada. The conference included presentations on a large variety of topics and a special track on Distributed Computing and Epistemic Logic arranged by Sergio Rajsbaum and Alexandru Baltag.

In preparing this volume, we invited all authors presenting at the conference to submit a manuscript for consideration. Among those that opted to submit, seven papers were selected after peer review. Beyond these selected papers, the volume also includes one invited contribution from each of the prize winners for IACAP's two annual awards: *The Covey Award* and *The Herbert A. Simon Award.*

Since 2009, the Covey Award "recognizes senior scholars with a substantial record of innovative research in the field of computing and philosophy broadly

[1] See: https://www.iacap.org/conferences/. At the time of writing this Introduction, the next IACAP conference, in 2021, is currently planned to be arranged *online* by the *University of Hamburg*. It is a collaboration with *CEPE* (*Computer Ethics—Philosophical Enquiry*).

conceived."[2] In 2019, the Covey Award went to John Weckert (*Charles Sturt University*) for his "broad and deep influence on the ethics of information technology" and "the significant impact John has had fostering by both example and personal influence the careers of many scholars who have continued following the fruitful paths of research his work brought to light" (Berkich 2019a).[3]

Since 2010, the Herbert A. Simon Award for Outstanding Research in Computing and Philosophy "recognizes scholars at an early stage of their academic career who are likely to reshape debates at the nexus of Computing and Philosophy by their original research."[4] In 2019, the Simon Award went to Juan M. Durán (Delft University of Technology) "for his emphasis on the epistemic and moral normative dimensions of computational methods in the sciences, representing transformational and seminal scholarship in the philosophy of science." (Berkich 2019b).[5]

The themes of the nine chapters of this book are diverse. The themes include epistemology, epistemic logic, computational logic, ethics, applied ethics, algorithmic fairness, philosophy of technology, human-robot interaction, philosophy of science, and pedagogy, which illustrate the breadth of the topics that are commonly dealt with at an IACAP conference. Below we briefly introduce each of the chapters and raise some questions.

Current technological developments show that it is more pressing than ever to unify philosophers and computer scientists' efforts to understand knowledge in its full complexity. In Chap. 1, "Knowledge and Simplicial Complexes," Hans van Ditmarsch, Eric Goubault, Jérémy Ledent, and Sergio Rajsbaum offer an interesting example of how bringing together philosophical developments on dynamic epistemic logic with topological insights from simplicial complexes and distributed computing sheds new light on core questions about knowledge and belief, like the notion of consensus.

van Ditmarsch et al. offer alternative ways to model epistemic notions such as knowledge and belief, as well as group notions such as mutual, common,

[2] http://www.iacap.org/awards/

[3] Previous winners are (chronologically ordered, most recent winner first): Deborah G. Johnson (*University of Virginia*), Raymond Turner (*University of Essex*), Jack Copeland (*University of Canterbury*), William J. Rapaport (*University at Buffalo, The State University of New York*), Selmer Bringsjord (*Rensselaer Polytechnic Institute*), Margaret Boden (*University of Sussex*), Luciano Floridi (*University of Hertfordshire*), Terrell Bynum (*Southern Connecticut State University*), John R. Searle (*University of California, Berkeley*), and Edward N. Zalta (*Stanford University*). See: http://www.iacap.org/awards/

[4] http://www.iacap.org/awards/

[5] Previous winners are (chronologically ordered, most recent winner first): Thomas C. King (*Oxford Internet Institute*), Andrea Scarantino (*Georgia State University*), Marcin Milkowski (*The Institute of Philosophy and Sociology of the Polish Academy of Sciences*), Michael Rescorla (*University of California-Santa Barbara*), Gualterio Piccinini (*University of Missouri–St. Louis*), Judith Simon (*University of Vienna*), Patrick Allo (*Vrije Universiteit Brussels*), John Sullins (*Sonoma State University*), and Mariarosaria Taddeo (*University of Hertfordshire*; *University of Oxford*). See: http://www.iacap.org/awards/

and distributed knowledge on simplicial complexes.[6] The appeal to algebraic topological tools like simplicial complexes has shown fertile explanatory advantages in diverse research areas, like graph theory (Lovász 1978), neuroscience (Reimann et al. 2017), and semantic networks (Salnikov et al. 2018; Christianson et al. 2020), among many others. For philosophers unfamiliar with algebraic topology, simplicial complexes' potential applications to epistemology may go unnoticed until the duality between simplicial complexes and Kripke models is shown, which is precisely the departure point of this chapter. Roughly speaking, if a Kripke model is viewed as a simplicial complex, we can use algebraic topology tools to study that structure's higher dimensional properties. For instance, the authors state that topological invariants are preserved when the system evolves after communication, and these topological invariants determine the problems that agents can solve. To sum up, the chapter is a contribution that provides novel insights for computer scientists interested in distributed computing and philosophers interested in dynamic epistemic logic, group knowledge, and the development of more accurate formal tools to model epistemic phenomena.

The development of computational technology has allowed us to process large quantities of data, to the point that big data, machine learning, and other data-driven methods have become essential to many scientific enterprises. This achievement comes, perhaps, at a cost: it has been claimed that data per se does not contribute to developing science, and that scientific theories should guide experimental design to efficiently collect data and produce reliable predictive models and conceptual knowledge. In Chap. 2, "Meta-Abduction: Inference to the Probabilistically Best Prediction," Christian J. Feldbacher-Escamilla suggests one way to help bridge the gap between vast troves of algorithmically massaged data and the predictive and explanatory goals of scientific theory by offering a characterization and justification for abduction as an inference to the probabilistically best prediction.

Feldbacher-Escamilla presents a taxonomy of abductive inferences to focus on a species of selective abduction, which aims at inferring the probabilistically best hypothesis, explanation, or theory based on data. Moreover, Feldbacher-Escamilla develops an account of meta-abduction as inference to the best probabilistic explanation, applying Schurz's theory of meta-induction and Feldbacher-Escamilla's framework of prediction games to show how such species of abduction can be justified. This theoretical contribution is especially relevant given the current "need to turn data into true predictions, that is, predictions of events in novel circumstances, or predictions of events before they occur (not *post hoc* explanations)." (Coveney et al. 2016)

[6] Simplicial complexes are made of simplices. Roughly speaking, a simplex can be understood as a generalized triangle (a triangle in any dimension): a 0 simplex is a single vertex, a 1 simplex is two vertices connected by an edge, a 2 simplex is three vertices connected pairwise by edges with a single face (a triangle), and a 3 simplex is four vertices connected pairwise by edges joint by four faces which are filled in to form a solid (a tetrahedron). Hence the duality between simplicial complexes and Kripke models which tend to be graphically represented through vertices connected by edges.

Big data methods in science raise further concerns. There are research ethical worries (such as about P-hacking), which often also raises epistemic worries (can we trust the results?).[7] In Chap. 3, "Is There Anything Special About the Ignorance Involved in Big Data Practices?", María del Rosario Martínez-Ordaz addresses epistemic challenges of big data. Martínez-Ordaz starts off by discussing some preliminary methodological worries relating to how big data is often used to find correlations and the worry about determining whether we can "rationally trust correlations as legitimate instances of scientific knowledge" (p. 117). She argues that since it is impossible for a human agent to process a big data set, we become "necessarily constrained by specific technological resources" (p. 117). Moreover, she argues that technological dependence is a source of epistemic opacity. Next, she connects these methodological worries to related worries about knowledge, before she turns to the main issue of the chapter: ignorance.

Martínez-Ordaz supplies a taxonomy of different types of or sources of ignorance (factual ignorance, objectual ignorance, procedural ignorance, and ignorance of theoretical structures; pp. 123–124) and argues that the main problem with big data has to do with ignorance of theoretical structure, but that there are two sources of reliable knowledge: objectual knowledge (i.e., knowledge of particular objects) and modal knowledge (which we take here to be knowledge of the possibility space) of "how these objects (could) behave and relate to one another" (p. 137). Based on this analysis, Martínez-Ordaz ends her chapter with a case-study on cosmology and big data.

Some of her negative arguments may meet resistance. For example, technological dependence is arguably not something that is special for big data. Most, if not all, empirical studies depend on instruments, from the mundane (a magnifying glass, say) to the technically more complex (an electron microscope, for example), and it would be challenging to find an empirical study in which there is no related dependence. Here it is worth to point out that Jumbly Grindrod (2019) has recently argued that algorithms function like "instrumental knowledge" (cf. Sosa 2006). Moreover, it is possible to construct big data sets with proven validity, so that while the scientist cannot process the data set themself, they can still know that the data set is valid.

As a final point, we believe that a fruitful way to develop her main points would be to introduce a measurement of ignorance and opacity, since the problem arguably is not whether there is or is not some ignorance and opacity, but how pervasive ignorance and opacity are.

Going back at least to the Four Color Theorem, mathematicians have wrestled with whether computer assisted proofs, which may have no traditional "paper-and-pencil" equivalent, can be counted as proofs. Favio E. Miranda-Perea and Lourdes

[7] P-hacking is the process of looking for correlations in a data set to find data points that would manifest as statistically significant. Since a large data set will—due to random variations—include correlations, we cannot apply a significance test to see if it is likely that those correlations are due to random variation or a real connection.

del Carmen González Huesca discuss this topic in Chap. 4, "On the Conciliation of Traditional and Computer-Assisted Proofs." In this work, Miranda-Perea and González Huesca elaborate on the concept of transitional proofs, which intends to bridge computer-assisted proofs and traditional paper-and-pencil proofs. This proposal is instantiated through the use of Coq, a powerful and popular proof assistant. Those unfamiliar with computer-assisted proofs can benefit from the detailed explanations in this chapter; however, various mathematicians, computer scientists, and software engineers specialized in automated reasoning may find some claims controversial. Particularly, novelty claims on this work's contributions might be contested, for example, the formalization of backward proofs, induction on assumption contexts, or the discovery of superior correctness proofs for specific algorithms.

Implicit bias is a well-known problem for algorithmic decision-making. In various ethical guidelines for AI (see, e.g., AI HLEG 2019; Brey et al. 2019) fairness is a key requirement for ethical AI. In Chap. 5, "Predictive Fairness," Anders Herlitz discusses the issue of algorithmic fairness. He starts by presenting a discussion about the fairness of the algorithm COMPAS (the Correctional Offender Management Profiling for Alternative Sanctions)—a tool used by US courts to determine an offender's recidivism rate. Herlitz notes that there has been disagreement about whether COMPAS is biased against blacks or not and that this disagreement is rooted in different conceptions of fairness, what Herlitz calls "Equality in accuracy" and "Equality in inaccuracy."

Kleinberg et al. (2017) have shown that these two conceptions of fairness—in Kleinberg et al. one of these conceptions corresponds to two separate conditions—cannot both be satisfied if the recidivism rate differs between two salient social groups. As Herlitz sums it up: "Either (i) the decision method correctly identifies the relevant property (e.g. recidivism) more often in one subgroup (e.g. black defendants) than another (e.g. white defendants); or (ii) the decision method systematically ascribes higher probabilities to individuals who have and/or individuals who lack the property in one group compared to the probabilities ascribed to individuals who have and/or individuals who lack the property in another group" (p. 141). In the second section, Herlitz turns to discuss three possible responses: dismiss one of the conditions or conceptions of fairness, address the unfairness at another stage of the process, or reject algorithmic decision-tools altogether. Herlitz finds all three options problematic. Thus, in the third section, he suggests three principles that ought to "guide policy that rely on systematic decision methods that generate probability predictions: *Dominance, Transparency* and *Priority to the worse off*" (p. 156). Simply put, Dominance is satisfied if one method is better than another relative to all relevant (fairness) factors and worse relative to none; Transparency demands awareness and transparency from the user of the unfair decision-method; and Priority to the worse off says that unfairness should benefit the group that is worst off.

We find these suggestions sound and helpful, but in practice they may require further operationalization. For example, as Herlitz indicates, Dominance implies a goal to minimize unfairness. However, we would have to specify further how to choose between methods that violate some fairness principles very little and rarely, and those that violate only one principle but do so more extremely and very often.

Robotic applications are breaking new grounds. Techniques are developing for robots in such diverse areas as health care, child care, elderly care, and the transport industry, well beyond their traditional domain in industrial manufacturing. This raises various issues of the role of robots in the society and especially when robots are asked not only to perform a limited task with objects, but to engage in a social environment. In Chap. 6, "Castigation by Robot: Should Robots Be Allowed to Punish Us?", Alan R. Wagner and Himavath Jois offer an introductory overview to the topic of robot punishment. Wagner and Jois apply a broad definition of punishment, according to which it is a behavior "selected for the purpose of adding cost to one's action" (p. 164). Under this definition, the idea that robots should be allowed and need to administer punishment is not surprising, as they note even a warning sound may be a punishment. Wagner and Jois argue that if robots perform substantial roles and tasks in society, the ability to punish (even if only minor) may be central to maintain the relevant authority needed to perform those roles and tasks. However, Wagner and Jois also argue that it is important that robots "support the existing system of social norms," even bad norms—because it is important to maintain consistency, by which we suppose they mean social order (p. 168). We find this problematic. First, we question the empirical claim that maintaining social norms maintains social order or consistency. Indeed, because social norms depend on second-order beliefs (i.e., beliefs about others' normative preferences), it follows that there are social norms that are upheld against a super majority's will (see, e.g., Bicchieri 2017 for empirical examples). Thus, some social norms are upheld because many or even most people hold false second-order beliefs. Hence, it is not clear that maintaining such norms would maintain consistency. Second, and more importantly, we worry that norm adherence can be abused to maintain oppressive norms (e.g., gender mutilation) or that claims about the necessity of social norms can be used to retain undemocratic powers and deny people their human rights. Perhaps Wagner and Jois can temper their claim about norms by reformulating their principle of norm adherence in terms of a *prima facie* or *pro tanto* principle focusing on good or generally accepted norms.

In Chap. 7, "Implementing Algorithmic and Computational Design in Philosophical Pedagogy," Rocco Gangle argues that algorithmic reasoning—or, what he calls "Diagrammatic Methods"—can be used to teach introductory philosophy courses. The pedagogical tools he envisions using might best be described as diagrams illustrating the different relations between arguments and premises. The idea here is simply to use graphical diagrammatic structures to show and analyze these relations. It is an interesting idea that may help students struggling with more formal expressions. However, we wonder to what extent the diagrams can be applied to all arguments even in an introductory course. While that may be a learning experience, because of the—as Gangle notes— "the self-reflexive logical aspect of philosophy" (p. 181), the value of such a learning experience arguably diminishes over time. Nevertheless, the manuscript contains several interesting examples (with detailed illustrations), which many teachers likely will find interesting to try out themselves.

Gangle also suggests that the diagrammatic representation may also be used for more sophisticated philosophical reasoning and research. Ultimately, the proof is in the pudding (or, for the pedagogical claims, the proof is in a well-designed experiment).

According to Moore's law the number of transistors in a microchip will double every second year. This forecast has held steady since the 1970s, and even if some expect the number of transistors to flatten (Shalf 2020; Waldrop 2016), it is fair to say that this development has had tremendous implications. With more and more devices becoming "smart" (phones, TVs, home appliances, etcetera), technology is developing in new ways. But is this development sound?

In Chap. 8, "Our Technology Fetish," John Weckert argues that we (humans) have a technological fetish, which he argues is bad, and we ought to slow down and reflect on technological development, rather than push to the next frontier. Weckert discusses many aspects and problems relating to technology, including surveillance, control, and risk. One might ask, however, is the supposed fetish really a fetish about the technological device as a technological artifact or an accessory? It seems that many desire the latest devices not necessarily for what they offer, as technological artifacts, but because of what they signify—in the same way as a new pair of sneakers or a handbag of the right brand signifies status in the relevant reference frame. If that is the case, then is that really a *technological* fetish?

The nature and understanding of computer simulations have been the subject of great interest in a wide range of fields: software engineering, computer science, mathematics, and philosophy of science—including philosophy of computer science and philosophy of mathematics. In Chap. 9, "Models, Explanation, Representation, and the Philosophy of Computer Simulations," Juan M. Durán offers a survey of the central positions within the interdisciplinary debate on computer simulations and argues that these simulations are units of analysis on their own and cannot be reduced to mathematical models. Thus, Durán's proposal approaches simulations as a branch of computer science and software engineering.

To underpin his defense of computer simulations as distinctive units of analysis, Durán relies on examples that aim to show how the representation of a target system through a computer simulation differs from its representation through a set of equations; the syntax of simulation models uncovers these differences. However, these claims are controversial because, so far, it is unclear that computer simulations actually exceed the expressive power of mathematical languages and models. The controversial nature of Durán's central claims serves as an invitation to discuss computer simulations' role and value within scientific theories and practices.

Lastly, this book would not have been possible without the help of a large number of people. We want to thank Deborah G. Johnson for writing the foreword. We also want to thank all reviewers for their help. While the reviewing was double-blinded, we publish the names of the reviewers consenting to be named below (some reviewers preferred to retain their anonymity also at this stage): Adam Chlipala, Alan Bundy, Aybüke Özgün, Barbro Fröding, Chantal Keller, Charles Rathkopf, Claus Beisbart, Cosmo Grant, Daniel Lim, Don Berkich, Dorna Behdadi, Eckhart Arnold, Gordana Dodig-Crnkovic, Göran Collste, John Symons, Jorge Manero,

Juan M. Durán, Kye Palider, Kyrill Winkler, Lenin Vazquez-Toledo, Marc Jiménez-Rolland, Martin Frické, Matthieu Fontaine, Michael Weisberg, Nicola Angius, Orlin Vakarelov, Otávio Bueno, Pak-Hang Wong, Patricia Mirabile, Paul Humphreys, Rasmus K. Rendsvig, Ryan Jenkins, Soraj Hongladarom, Stefan Buijsman, Stefano Canali, Stefano Vincini, Steve McKinlay, and Yoav Meyrav.

Mexico City, Mexico — Nancy Abigail Nuñez Hernández
Stockholm, Sweden — Björn Lundgren

References

AI HLEG (High-Level Expert Group on Artificial Intelligence). 2019. *Ethics guidelines for trustworthy AI.* https://ec.europa.eu/futurium/en/ai-alliance-consultation.

Berkich, D. 2019a. 2019 Covey Award: Emeritus Professor John Weckert. *IACAP announcement on PHILOS-L email list.* https://listserv.liv.ac.uk/cgi-bin/wa?A2=PHILOS-L;f0e8d75e.1906.

Berkich, D. 2019b. 2019 Simon Award: Dr. Juan M. Durán. *IACAP announcement on PHILOS-L email list.* https://listserv.liv.ac.uk/cgi-bin/wa?A2=PHILOS-L;5222a460.1906.

Bicchieri, C. 2017. Norms in the wild: how to diagnose, measure, and change social norms. New York: Oxford University Press.

Brey, P., B. Lundgren, K. Macnish, and M. Ryan. 2019. Guidelines for the development and the use of SIS. *Deliverable D3.2 of the SHERPA project.* https://doi.org/10.21253/DMU.11316833.v2.

Christianson, N.H., A. Sizemore Blevins, and D.S. Bassett. 2020. Architecture and evolution of semantic networks in mathematics texts. *Proceedings of the Royal Society A* 476(2239): 20190741. https://doi.org/10.1098/rspa.2019.0741.

Coveney, P.V., E.R. Dougherty, & R.R. Highfield. 2016. Big data need big theory too. *Philosophical Transactions of the Royal Society A: Mathematical, Physical and Engineering Sciences* 374(2080): 20160153. https://doi.org/10.1098/rsta.2016.0153.

Grindrod, J. 2019. Computational beliefs. *Inquiry.* https://doi.org/10.1080/0020174X.2019.1688178.

Kleinberg, Joel, Sendhil Mullainathan, and Manish Raghavan. 2017. *Inherent Trade-Offs in the Fair Determination of Risk Scores.* arXiv:1609.05807v2.

Lovász, L. 1978. Kneser's conjecture, chromatic number, and homotopy. *Journal of Combinatorial Theory*, Series A, 25(3): 319–324. https://doi.org/10.1016/0097-3165(78)90022-5.

Reimann, M.W., M. Nolte, M. Scolamiero, K. Turner, R. Perin, G. Chindemi, P. Dłotko, R. Levi, K. Hess, and H. Markram. 2017. Cliques of neurons bound into cavities provide a missing link between structure and function. *Frontiers in computational neuroscience*, 11, 48. https://doi.org/10.3389/fncom.2017.00048.

Salnikov, V., D. Cassese, R. Lambiotte, and N.S. Jones. 2018. Co-occurrence simplicial complexes in mathematics: Identifying the holes of knowledge. *Applied Network Science* 3, 37. https://doi.org/10.1007/s41109-018-0074-3.
Shalf, J. 2020. The future of computing beyond Moore's Law. *Philosophical Transactions of Royal Society A* 37820190061. https://doi.org/10.1098/rsta.2019.0061.
Sosa, E. 2006. Knowledge: Instrumental and testimonial. In *The epistemology of testimony,* ed. J. Lackey and E. Sosa, 116–123. Oxford: Oxford University Press.
Turing, A.M. 1936. On computable numbers, with an application to the Entscheidungsproblem. In *Proceedings of the London Mathematical Society* S2–42: 230–265. [Published 1937]. https://doi.org/10.1112/plms/s2-42.1.230.
Turing, A.M. 1950. Computing machinery and intelligence. *Mind* LIX(236): 433–460. https://doi.org/10.1093/mind/LIX.236.433.
Waldrop, M.M. 2016. The chips are down for Moore's law. *Nature* News feature 530: 144–147. https://doi.org/10.1038/530144a.

Contents

Chapter 1
Knowledge and Simplicial Complexes

Hans van Ditmarsch, Éric Goubault, Jérémy Ledent, and Sergio Rajsbaum

Abstract Simplicial complexes are a versatile and convenient paradigm on which to build all the tools and techniques of the logic of knowledge, on the assumption that initial epistemic models can be described in a distributed fashion. Thus, we can define: knowledge, belief, bisimulation, the group notions of mutual, distributed and common knowledge, and also dynamics in the shape of simplicial action models. We give a survey on how to interpret all such notions on simplicial complexes, building upon the foundations laid in Goubault et al. (Inf Comput 278:104597, 2021).

1.1 Introduction

Epistemic logic investigates knowledge and belief, and change of knowledge and belief, in multi-agent systems. Since Hintikka (1962) it has developed into multiple directions. Knowledge change was extensively modelled in temporal epistemic logics (Alur et al. 2002; Halpern and Moses 1990; Pnueli 1977) and more recently in dynamic epistemic logics (van Ditmarsch et al. 2008), where action model logic was particularly successful (Baltag et al. 1998). Modelling asynchrony and concurrency

H. van Ditmarsch (✉)
Open University of the Netherlands, Heerlen, Netherlands
e-mail: hans.vanditmarsch@ou.nl

É. Goubault
LIX, École Polytechnique, CNRS, Institut Polytechnique de Paris, Palaiseau, France
e-mail: goubault@lix.polytechnique.fr

J. Ledent
University of Strathclyde, Glasgow, Scotland
e-mail: jeremy.ledent@strath.ac.uk

S. Rajsbaum
Instituto de Matemáticas, Universidad Nacional Autónoma de México (UNAM), Mexico D.F., Mexico
e-mail: rajsbaum@im.unam.mx

B. Lundgren, N. A. Nuñez Hernández (eds.), *Philosophy of Computing*,
Philosophical Studies Series 143, https://doi.org/10.1007/978-3-030-75267-5_1

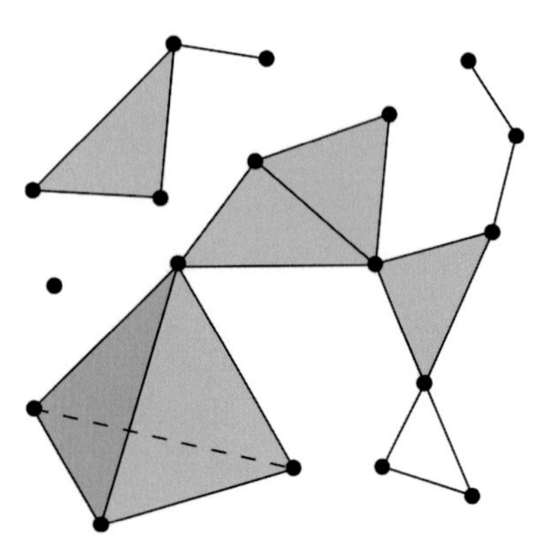

Fig. 1.1 A simplicial complex

in epistemic logic has been investigated in many works in a temporal or dynamic setting (Dixon et al. 2015; Halpern and Moses 1990; Harel et al. 2000; Knight 2013; Panangaden and Taylor 1992; Peleg 1987; Pnueli 1977) and more recently in dynamic epistemic logic (Balbiani et al. 2019; Degremont et al. 2011; Knight et al. 2019).

Combinatorial topology has been used to great effect in distributed computing to model concurrency and asynchrony since the early works (Biran et al. 1990; Fischer et al. 1985; Loui and Abu-Amara 1987) that identified one-dimensional connectivity invariants, and the higher dimensional topological properties discovered in the early 1990s e.g. Herlihy and Shavit (1993, 1999). A recent overview is Herlihy et al. (2013). The basic structure in combinatorial topology is the *simplicial complex,* a collection of subsets of a set of vertices closed under containment. The subsets, called *simplices*, can be vertices, edges, triangles, tetrahedrons, etc. (Fig. 1.1). Geometric manipulations that preserve some of the topology of an initial simplicial complex, such as subdivision, correspond in a natural and transparent way to the evolution of a distributed computation by a set of agents. In turn, the topological invariants preserved, determine the computational power of the distributed computing system (defined by the possible failures, asynchrony, and communication media).

Recently, an epistemic logic interpreted on simplicial complexes has been proposed (van Ditmarsch et al. 2021; Goubault et al. 2021; Ledent 2019). There is a categorical correspondence between the usual Kripke frames to interpret epistemic logic and simplicial complexes. The epistemic logic of Goubault et al. (2021) is also a dynamic epistemic logic: the action models of Baltag et al. (1998) correspond to simplicial complexes in the same way as Kripke models, but then with a dynamic interpretation, such that action execution is a product operation on simplicial complexes. In this way, Goubault et al. (2021) models distributed computing tasks and algorithms in their combinatorial topological incarnations as simplicial complexes. As their knowledge operators can be interpreted for typical asynchronous algorithms (e.g. when processes communicate by writing and reading shared variables), this is a relevant notion of asynchronous knowledge, as it reasons over indistinguishable asynchronous sequences of events, although differently so than Degremont et al. (2011). Predating Goubault et al. (2021) the work of Porter (2002) already established a relation between multi-agent epistemic models and simplicial complexes by way of the logic of knowledge. In that work the dynamics were based on the runs-and-systems approach of Halpern and Moses (1990), Fagin et al. (1995) and not on a dynamic epistemic logic as in Goubault et al. (2021).

In this work we further explore using simplicial complexes as models for epistemic notions, and survey what the logic of knowledge can contribute to the description of simplicial complexes and their dynamic evolution. Our motivation begins in the same way as in Goubault et al. (2021): the realisation discovered by distributed computing research, that the usual Kripke semantics for multi-agent epistemic logic has implicit a higher dimensional structure, that gets exposed using simplicial complexes. That there are topological invariants preserved when the system evolves after communication, and that this topological invariants in turn determine the problems that the agents can solve. In this paper the goal is to go beyond this initial motivation, and present an introduction to a variety of issues that arise when using simplicial complexes in epistemic logic. We stress that each one can be studied in more detail, we hope only to show that many interesting issues arise when moving from the usual one dimensional Kripke structure, to its higher dimensional simplicial complex dual. Many problems remain open.

Organization and Results This is an overview of our contribution and of our results. Section 1.2 introduces informally, and motivates the simplicial complex approach to epistemic logic. Section 1.3 presents the logical and topological tools and techniques that we build upon in this work and also reviews the results of Goubault et al. (2021), such as the categorical correspondence between simplicial models and the frames of Kripke models satisfying certain structural conditions. Section 1.4 defines bisimulation for simplicial complexes, and gives the obvious adequacy results for the notion: bisimulation invariance. Bisimulation allows us to verify whether different complexes contain the same information. Section 1.5 presents a local semantics for the logic of knowledge on simplicial complexes. In Goubault et al. (2021) the designated object for interpreting formulas always is a facet of the simplicial model under consideration. By local semantics we mean that the designated object can be any simplex. Local semantics therefore allow us to check the truth even in a vertex of a complex, which represents the local state of an agent. If the logic of knowledge were ever to become an acceptable tool for combinatorial topologists, this flexibility seems an important requirement. Section 1.6 presents semantics on simplicial complexes for the well-known group epistemic notions of mutual, common, and distributed knowledge. In that section we also explore novel group epistemic notions describing higher-dimensional features of complexes. We propose a notion of common distributed knowledge that can model check truth on a manifold. Section 1.7 defines belief for simplicial complexes. Unlike knowledge of propositions, belief of propositions does not entail that these propositions are true. Beliefs may be false. This is useful to model distributed systems wherein agents may be mistaken about what other agents know or believe, or even about their own local state. Section 1.8 defines simplicial action models. Unlike Goubault et al. (2021), we define preconditions on vertices (from which preconditions for facets can be derived, and vice versa) and we also model factual change (change of the value of variables). As an example of factual change we then model binary consensus in distributed computing. Finally, Sect. 1.9 investigates when epistemic models (Kripke models wherein all accessibility relations are equivalence relations) can be transformed into local epistemic models with the same information content. An epistemic model is local if for each variable there is an agent such that the value of that variable is uniform in any of her equivalence classes. We formulate some as yet unanswered conjectures there on the existence of such transformation. An answer is relevant to widen the scope of the categorical correspondence results of Goubault et al. (2021), as they are restricted to local epistemic models. A short concluding section describes even further avenues for exploration, such as the interaction between knowledge and awareness on impure complexes.

1.2 An Informal Introduction to Epistemic Logic on Simplicial Complexes

Epistemic logic, also known as the logic of knowledge, has been interpreted on a wide variety of model classes, including Kripke models where all relations are equivalence relations (a.k.a. epistemic models) (Fagin et al. 1995), rough sets (Banerjee and Khan 2007), neighbourhood structures (Chellas 1980; Pacuit 2017), subset space models (Dabrowski et al. 1996) (not necessarily discrete) topological spaces (Parikh et al. 2007; Özgün 2017), and the central topic of this paper, simplicial complexes (Goubault et al. 2021; Ledent 2019). Simplicial complexes provide a fascinating new point of view on epistemic logic. In this section, we give an intuitive explanation of the notion of simplicial model, its relation with the more traditional epistemic models, and the new questions that arise from this new simplicial point of view. Precise definitions of all the concepts discussed here will be given in Sect. 1.3.

In epistemic logic, the construction $K_a\varphi$ stands for "agent a knows that proposition φ is true". The usual Kripke semantics for this employs an *epistemic model*, which consists of a set S of *global states*, and for each agent, an equivalence relation on S called the *indistinguishability* relation. Formula $K_a\varphi$ is *true* in global state s whenever φ is true in every state t that is indistinguishable from s for the agent a.

In the simplicial complex point of view, each global state is represented by a basic geometric shape called a *simplex*. The dimension of these simplices depends on the total number of agents that we consider: they are *edges* for 2 agents, *triangles* for 3 agents, tetrahedrons for 4 agents, and more generally, n-simplices for $n + 1$ agents. The notion of indistinguishability between global states is encoded geometrically by gluing these basic blocks together. Thus, simplicial models are of a geometric nature. Indeed, in the field of combinatorial topology, simplicial complexes are the primary representation of a topological space, see for example Edelsbrunner and Harer (2010) and Kozlov (2008).

Epistemic Models Based on Local States Global states are the main elements of epistemic models, which have served as the basis of the most widely used semantics for all varieties of modal logic, since the 1960s. However, in many situations, the focus is on local states. This is for example the case in distributed computing, where the central role of topology for multi-agent computation was discovered. Here, the agents are parallel processes that communicate with each other via message-passing or shared memory. The local state of an agent is defined by its local memory, program counter, etc. Indeed, we take most of our examples and motivations from this area.

Another motivation for focusing on local states is the relativistic perspective, where at any given moment an agent's *local state* exists and is well-defined but not necessarily so the (global) state. In fact, there may be more than one state which is compatible with a set of local states of the agents. Indeed, even talking about absolute real time is impossible according to relativity theory, so it may not make sense to talk about the state at some real time t. This relativistic approach has been considered since early on in computer science (Lamport 1978).

Simplicial Models We thus move away from defining an epistemic model in terms of global states, to one defined in terms of local states. Then, a global state s for $n+1$ agents can be decomposed as a tuple $(\ell_0, \dots, \ell_n)$, consisting of one local state for each agent. In a simplicial model, each of these local states is modeled separately as a *vertex* of a simplicial complex. With this point of view, a global state corresponds to a set of $n + 1$ vertices, that is, an n-dimensional *simplex*. Two global states are indistinguishable by some agent when it has the same local state in both of them. In a simplicial model, this fact is represented by a situation where one vertex (the local state of the agent) belongs to two different simplices (the two global states).

To give an intuitive understanding of how simplicial complexes are used to model epistemic situations, we present a few examples adapted from Goubault et al. (2021), Herlihy et al. (2013), and Ledent (2019). In all examples, we work with either two or three agents in order to be able to draw low-dimensional pictures. The agents are often represented as colours, black, gray and white. The local state of an agent is written inside or besides a vertex of the corresponding colour. Global states are represented as edges (for 2 agents) or triangles (for 3 agents).

Example 1 For readers familiar with the muddy children puzzle, the picture below represents the initial knowledge for three children $\{a, b, c\}$ (represented in gray, white, black, respectively), viewed as a simplicial model. The local view of a child is written as a vector of three symbols among $\{0, 1, \bot\}$, where each symbol

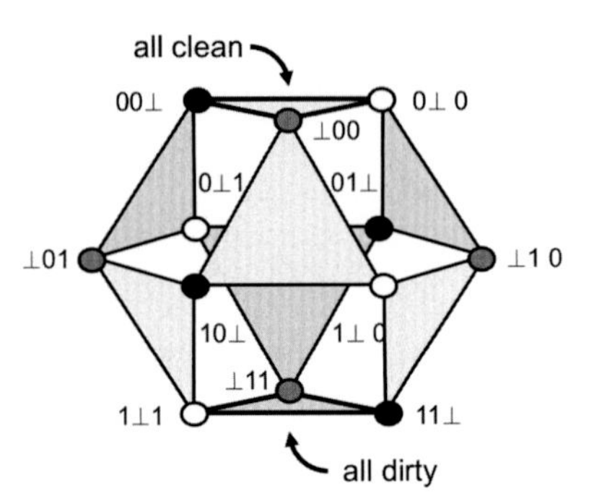

Fig. 1.2 Example 1

corresponds to a, b, c, in that order. The '0' symbol means clean, '1' means muddy, and '$\bot$' means that the child cannot see her own forehead (Fig. 1.2).
For instance, the black vertex labeled by $00\bot$ corresponds to the local state of the third child ("c"), which is able to see that the other two children are not muddy. Since she cannot see her own forehead, this vertex belongs to two possible global states: one where all children are clean, and one where only the third child is muddy. On the picture, global states are represented by triangles, and indeed the black vertex labeled by $00\bot$ belongs to two of those triangles. $\dashv$

Example 2 Each agent is secretly given a binary value 0 or 1, but does not know which value has been received by the other agents. So, every possible combination of 0's and 1's is a possible state. In contrast with the previous example, here each agent knows only its own value and not the other agents' values. The figures below depict this situation for 2 and 3 agents, as epistemic models and as simplicial models.

In the epistemic models, the agents are called a, b, c, and the global state is represented as a vector of values, e.g., 101, representing the values given to the agents a, b, c (in that order). In the 3-agents case, the labels of the dotted edges have been omitted to avoid overloading the picture, as well as other edges that can be deduced by transitivity.

In the simplicial model, agents are represented as colours (gray, white and black). The local state is represented as a single value in a vertex, e.g., "1" in a gray vertex means that agent a has been given input value 1. The global states correspond to edges in the 2-agents case, and triangles in the 3-agents case (Fig. 1.3).

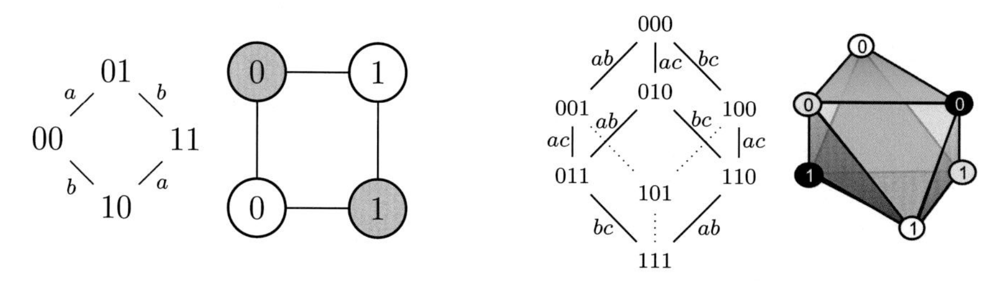

Fig. 1.3 Example 2

Notice that the simplicial complex on the right is an octahedron, which is a triangulated sphere. More generally, it is known that the binary input simplicial complex for $n + 1$ agents is topologically equivalent to an n-dimensional sphere (Herlihy et al. 2013). ⊣

Example 3 The toy example below illustrates the equivalence between epistemic models and simplicial models. The picture shows an epistemic model (left) and its associated simplicial model (right). No variable labelings are shown. The three agents, named a, b, c, are represented as colours gray, white and black on the vertices of the simplicial complex. The three global states of the epistemic model correspond to the three triangles (i.e., two-dimensional facets) of the simplicial complex. The two states indistinguishable by agent c, are glued along their black vertex; while the two states indistinguishable by agents a and b are glued along the gray-and-white edge. The two functors κ and σ translating back and forth between the two kinds of models are defined formally in Sect. 1.3.3.

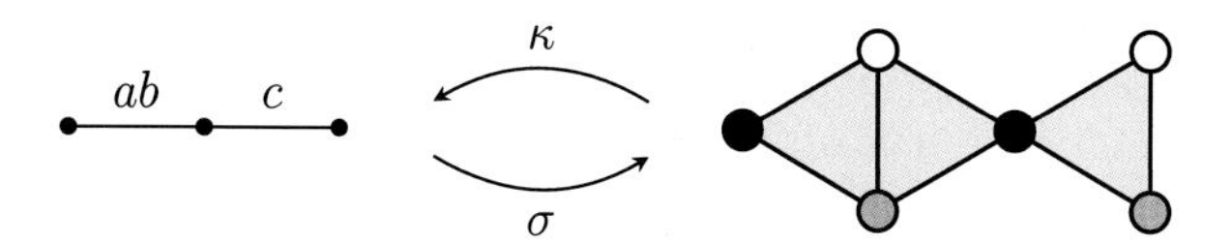

⊣

Example 4 (Asynchronous Computation and Subdivisions) We now give an example that illustrates how a simplicial model can change after communication. In Sect. 1.8 we explore change of information using action models. We describe here an important geometric operation on simplicial complexes called the *chromatic subdivision*. We describe it for 3 agents, but it generalizes to any number of agents. It appears in various distributed computing situations, where agents communicate by shared memory or message passing (Herlihy et al. 2013).

Consider three agents a, b, c, represented with colours gray, white and black, and suppose that agent a (gray) may receive as input value either 0 or 1, while the two other agents receive input 0. The two triangles on the left represent two facets of the input model, with input values 000 (green) and 100 (yellow). Now, suppose

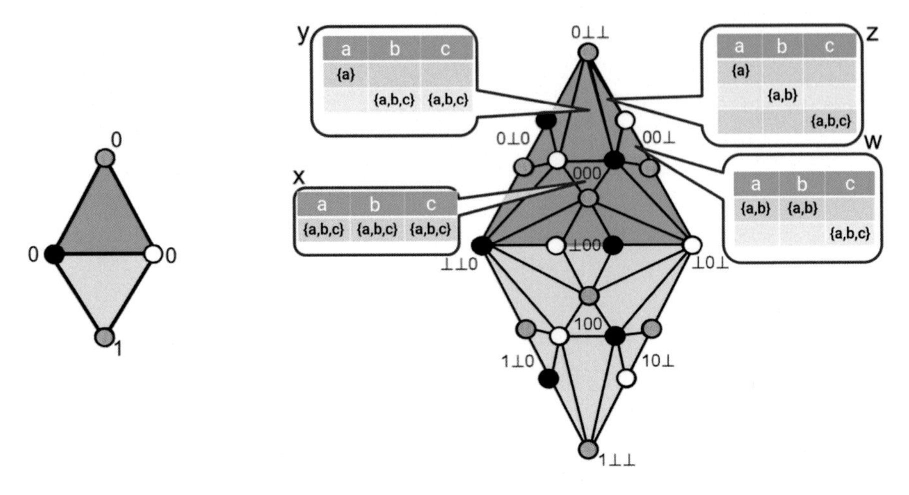

Fig. 1.4 Example 4

they communicate to each other their inputs, but in the following unreliable way. After communication, it is possible that an agent missed hearing the input value of one or two of the other agents. There are four possible patterns of communication, x, y, z, w, illustrated in the figure. Each pattern is defined by a table, with one column per agent, and where a row represents what agents hear, consecutively from top to bottom. An agent i receives the value from agent j if and only if agent j occurs in the column for agent i. After communication, we have the simplicial model on the right. Each vertex is labeled with a vector indicating which values it received (Fig. 1.4).

The triangle x represents the situation where all agents heard their inputs (all 0 in this case) from each other, and intersects in an edge with the triangle y, where agent a (gray) did not hear from the other two. Triangle x intersects only in a vertex with triangle w, where both a and b heard from each other, but not from c. Triangle x also intersects in only one vertex with triangle z, where agent a heard from no-one, agent b heard only from a, and agent c heard from everybody.

In Goubault et al. (2021) this chromatic subdivision is described as an action model. This fairly complex description is not repeated in this work, where we restrict ourselves to simpler action models. ⊣

Example 5 (Synchronous Computation) Consider again black, white, and gray agents. They get inputs 012, respectively. Suppose they communicate to each other their inputs in a reliable way, but now, at most one of the agents may crash. Thus, when they communicate, it is possible that an agent missed hearing the input value of another agent, and in that case, it knows that this agent crashed. After communication, we have the simplicial model below. The center triangle represents the state where no one crashed, and each agent learned the inputs of every agent. The other edges represent the cases where one agent crashed, and that is why the

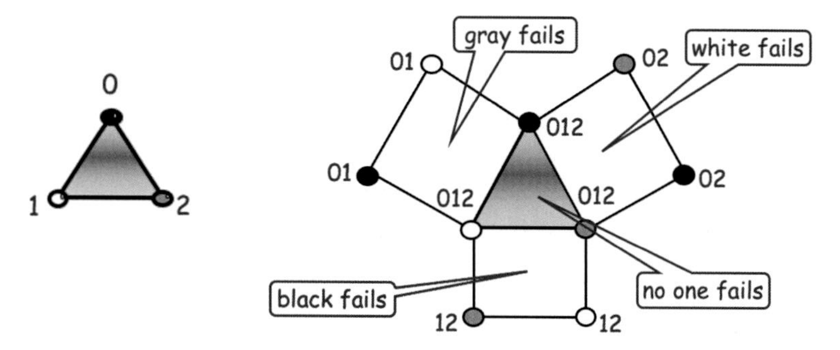

Fig. 1.5 Example 5

dimension goes down. Thus, such global states are represented as edges instead of triangles. The local state of the crashed agent is no longer relevant (Fig. 1.5). This is an example of synchronous computation, which has been thoroughly studied in distributed computing, see e.g. Herlihy et al. (2013, chapter 13.5.2). ⊣

The correspondence between epistemic models and simplicial models exhibited in Goubault et al. (2021) (and depicted in Example 3) can be extended to model various epistemic notions. We conclude this motivating section with an overview of novel ways presented in this contribution in which epistemic logic can be relevant to describe simplicial complexes, with pointers to further sections giving details.

- **Locality.** A crucial property that allows one to translate an epistemic model into a simplicial one is called *locality* (see Sect. 1.3.1). Intuitively, an epistemic model is local when each propositional variable talks only about the local state of one of the agents. This condition can be lifted in the case of *finite* epistemic models as demonstrated in Ledent (2019). In Sect. 1.9 we investigate the consequences of this solution for the information content of models, and what happens in the infinite case.
- **Dynamics.** A simplicial complex version of the *action models* found in dynamic epistemic logic was defined in Goubault et al. (2021). In Sect. 1.8, we build on this idea in order to include the notion of *factual change*, which allows the values of propositional variables to be modified by the action model.
- **Bisimulation.** A central topic in modal logics is that of bisimulation between epistemic models. Bisimulation of simplicial models has been proposed in Goubault et al. (2019) and Ledent (2019). We further discuss simplicial bisimulation in Sect. 1.4, and relate it with simplicial maps and covering spaces.
- **Maps.** In topology maps from one space to another are continuous functions, which in the case of simplicial complexes are *simplicial maps*. The simulations and bisimulations introduced in Sect. 1.4 are natural generalizations of simplicial maps, on condition that they are also value preserving. Then, in Sect. 1.7 we model the notion of belief (knowledge that may be incorrect) on simplicial complexes, by way of so-called *belief functions* that induce simplicial maps in an

obvious way, however in that case they need not be value preserving, as beliefs may be incorrect.

- **Topology.** The most evident new feature of simplicial models is their geometric nature. Hence, one can analyze the topological structure of the underlying space, such as its high-dimensional connectivity (e.g. homotopy groups, representing its "holes"). Examples 1 and 2 make the geometric space evident that would be hard to see in an epistemic model.
  The basic knowledge operator $K_a\varphi$ is intrinsically one-dimensional. However, higher dimensional structure in complexes can be described by, for example, group epistemic operators, such as the common distributed knowledge presented in Sect. 1.6 that can check truth on manifolds.
- **Local semantics.** In epistemic models the typical unit of interpretation is the (global) state, as in the semantics of epistemic logic on simplicial complexes presented in Goubault et al. (2021) and Ledent (2019). However, there are other relevant units of interpretation on simplicial complexes, as we can determine truth in facets but also in simplices of any dimension. Already in Example 3, we see that there are 3 states, corresponding to three facets, while the total number of simplices is 17. Local interpretation on simplices is addressed in Sect. 1.5.
- **Hierarchical structure.** In distributed computing, simplicial models often have a hierarchical structure, meaning that they are composed of submodels representing the behavior of systems with less agents, and hence of lower dimension. In Example 4, notice that the simplicial model on the right is a subdivision of the initial simplicial model, and furthermore, the green sub-complex is a subdivision of the initial green triangle, while the yellow sub-complex is a subdivision of the initial yellow triangle. Notice that whenever the white and gray agents do not hear from the black one, they cannot tell what its input was. Thus, the intersection C' of both the green and the yellow complexes consists of three edges (and their vertices). This complex C' is a subdivision of the initial edge, which is the intersection of the green and the yellow triangle. Hence, we can analyze a subcomplex C' of C, of dimension 1, with an interesting semantic meaning: it represents a subsystem consisting of only two agents, white and gray. One can ask what the white and gray agents know on vertices or edges of C', as if the larger system of three agents did not exist.
  This sort of semantics for edges instead of triangles is addressed in full generality in Sect. 1.6.1, as language-restricted local semantics.
- **Variable number of agents.** In distributed computing, whenever crashes are detectable, facets of different size may occur (see Example 5). In epistemic logic such phenomena are modelled in logics of awareness and knowledge, and in particular the kind of unawareness known as *unawareness of propositional variables*, as in Fagin and Halpern (1988) and van Ditmarsch and French (2009). A thorough treatment of this topic is deferred to future research. Another direction for future investigation are adjunctions between epistemic frames and simplicial complexes, that generalize the (equivalent) correspondence between frames and complexes of Goubault et al. (2021).

For the sake of completeness, we also mention below some of the important features of simplicial models, even though we do not address them in the rest of the paper. They are not specific to the epistemic logic perspective, and have been thoroughly studied in distributed computing. See Herlihy et al. (2013) for more detail.

- **Topological invariants.** As it turns out, there are topological invariants that are maintained starting from an initial simplicial model, after the agents communicate with each other. In the most basic case, illustrated in Example 4, the initial simplicial model is subdivided, but in other situations where the communication is more reliable "holes" may be introduced, as in Example 5. Many other examples are described in Herlihy et al. (2013), and the notion of "holes" is formalized in terms of homotopy groups.
- **Power of a computational model.** The topological invariants preserved by the distributed model (number of failures, communication media, etc.) in turn determine what the agents can distributively compute. After all, whatever computation an agent is performing, the outcome is a function of its local state. Thus, decisions induce a simplicial map from the simplicial complex representing what the agents know about the inputs after communication, to an output simplicial complex model representing what they are expected to compute based on that knowledge. This approach is developed in Goubault et al. (2021).
- **Specifications.** In Goubault et al. (2021) it is explained how a simplicial model is used as knowledge that is supposed to be gained by a set of agents communicating with each other, via a simplicial map from the simplicial model at the end of the agent's communications.

1.3 Logical and Topological Tools

1.3.1 Logical Tools

Given are a set A of $n+1$ *agents* $a_0, \dots, a_n$ (or $a, b, \dots$) and a countable set P of *global variables* $p, q, p', q', \dots$ (possibly indexed).

Language The *language of epistemic logic* $\mathcal{L}_K(A, P)$ is defined as

$$\varphi ::= p \mid \neg\varphi \mid \varphi \wedge \varphi \mid K_a\varphi$$

where $p \in P$ and $a \in A$. We will write $\mathcal{L}_K(P)$ if A is clear from the context, and $\mathcal{L}_K$ if A and P are clear from the context. For $\neg p$ we may write $\overline{p}$. Expression $K_a\varphi$ stands for 'agent a knows (that) φ.' The fragment without inductive construct $K_a\varphi$ is the *language of propositional logic* (also known as the Booleans) denoted $\mathcal{L}_\emptyset(A, P)$. For $\neg K_a \neg\varphi$ we may write $\widehat{K}_a\varphi$, for 'agent a considers it possible that φ.'

Epistemic Models *Epistemic frames* are pairs $\mathcal{M} = (S, \sim)$ where S is the domain of *(global) states*, and $\sim$ is a function from the set of agents A to (binary) *accessibility relations* on S, that are required to be equivalence relations. For $\sim(a)$ we write $\sim_a$. *Epistemic models* are triples $\mathcal{M} = (S, \sim, L)$, where $(S, \sim)$ is an epistemic frame and where *valuation* L is a function from S to $\mathcal{P}(P)$. A pair $(\mathcal{M}, s)$ where $s \in S$ is a *pointed* epistemic model. We also distinguish the *multi-pointed* epistemic model $(\mathcal{M}, S')$ where $S' \subseteq S$. For $(s,t) \in \sim_a$ we write $s \sim_a t$, and for $\{t \mid s \sim_a t\}$ we write $[s]_a$: this is an equivalence class of the relation $\sim_a$.

We now introduce terminology for epistemic models for distributed systems.

Given an epistemic model $\mathcal{M} = (S, \sim, L)$, a variable $p \in P$ is *local for agent* $a \in A$ iff for all $s, t \in S$, if $s \sim_a t$ then $p \in L(s)$ iff $p \in L(t)$. Variable p is *local* iff it is local for some agent a, and a model $\mathcal{M}$ is local iff all variables are local. We write P_a for the set of local variables for agent a, i.e., $P_a = \{p \in P \mid \text{for } s, t \in S, \text{ if } s \sim_a t, \text{ then } p \in L(s) \text{ iff } p \in L(t)\}$. If an epistemic model is local, then $\bigcup_{a \in A} P_a = P$. Without loss of generality we may assume that all members of $\{P_{a_0}, \dots, P_{a_n}\}$ are mutually disjoint, and that members of each P_a are subindexed with the agent name as in: $p_a, p'_a, q_a, \dots$.[1]

We can also approach locality from the other direction. Let mutually disjoint sets $S_{a_0} \dots S_{a_n}$ of *local states* and mutually disjoint sets $P_{a_0}, \dots, P_{a_n}$ of *local variables* (for resp. agents $a_0, \dots, a_n$) be given. Let for each $a \in A$, $L_a : S_a \to \mathcal{P}(P_a)$ be a valuation for the local states of agent a. A *distributed model* is an epistemic model $(S, \sim, L)$ such that:

- $S \subseteq S_{a_0} \times \dots \times S_{a_n}$;
- for all $a \in A$ and $(s_0, \dots, s_n), (t_0, \dots, t_n) \in S$: $(s_0, \dots, s_n) \sim_a (t_0, \dots, t_n)$ iff $s_a = t_a$;
- for all $a \in A$ and $(s_0, \dots, s_n) \in S$: $L((s_0, \dots, s_n)) = \{p \in P \mid \exists a \in A : p \in L_a(s_a)\}$.

In other words, each global state $s \in S$ is an $(n+1)$-tuple $s = (s_0, \dots, s_n)$ of local states for the agents $0, \dots, n$ respectively.

An epistemic frame $(S, \sim)$ is *proper* if $\bigcap_{a \in A} \sim_a = Id$, where the identity relation $Id := \{(s,s) \mid s \in S\}$. An epistemic model is proper if its underlying frame is proper.

Clearly, a distributed model is a local proper epistemic model, and we can also see a local proper epistemic model as a distributed model: if $\mathcal{M} = (S, \sim, L)$ is proper, then the model wherein domain S is replaced by the domain $\{([s]_{a_0}, \dots, [s]_{a_n}) \mid s \in S\}$ (and wherein the valuation can be relativized to a in the obvious way) is a distributed model. The definition is adequate, as for all $s \in S$ it holds that $[s]_{a_0} \cap \dots \cap [s]_{a_n} = \{s\}$.

[1] Instead of atoms P, considers atoms $P \times A$, where we write p_a for (p, a). For each agent a, let P_a be the set of all p_a that are local for a. Now, all P_a are disjoint.

An epistemic model is *factual* if all global states have different valuations, i.e., if for all $s, t \in S$ there is a $p \in P$ such that $p \in L(s)$ and $p \notin L(t)$, or $p \notin L(s)$ and $p \in L(t)$.

Semantics on Epistemic Models The interpretation of a formula $\varphi \in \mathcal{L}_K$ in a global state of a given pointed model $(\mathcal{M}, s)$ is by induction on the structure of φ. The expression "$\mathcal{M}, s \models \varphi$" stands for "in global state s of epistemic model $\mathcal{M}$ it holds that (or: it is true that) φ."

$$
\begin{array}{ll}
\mathcal{M}, s \models p & \text{iff } p \in L(s) \\
\mathcal{M}, s \models \neg\varphi & \text{iff } \mathcal{M}, s \not\models \varphi \\
\mathcal{M}, s \models \varphi \wedge \psi & \text{iff } \mathcal{M}, s \models \varphi \text{ and } \mathcal{M}, s \models \psi \\
\mathcal{M}, s \models K_a\varphi & \text{iff } \mathcal{M}, t \models \varphi \text{ for all } t \text{ with } s \sim_a t
\end{array}
$$

Formula φ is *valid* iff for all $(\mathcal{M}, s)$, $\mathcal{M}, s \models \varphi$. Given $(\mathcal{M}, s)$ and $(\mathcal{M}', s')$, by $(\mathcal{M}, s) \equiv (\mathcal{M}', s')$, for "$(\mathcal{M}, s)$ and $(\mathcal{M}', s')$ are *modally equivalent*," we mean that for all $\varphi \in \mathcal{L}_K$: $\mathcal{M}, s \models \varphi$ iff $\mathcal{M}', s' \models \varphi$. Informally, this means that these pointed models contain the same information. We let $[\![\varphi]\!]_{\mathcal{M}}$ stand for $\{s \in S \mid \mathcal{M}, s \models \varphi\}$. This set is called the *denotation* of φ in $\mathcal{M}$.

Standard references for epistemic logic are Meyer and van der Hoek (1995), Fagin et al. (1995), van Ditmarsch et al. (2015). The typical example of proper factual local epistemic models are interpreted systems (Fagin et al. 1995).

1.3.2 Topological Tools

We continue by introducing simplicial complexes and other topological tools. A reference on algebraic topology is Hatcher (2002), on a more combinatorial approach there are Edelsbrunner and Harer (2010) and Kozlov (2008), and on the use of combinatorial topology in distributed computing it is Herlihy et al. (2013). Subsequently and in the next subsection we review results from Goubault et al. (2021), that are also reported in Ledent (2019).

Simplicial Complexes Given a set of *vertices* V (sometimes also called *local states*; singular form *vertex*), a *(simplicial) complex* C is a set of non-empty finite subsets of V, called *simplices* (singular form *simplex*), that is closed under subsets (such that for all $X \in C$, $Y \subseteq X$ implies $Y \in C$), and that contains all singleton subsets of V. If $Y \subseteq X$ we say that Y is a *face* of X. A maximal simplex in C is a *facet*. The facets of a complex C are denoted as $\mathcal{F}(C)$, and the vertices of a complex C are denoted as $\mathcal{V}(C)$. The dimension of a simplex X is $|X| - 1$, e.g., vertices are of dimension 0, while edges are of dimension 1. The dimension of a complex is the maximal dimension of its facets. A simplicial complex is *pure* if all facets have the same dimension. A *manifold* is a pure simplicial complex C of dimension n such that: (i) for any $X, Y \in \mathcal{F}(C)$ there are facets $X = X_0, \dots, X_m = Y$ such that for all $i < m$ the dimension of $X_i \cap X_{i+1}$ is $n - 1$ (this is a 'path' from X to Y in C), and (ii) any simplex $Z \in C$ of dimension $n - 1$ is a face of one or two facets of C (those that are faces of only one facet form the *boundary* of C).

Complex D is a *subcomplex* of complex C if $D \subseteq C$. A subcomplex of interest is the *m-skeleton* D of a pure n-dimensional complex C, i.e., the maximal subcomplex D of C of dimension m.

We decorate the vertices of simplicial complexes with agent's names, that we refer to as *colours*. A *chromatic map* $\chi : C \to A$ assigns colours to vertices such that different vertices of the same simplex are assigned different colours. Thus, $\chi(v) = a$ denotes that the local state or vertex v belongs to agent a. Dually, the vertex of a simplex X coloured with a is denoted X_a. A pair consisting of a simplicial complex and a colouring map χ is a *chromatic simplicial complex*.

From now on, all simplicial complexes will be chromatic pure simplicial complexes, and we will also always assume that the number of agents is the number of vertices in a facet, i.e., $|A| = n + 1$ where n is the dimension of the complex.

Simplicial Models We now also decorate the vertices of simplicial complexes with *local variables* $p_a \in P_a$ for $a \in A$, where we recall that $\bigcup_{a \in A} P_a = P$ (all sets of local variables are assumed to be disjoint). While chromatic maps are denoted $\chi, \chi', \ldots$, *valuations* (valuation maps) assigning sets of local variables to vertices are denoted $\ell, \ell', \ldots$ For any $X \in C$, $\ell(X)$ stands for $\bigcup_{v \in X} \ell(v)$. A *simplicial model* $\mathcal{C}$ is a triple (C, χ, ℓ) where C is a simplicial complex.

Simplicial Maps A *simplicial map* (*simplicial function*) between simplicial complexes C and C' is a simplex preserving function f between its vertices, i.e., $f : \mathcal{V}(C) \to \mathcal{V}(C')$ such that for all $X \in C$, $f(X) := \{f(v) \mid v \in X\} \in C'$. We let $f(C)$ stand for $\{f(X) \mid X \in C\}$. A simplicial map is *rigid* if it is dimension preserving, i.e., if for all $X \in C$, $|f(X)| = |X|$. We will abuse the language and also call f a simplicial map between chromatic simplicial complexes (C, χ) and (C', χ'), and between simplicial models (C, χ, ℓ) and (C', χ', ℓ').

A *chromatic simplicial map* is a *colour preserving* simplicial map between chromatic simplicial complexes, i.e., for all $v \in \mathcal{V}(C)$, $\chi'(f(v)) = \chi(v)$. We note that it is therefore also rigid. A simplicial map f between simplicial models is *value preserving* if for all $v \in \mathcal{V}(C)$, $\ell'(f(v)) = \ell(v)$. If f is not only colour preserving but also value preserving, and its inverse f^{-1} as well, then $\mathcal{C}$ and $\mathcal{C}'$ are *isomorphic*, notation $\mathcal{C} \simeq \mathcal{C}'$. It is customary to define isomorphy between simplicial complexes instead of between simplicial models. However, we will later define a bisimulation as a relation between simplicial models, such that isomorphy implies bisimilarity but not necessarily vice versa. We therefore defined isomorphy between simplicial models.

1.3.3 Epistemic Logic on Simplicial Models

Semantics on Simplicial Models The interpretation of a formula $\varphi \in \mathcal{L}_K(A, P)$ in a facet $X \in \mathcal{F}(C)$ of a given simplicial model $\mathcal{C} = (C, \chi, \ell)$ is by induction on the structure of φ. Merely the clause for knowledge looks different than that on epistemic models.

$$
\begin{array}{ll}
\mathcal{C}, X \models p_a & \text{iff } p_a \in \ell(X) \\
\mathcal{C}, X \models \neg\varphi & \text{iff } \mathcal{C}, X \not\models \varphi \\
\mathcal{C}, X \models \varphi \wedge \psi & \text{iff } \mathcal{C}, X \models \varphi \text{ and } \mathcal{C}, X \models \psi \\
\mathcal{C}, X \models K_a\varphi & \text{iff } \mathcal{C}, Y \models \varphi \text{ for all } Y \in \mathcal{F}(C) \text{ with } a \in \chi(X \cap Y)
\end{array}
$$

Validity and modal equivalence are also defined similarly. Formula φ is *valid* iff for all $(\mathcal{C}, X)$, $\mathcal{C}, X \models \varphi$; and given $(\mathcal{C}, X)$ and $(\mathcal{C}', X')$, by $(\mathcal{C}, X) \equiv (\mathcal{C}', X')$ we mean that for all $\varphi \in \mathcal{L}_K(A, P)$: $\mathcal{C}, X \models \varphi$ iff $\mathcal{C}', X' \models \varphi$.

In the clause $K_a\varphi$ for knowledge, the part $a \in \chi(X \cap Y)$ says in other words that agent a cannot distinguish facet X from facet Y.

Example 6 Consider the simplicial model $\mathcal{C}$ depicted below. For the convenience of the exposition the labels of the vertices are pairs $(i, 0)$ or $(i, 1)$, written as $i0$ and $i1$ respectively, where i is colour or agent and where each agent i has a single local proposition p_i such that 0 means that p_i is false and 1 means that p_i is true. These conventions will be used from now on throughout the paper.

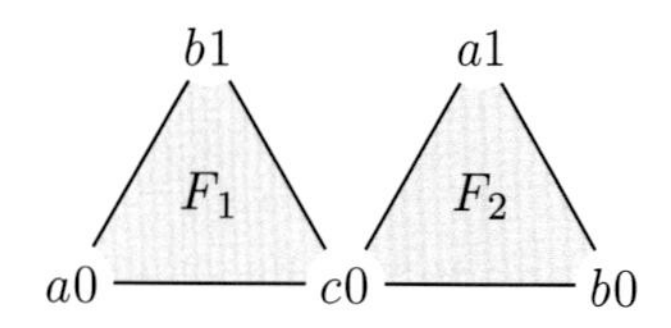

For example, $\mathcal{C}, F_1 \models K_a p_b$, because agent a only considers facet F_1 possible and $\mathcal{C}, F_1 \models p_b$, because $p_b \in \ell(F_1)$. Whereas $\mathcal{C}, F_1 \models \neg K_c p_b$, because agent c is uncertain between facets F_1 and F_2 (the c vertex is in the intersection of F_1 and F_2), and $\mathcal{C}, F_2 \models \neg p_b$. ⊣

Correspondence Between Simplicial Models and Epistemic Models Given agents A and variables P, let $\mathcal{K}$ be the class of local proper epistemic models and let $\mathcal{S}$ be the class of simplicial models. In Goubault et al. (2021), simplicial models are shown to correspond to local proper epistemic models by showing that the underlying proper epistemic frames and simplicial complexes correspond (categorically) via functions (functors) $\sigma : \mathcal{K} \to \mathcal{S}$ (σ for Simplicial) and $\kappa : \mathcal{S} \to \mathcal{K}$ (κ for Kripke), and then extending σ and κ in the obvious way, such that they are also valuation preserving, to models. The correspondence uses the observed relation between local proper epistemic models and distributed models. These are the details.

Given a local proper epistemic model $\mathcal{M} = (S, \sim, L)$, we define $\sigma(\mathcal{M}) = (C, \chi, \ell)$ as follows. Its vertices are $\mathcal{V}(\sigma(\mathcal{M})) = \{[s]_a \mid a \in A, s \in S\}$, where $\chi([s]_a) = a$. Its facets are $\mathcal{F}(\sigma(\mathcal{M})) = \{\{[s]_{a_0}, \ldots, [s]_{a_n}\} \mid s \in S\}$. For such a facet $\{[s]_{a_0}, \ldots, [s]_{a_n}\}$ we write $\sigma(s)$. Then, $[s]_a \in \sigma(s) \cap \sigma(t)$ iff $s \sim_a t$ (i.e., such that $[s]_a = [t]_a$), and for all $p_a \in P$, $p_a \in \ell([s]_a)$ iff $p_a \in L(s)$. We note that $\sigma(\mathcal{M})$ has dimension n.

Given a simplicial model $\mathcal{C} = (C, \chi, \ell)$, we define an epistemic model $\kappa(\mathcal{M}) = (S, \sim, L)$ as follows. Its domain S consists of states $\kappa(X)$ for all $X \in \mathcal{F}(C)$. We define $\kappa(X) \sim_a \kappa(Y)$ iff $a \in X \cap Y$, and $p_a \in L(\kappa(X))$ iff $p_a \in \ell(X)$. We note that $\kappa(\mathcal{M})$ is local and proper.

One can now show that:

Proposition 7 (Goubault et al. (2021)) *For all $\mathcal{M} \in \mathcal{K}$, $\kappa(\sigma(\mathcal{M})) \simeq \mathcal{M}$, and for all $\mathcal{C} \in \mathcal{S}$, $\sigma(\kappa(\mathcal{C})) \simeq \mathcal{C}$.* ⊣

Proposition 8 (Goubault et al. (2021)) *For all $\varphi \in \mathcal{L}_K$, $\mathcal{M}, s \models \varphi$ iff $\sigma(\mathcal{M}), \sigma(s) \models \varphi$, and $\mathcal{C}, X \models \varphi$ iff $\kappa(\mathcal{C}), \kappa(X) \models \varphi$.* ⊣

The logic **S5** augmented with so-called *locality axioms* $K_a p_a \vee K_a \neg p_a$ formalizing that all variables are local, is sound and complete with respect to the class of simplicial models (Goubault et al. 2021, Cor. 1).

1.4 Bisimulation for Simplicial Complexes

1.4.1 Bisimulation

If two models are isomorphic, they make the same formulas true, i.e., they contain the same information (what we also called *modally equivalent*). In modal logic and concurrency theory, a weaker form of sameness than isomorphy already guarantees that, *bisimilarity* (Baier and Katoen 2008; Blackburn et al. 2001; Sangiorgi 2011; van Benthem 2010; van Glabbeek 1990). As truth is invariant under bisimulation, it is an important way to determine whether structures contain the same information. In distributed computing by way of combinatorial topology, bisimulation between simplicial models has been proposed in van Ditmarsch et al. (2021) and Ledent (2019). We present and further explore this notion, and demonstrate its adequacy with respect to bisimulation for epistemic models. This may be considered of interest, for example, if different subdivisions of complexes are shown to be bisimilar, then they contain the same information after all.

We first define the standard notion of bisimulation between epistemic models, and then bisimulation between simplicial models.

Bisimulation Between Epistemic Models A *bisimulation* between $\mathcal{M} = (S, \sim, L)$ and $\mathcal{M}' = (S', \sim', L')$, notation $\mathfrak{R} : \mathcal{M} \underline{\leftrightarrow} \mathcal{M}'$, is a non-empty relation

$\mathfrak{R} \subseteq S \times S'$ such that for all $s \in S$, $s' \in S'$ with $\mathfrak{R}ss'$ the following three conditions are satisfied:

- **atoms**: for all $p \in P$, $p \in L(s)$ iff $p \in L'(s')$.
- **forth**: for all $a \in A$, for all t with $s \sim_a t$, there is a t' with $s' \sim'_a t'$ such that $\mathfrak{R}tt'$.
- **back**: for all $a \in A$, for all t' with $s' \sim'_a t'$, there is a t with $s \sim_a t$ such that $\mathfrak{R}tt'$.

A bisimulation $\mathfrak{R}$ such that for all $s \in S$ there is a $s' \in S'$ such that $\mathfrak{R}ss'$ and for all $s' \in S'$ there is a $s \in S$ such that $\mathfrak{R}ss'$ is a *total bisimulation*. If there is a bisimulation $\mathfrak{R}$ between $\mathcal{M}$ and $\mathcal{M}'$ we say that $\mathcal{M}$ and $\mathcal{M}'$ are *bisimilar*, denoted $\mathcal{M} \underline{\leftrightarrow} \mathcal{M}'$. A bisimulation between pointed models $(\mathcal{M}, s)$ and $(\mathcal{M}', s')$ is a bisimulation $\mathfrak{R}$ such that $\mathfrak{R}ss'$, notation $\mathfrak{R} : (\mathcal{M}, s) \underline{\leftrightarrow} (\mathcal{M}', s')$, and if there is such a bisimulation we write $(\mathcal{M}, s) \underline{\leftrightarrow} (\mathcal{M}', s')$. A model is *bisimulation minimal* if the only total bisimulation on that model (i.e., between the model and itself) is the identity relation.

Bisimulation Between Simplicial Models Let simplicial models $\mathcal{C} = (C, \chi, \ell)$ and $\mathcal{C}' = (C', \chi', \ell')$ be given. A non-empty relation $\mathfrak{R}$ between $\mathcal{F}(C)$ and $\mathcal{F}(C')$ is a *(simplicial) bisimulation* between $\mathcal{C}$ and $\mathcal{C}'$, notation $\mathfrak{R} : \mathcal{C} \underline{\leftrightarrow} \mathcal{C}'$, iff for all $Y \in \mathcal{F}(C)$ and $Y' \in \mathcal{F}(C')$ with $\mathfrak{R}YY'$ the following three conditions are satisfied:

- **atoms**: for all $a \in A$ and $p_a \in P_a$, $p_a \in \ell(Y)$ iff $p_a \in \ell(Y')$.
- **forth**: for all $a \in A$, if $Z \in \mathcal{F}(C)$ and $a \in \chi(Y \cap Z)$ there is a $Z' \in \mathcal{F}(C')$ with $a \in \chi(Y' \cap Z')$ such that $\mathfrak{R}ZZ'$.
- **back**: for all $a \in A$, if $Z' \in \mathcal{F}(C')$ and $a \in \chi(Y' \cap Z')$ there is a $Z \in \mathcal{F}(C)$ with $a \in \chi(Y \cap Z)$ such that $\mathfrak{R}ZZ'$.

A *total simplicial bisimulation* $\mathfrak{R}$ is a simplicial bisimulation such that for all $X \in \mathcal{F}(C)$ there is a $X' \in \mathcal{V}(C')$ with $\mathfrak{R}XX'$ and for all $X' \in \mathcal{F}(C')$ there is a $X \in \mathcal{F}(C)$ with $\mathfrak{R}XX'$. If there is a bisimulation between $\mathcal{C}$ and $\mathcal{C}'$ we write $\mathcal{C} \underline{\leftrightarrow} \mathcal{C}'$. A bisimulation between pointed simplicial models $(\mathcal{C}, X)$ and $(\mathcal{C}', X')$, where $X \in \mathcal{F}(C)$ and $X' \in \mathcal{F}(C')$, is a bisimulation $\mathfrak{R}$ such that $\mathfrak{R}XX'$, notation $\mathfrak{R} : (\mathcal{C}, X) \underline{\leftrightarrow} (\mathcal{C}', X')$, and if there is such a bisimulation we write $(\mathcal{C}, X) \underline{\leftrightarrow} (\mathcal{C}', X')$. Relation $\mathfrak{R}$ is a *simulation* if it satisfies the **atoms** and the **forth** conditions.

Intuitively, the **forth**-clause preserves ignorance when going from $\mathcal{C}$ to $\mathcal{C}'$ (typically, each facet in $\mathcal{C}$ is required to have an $\mathfrak{R}$-image in $\mathcal{C}'$, so that $\mathcal{C}$ can be viewed as a substructure of $\mathcal{C}'$, introducing more uncertainty there) and the **back**-clause preserves knowledge when going from $\mathcal{C}$ to $\mathcal{C}'$ (now, typically, each facet in $\mathcal{C}'$ is required to have an $\mathfrak{R}$-origin in $\mathcal{C}$, so that $\mathcal{C}'$ can be viewed as a substructure of $\mathcal{C}$, representing reduction of uncertainty in $\mathcal{C}'$).

A relation $\mathfrak{R} : \mathcal{F}(C) \to \mathcal{F}(C')$ between facets *induces* a (similarly denoted) colour preserving relation $\mathfrak{R} : \mathcal{V}(C) \to \mathcal{V}(C')$ between vertices by way of: if $\mathfrak{R}XX'$ then for all $a \in A$, $\mathfrak{R}X_aX'_a$. Dually, given a relation $\mathfrak{R} : \mathcal{V}(C) \to \mathcal{V}(C')$ between vertices that is *chromatic*, i.e., colour preserving (if $\mathfrak{R}vv'$ then $\chi(v) = \chi(v')$), then for any simplices X, X' with $\chi(X) = \chi(X')$, $\mathfrak{R}XX'$ denotes that for all $a \in \chi(X)$, $\mathfrak{R}X_aX'_a$.

As bisimulations between simplicial models are relations between facets, and as facets correspond to global states in epistemic models, such bisimulations are very much like bisimulations between epistemic models.

Example 9 Consider the simplicial models depicted below. (We recall the notational conventions made explicit in Example 6.) The two simplicial models on the left are (totally) bisimilar: all four facets on the far left are related to the single facet on the near left. The two simplicial models on the right are also (totally) bisimilar: the two opposed $\{a0, b0, c0\}$ facets on the near right are related to the facet $\{a0, b0, c0\}$ of the complex on the far right, and the $\{a1, b0, c0\}$ facet on the near right is related to the facet $\{a1, b0, c0\}$ on the far right.

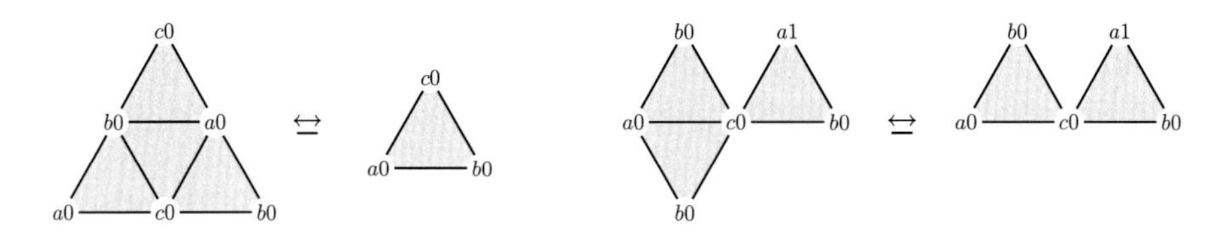

⊣

1.4.2 *Why Defining Bisimulation Between Vertices Does Not Work*

The reader may wonder why bisimulations between simplicial models are not defined between vertices instead of between facets, and with the **back** and **forth** conditions only required between vertices. This is undesirable, because if we were to do so, then different simplicial models with the same one-dimensional skeleton could incorrectly become bisimilar. Example 10 illustrates how this can go wrong. Similar counterexamples can easily be found if we merely require **back** and **forth** for simplices of dimension m smaller than the dimension n of the complex. This demonstrates that **back** and **forth** need to be required for facets.

More interestingly, now suppose bisimulations between simplicial models were defined as relations between vertices but with the **back** and **forth** conditions between facets. This is also undesirable: although a relation between facets induces a (unique) relation between vertices, different relations between facets may induce the same relation between vertices. Therefore, only specifying the relation between vertices does not determine a relation between facets. However, there is a unique *maximal* such relation, defined as: for all facets X and X', $\mathfrak{R}XX'$ iff for all $a \in A$, $\mathfrak{R}X_aX'_a$. Example 11 illustrates this.

Given a bisimulation, its induced chromatic relation between vertices interestingly compares to chromatic simplicial maps. We recall the notion of chromatic simplicial map as a colour and simplex preserving function between the vertices of a complex. Given $\mathcal{C} = (C, \chi, \ell)$ and $\mathcal{C}' = (C', \chi', \ell')$, we will call a relation $\mathfrak{R}$ between vertices *simplex preserving* if for all $X \in C$ with X in the domain of $\mathfrak{R}$ there is a $X' \in C'$ such that $\mathfrak{R}XX'$. It is easy to see that a bisimulation $\mathfrak{R}$ between simplicial models induces a relation $\mathfrak{R}$ between vertices that is simplex preserving and such that its converse relation $\mathfrak{R}^{-1} := \{(v', v) \mid (v, v') \in \mathfrak{R}\}$ is also simplex preserving:

Let $X \in C$ be such that X is in the domain of $\mathfrak{R}$, then, as C is pure, there is a $Y \in \mathcal{F}(C)$ with $X \subseteq Y$. Let now $v \in X$. As v is in the domain of $\mathfrak{R}$, there is a $Z \in \mathcal{F}(C)$ with $v \in Z$ and a $Z' \in \mathcal{F}(C')$ such that $\mathfrak{R}ZZ'$. As $v \in Z$ and $v \in Y$, $v \in Z \cap Y$. Let now $\chi(v) = a$, then from $\mathfrak{R}ZZ'$, $v \in Z \cap Y$ and **forth** it follows that there is a Y' such that $\mathfrak{R}YY'$. The set $X' \subseteq \mathcal{V}(C')$ with $\mathfrak{R}XX'$ is therefore a face of Y' and thus a simplex. Similarly, using that **back** is satisfied for $\mathfrak{R}$ we show that $\mathfrak{R}^{-1}$ is simplex preserving. It seems interesting to explore how a bisimulation can be seen as a generalization of a chromatic simplicial map. Note that the **atoms** requirement corresponds to the value preservation of the simplicial map. In Propositions 14 and 15, later, we also address the relation between bisimulations (and simulations) and simplicial maps.

Example 10 It is not sufficient only to require **forth** and **back** for lower dimensional simplices, e.g., for edges only. For a counterexample, consider the following simplicial models $\mathcal{C}$ and $\mathcal{C}'$ (and with valuations as depicted). Simplicial model $\mathcal{C}$ consists of four facets/triangles, whereas simplicial model $\mathcal{C}'$ consists of three facets (the middle triangle $\{a1, b1, c1\}$ is not a facet). The 1-skeletons of $\mathcal{C}$ and $\mathcal{C}'$ (checking **forth** and **back** only for edges of triangles) are bisimilar by way of the relation $\mathfrak{R}$ mapping vertices with the same colour and value. However, $\mathcal{C}$ and $\mathcal{C}'$ are not bisimilar, because no facet of $\mathcal{C}'$ is bisimilar to F_4 in $\mathcal{C}$.

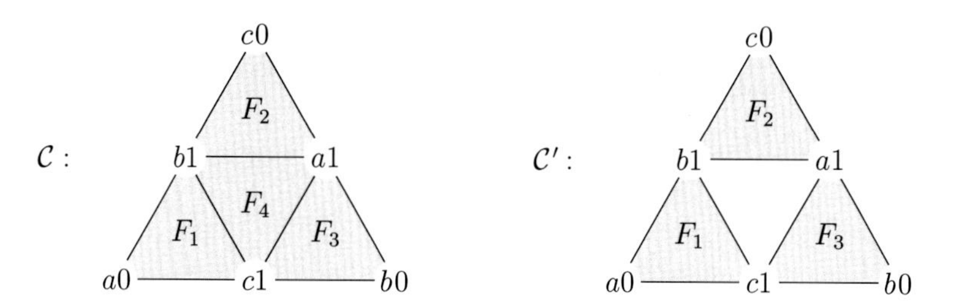

⊣

Example 11 Consider three one-dimensional complexes $\mathcal{C} = (C, \chi, \ell)$, $\mathcal{C}' = (C', \chi', \ell')$, and $\mathcal{C}'' = (C'', \chi'', \ell'')$ for two agents a, b, where we assume that a has the same value everywhere, and that b also has the same value everywhere.

Clearly, they are all bisimilar. Two different bisimulations between $\mathcal{C}$ and $\mathcal{C}'$ are:

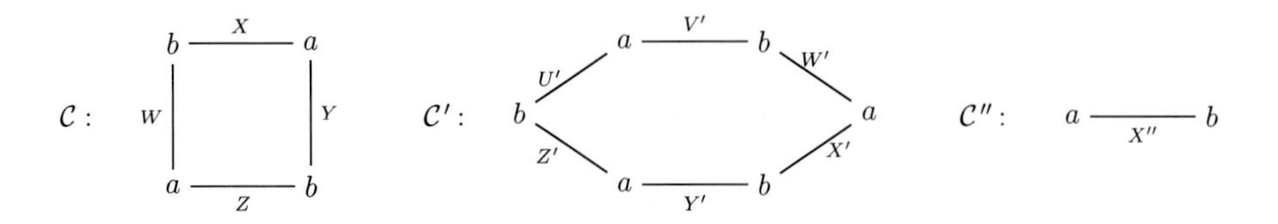

$$\begin{aligned}\mathfrak{R} &:= \{(X, V'), (X, X'), (X, Z'), (Y, U'), (Y, W'), (Y, Y'),\\ &\quad (Z, V'), (Z, X'), (Z, Z'), (W, U'), (W, W'), (W, Y')\}\\ \mathfrak{R}' &:= \mathcal{F}(C) \times \mathcal{F}(C')\end{aligned}$$

Relations $\mathfrak{R}$ and $\mathfrak{R}'$ induce the same relation between vertices, namely relating every a vertex in C to every a vertex in C', and every b vertex in C to every b vertex in C'. This demonstrates that the same relation between vertices may be consistent with different relations between facets. The relation $\mathfrak{R}'$ is the maximal bisimulation between C and C'. Assuming the relation by vertices as primitive, we can define also define $\mathfrak{R}'$ as: for all facets X and X', $\mathfrak{R}'XX'$ iff ($\mathfrak{R}'X_aX'_a$ and $\mathfrak{R}'X_bX'_b$).

Both C and C' are bisimilar to the single-edged complex C''. This complex is bisimulation minimal (C'' can be seen as the quotient of C, and of C', with respect to bisimilarity). ⊣

1.4.3 Elementary Results for Bisimulation

We continue with some elementary results for bisimulations between complexes. We recall the maps $\kappa : \mathcal{S} \to \mathcal{K}$ and $\sigma : \mathcal{K} \to \mathcal{S}$ between the simplicial models and the proper local epistemic models. We now have the following.

Proposition 12 *Let $\mathcal{M}, \mathcal{M}' \in \mathcal{K}$, let $C, C' \in \mathcal{S}$. Then:*

- *If $\mathcal{M} \underline{\leftrightarrow} \mathcal{M}'$, then $\sigma(\mathcal{M}) \underline{\leftrightarrow} \sigma(\mathcal{M}')$.*
- *If $C \underline{\leftrightarrow} C'$, then $\kappa(C) \underline{\leftrightarrow} \kappa(C')$.*
- *If $\mathcal{M} \underline{\leftrightarrow} \mathcal{M}'$, then $\kappa(\sigma(\mathcal{M})) \underline{\leftrightarrow} \kappa(\sigma(\mathcal{M}'))$.*
- *If $C \underline{\leftrightarrow} C'$, then $\sigma(\kappa(C)) \underline{\leftrightarrow} \sigma(\kappa(C'))$.*

⊣

Proof We show the first item. Let $\mathcal{M} = (S, \sim, L)$ and $\mathcal{M}' = (S', \sim', L')$ be local proper epistemic models.

Let $\mathfrak{R} : \mathcal{M} \underline{\leftrightarrow} \mathcal{M}'$. We define relation $\mathfrak{R}'$ between the facets $\sigma(s) = \{[s]_a \mid a \in A\}$ of $\sigma(\mathcal{M})$ and the facets $\sigma(s') = \{[s']_a \mid a \in A\}$ of $\sigma(\mathcal{M}')$ as: $\mathfrak{R}'\sigma(s)\sigma(s')$ iff $\mathfrak{R}ss'$. Note that this induces that for all $a \in A$, $\mathfrak{R}'[s]_a[s']_a$ iff $\mathfrak{R}ss'$.

We now show that $\mathfrak{R}'$ is a bisimulation. Let $\mathfrak{R}'\sigma(s)\sigma(s')$.

The clause **atoms** holds: For all $[s]_a \in \sigma(s)$, $[s']_a \in \sigma(s')$ and $p_a \in P$, we have that $p_a \in \ell([s]_a)$ iff $p_a \in L(s)$ iff (using **atoms** for $\mathfrak{R}$, given that $\mathfrak{R}ss'$ follows from $\mathfrak{R}'\sigma(s)\sigma(s')$) $p_a \in L'(s')$ iff $p_a \in \ell([s]_a)$.

The clause **forth** holds: Let $\sigma(t) \in \mathcal{F}(\sigma(\mathcal{M}))$ such that $a \in \chi(\sigma(s) \cap \sigma(t))$. From the definition of $\sigma(\mathcal{M})$ we get that $s \sim_a t$. Therefore, given that $\mathfrak{R}'\sigma(s)\sigma(s')$ so that also $\mathfrak{R}ss'$, and as $\mathfrak{R}$ is a bisimulation, there is $t' \in \mathcal{M}'$ such that $s' \sim_a t'$ and $\mathfrak{R}tt'$. We now choose $\sigma(t') \in \mathcal{F}(\sigma(\mathcal{M}'))$ to obtain $\mathfrak{R}'\sigma(t)\sigma(t')$ and $a \in \chi(\sigma(s') \cap \sigma(t'))$, as required.

The **back** step is similar.

The proof of the second item is also fairly similar.

The last two items follow from the first two items, but also directly from $\mathcal{M} \simeq \kappa(\sigma(\mathcal{M}))$ and $\mathcal{C} \simeq \sigma(\kappa(\mathcal{M}))$: isomorphy is stronger than bisimilarity, and bisimilarity is transitive. □

These results extend the known results that $\kappa(\sigma(\mathcal{M})) \simeq \mathcal{M}$ and that $\sigma(\kappa(\mathcal{C})) \simeq \mathcal{C}$ from Goubault et al. (2021) (Proposition 7) in the expected way. Clearly, we also have such correspondence between pointed simplicial complexes, for example, $\sigma(\kappa(\mathcal{C}, X)) \underline{\leftrightarrow} \sigma(\kappa(\mathcal{C}', X'))$, where X is a facet in the complex of $\mathcal{C}$ and X' is a facet in the complex of $\mathcal{C}'$.

Also for simplicial complexes, bisimilarity implies modal equivalence, and for finite complexes the implication also holds in the other direction, as is to be expected. It is often assumed in combinatorial topology and in distributed computing that complexes are finite, but not always. Interesting applications exist for the infinite case (Aguilera 2004).

Proposition 13 *Let* $(\mathcal{C}, X)$ *and* $(\mathcal{C}', X')$ *be given, with* $\mathcal{C} = (C, \chi, \ell)$ *and* $\mathcal{C}' = (C', \chi', \ell')$. *Then:*

- $(\mathcal{C}, X) \underline{\leftrightarrow} (\mathcal{C}', X')$ *implies* $(\mathcal{C}, X) \equiv (\mathcal{C}', X')$.

Let now, additionally, the sets of vertices of $\mathcal{C}$ *and* $\mathcal{C}'$ *be finite. Then:*

- $(\mathcal{C}, X) \equiv (\mathcal{C}', X')$ *implies* $(\mathcal{C}, X) \underline{\leftrightarrow} (\mathcal{C}', X')$. ⊣

Proof These proofs are elementary. Let us show the second item.

Define relation $\mathfrak{R} : \mathcal{F}(C) \times \mathcal{F}(C')$ as: for all $X \in \mathcal{F}(C)$ and $X' \in \mathcal{F}(C')$, $\mathfrak{R}XX'$ if $(\mathcal{C}, X) \equiv (\mathcal{C}', X')$. We show that $\mathfrak{R}$ is a bisimulation.

Let $\mathfrak{R}XX'$ be arbitrary.

First consider condition **atoms**. For all $p_a \in P$, $p_a \in \ell(X)$ iff $p_a \in \ell'(X')$, because $(\mathcal{C}, X) \equiv (\mathcal{C}', X')$ implies that $(\mathcal{C}, X) \models p_a$ iff $\mathcal{C}', X' \models p_a$.

Now consider **forth**. Let $Y \in \mathcal{F}(C)$ be such that $a \in \chi(X \cap Y)$. Consider $\boldsymbol{Y}' := \{Y' \in \mathcal{F}(C') \mid a \in \chi(X' \cap Y')\}$. A $\mathcal{C}'$ is finite, $\boldsymbol{Y}'$ is finite, let $\boldsymbol{Y}' = \{Y'_1, \dots, Y'_n\}$. Now suppose towards a contradiction that Y is not in the relation $\mathfrak{R}$ with any of the $Y'_1, \dots, Y'_n$, so that therefore $(\mathcal{C}, Y) \not\equiv (\mathcal{C}', Y'_1), \dots, (\mathcal{C}, Y) \not\equiv (\mathcal{C}', Y'_n)$. Then there are $\varphi_1, \dots, \varphi_n$ such that $\mathcal{C}', Y'_1 \models \varphi_1$ but $\mathcal{C}, Y \not\models \varphi_1, \dots, \mathcal{C}', Y'_n \models \varphi_n$ but $\mathcal{C}, Y \not\models \varphi_n$. Using the semantics of knowledge, it follows that $\mathcal{C}', X' \models K_a(\varphi_1 \vee \dots \vee \varphi_n)$

whereas $\mathcal{C}, X \not\models K_a(\varphi_1 \vee \cdots \vee \varphi_n)$. Therefore it is not the case that $(\mathcal{C}, X) \equiv (\mathcal{C}', X')$. This contradicts the assumption $\mathfrak{R}XX'$.

The condition **back** is shown similarly. □

Given complexes C and C', let $f : \mathcal{V}(C) \to \mathcal{V}(C')$ be a chromatic simplicial map. The relation $\mathfrak{R} \subseteq \mathcal{V}(C) \times \mathcal{V}(C')$ defined as, for all $x \in \mathcal{V}(C)$ and $x' \in \mathcal{V}(C')$, $\mathfrak{R}xy$ iff $f(x) = y$ is called the *induced* relation. Also, as $\mathfrak{R}$ is functional, this, in its turn, induces a unique relation $\mathfrak{R}$ between facets (in the usual way). We recall that a simplicial map is value preserving between simplicial models iff for all $v \in \mathcal{V}(C)$, $\ell'(f(v)) = \ell(v)$ (where ℓ is the valuation on C and ℓ' that on C'). We now have that:

Proposition 14 *The induced relation $\mathfrak{R}$ of a value preserving chromatic simplicial map f between simplicial models $\mathcal{C}$ and $\mathcal{C}'$ is a simulation.* ⊣

Proof The proof is elementary. □

As a direct consequence of Proposition 14, if both f and f^{-1} are value preserving chromatic simplicial maps, then the induced relation $\mathfrak{R}$, that determines a unique relation between facets as it is functional, is a bisimulation between $\mathcal{C}$ and $\mathcal{C}'$. This bisimulation is an isomorphism.

1.4.4 Subdivision and Bisimulation

Subdivisions typically do not result in bisimilar complexes.[2]Consider the two simplicial models for agents a, b below, where the edge with labels 1 is subdivided into three edges, while the other two edges of this simplicial model are not subdivided. These two simplicial models are not bisimilar. Clearly, because of **atoms**, the facet $a1$——$b1$ on the left must be in relation with a facet $a1$——$b1$ on the right. Which one? If we choose $a1$——$\boldsymbol{b1}$ on the right, then **forth** fails because the adjoining $b1$——$a2$ on the left cannot be related to $\boldsymbol{b1}$——$\boldsymbol{a1}$ on the right, as the values of the a vertices are different. If we choose $\boldsymbol{b1}$——$\boldsymbol{a1}$ on the right, then **forth** fails because the adjoining $b0$——$a1$ on the left cannot be related to $\boldsymbol{a1}$——$b1$ on the right, as the values of the b vertices are different. Finally and again similarly, if we choose $\boldsymbol{a1}$——$b1$ on the right, then **back** fails, as $b0$——$a1$ and $\boldsymbol{b1}$——$a1$ are not related.

We can also come to this conclusion in a different way: if the respective complexes have different properties of knowledge, then they are not bisimilar (Proposition 13, later). We now note, for example, that in the complex on the right in vertex $\boldsymbol{b1}$ agent b knows that the value of a is 1, whereas on the left agent b

[2] In a subdivision of a (pure chromatic) simplicial complex we replace simplexes by sets of simplexes of the same dimension. For general definitions see Herlihy et al. (2013).

is uncertain about the value of a if his value is 1. Similarly, $\boldsymbol{a1}$ on the right is the unique vertex where a knows that the value of b is 1.

$$b0 \;\text{———}\; a1 \;\text{———}\; b1 \;\text{———}\; a2 \qquad \not\underline{\leftrightarrow} \qquad b0 \;\text{———}\; a1 \;\text{———}\; \boldsymbol{b1} \;\text{———}\; \boldsymbol{a1} \;\text{———}\; b1 \;\text{———}\; a2$$

However, sometimes subdivisions result in bisimilar complexes. For example, if, instead, $a1$——$b1$ were the unique edge of a simplicial model, then its subdivision must be bisimilar, as below:

$$a1 \;\text{———}\; b1 \qquad \underline{\leftrightarrow} \qquad a1 \;\text{———}\; \boldsymbol{b1} \;\text{———}\; \boldsymbol{a1} \;\text{———}\; b1$$

Similarly, for dimension 2, the left complex below consists of a single facet, and it is bisimilar to its subdivision on the right (by connecting vertices with the same colour), where we assume that all vertices of the same colour satisfy the same local variable(s).

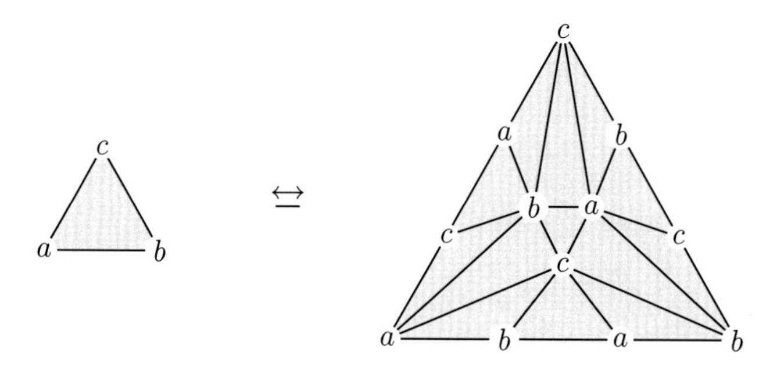

Bisimulation may be a useful tool to determine whether seemingly similar simplicial complexes contain the same information. Algorithms determining whether given models are bisimilar are presented in for example Paige and Tarjan (1987) and Fisler and Vardi (2002). Another use may be to determine whether, given some initial complex, clearly different subdivisions (where agents have different knowledge, such as a knows p in one but never in the other, and b knows q in the other but never in the one) both have subsequent subdivisions resulting, after all, in bisimilar complexes again. This is the property known as confluence, or Church-Rosser. Isomorphy is then often a bridge too far, but bisimilarity may be all we need.

1.4.5 Covering Complex and Bisimulation

As explained by Hatcher (2002), algebraic topology can be roughly defined as the study of techniques for forming algebraic images of topological spaces. Thus, continuous maps between spaces are projected onto homomorphisms between their algebraic images, so topologically related spaces have algebraically related images. One of the most simple and important functors in algebraic topology is the *fundamental group*. It creates an algebraic image of a space from the loops in the space (i.e., paths starting and ending at the same point). The fundamental group of a space $\boldsymbol{X}$ is defined so that its elements are loops in $\boldsymbol{X}$ starting and ending at a fixed basepoint $x_0 \in \boldsymbol{X}$, but two such loops are regarded as determining the same element of the fundamental group if one loop can be continuously deformed into the other within the space $\boldsymbol{X}$. There is a very deep connection between the fundamental group and *covering spaces*, and they can be regarded as two viewpoints toward the same thing (Rotman 1973). This means that algebraic features of the fundamental group can often be translated into the geometric language of covering spaces. Roughly speaking, a space $\boldsymbol{Y}$ is called a covering space of $\boldsymbol{X}$ if $\boldsymbol{Y}$ maps onto $\boldsymbol{X}$ in a locally homeomorphic way, so that the pre-image of every point in $\boldsymbol{X}$ has the same cardinality.

We consider the combinatorial version of covering complex (Rotman 1973), extended to the epistemic setting. Let $\mathcal{C} = (C, \chi, \ell)$ and $\mathcal{C}' = (C', \chi', \ell')$ be simplicial models. An *epistemic covering complex* of $\mathcal{C}'$ is a pair $(\mathcal{C}, f)$ where C is a connected complex and $f : \mathcal{C} \to \mathcal{C}'$ is a value preserving simplicial map such that for every simplex X in C', $f^{-1}(X)$ is a union of pairwise disjoint simplices. We observe that C' is the image of f. Hence, C' is also connected.

Applications of covering spaces to distributed computing are explored in Fraigniaud et al. (2013), and this is an example. In the figure we assume that a vertex v and $f(v)$ have the same valuations, and an edge in C is sent by f to the edge of the same label in C'. Note that in the picture below, agent names are represented as colors (black and white), and labels on the edges are just here to specify the map f (Fig. 1.6).

Proposition 15 *If $(\mathcal{C}, f)$ is an epistemic covering complex of $\mathcal{C}'$ then there is a total bisimulation between $\mathcal{C}$ and $\mathcal{C}'$.* ⊣

Proof By Proposition 14 the induced relation $\mathfrak{R}$ of $f : \mathcal{C} \to \mathcal{C}'$ is a simulation from $\mathcal{C}$ to $\mathcal{C}'$. It is easy to see that the **back** property holds as well. □

1.5 Local Semantics for Simplicial Complexes

In Goubault et al. (2021) the semantics of knowledge on simplicial complexes employs a satisfaction relation between a pair consisting of a simplicial model and a facet of that model, and a formula. However, in combinatorial topology we not only

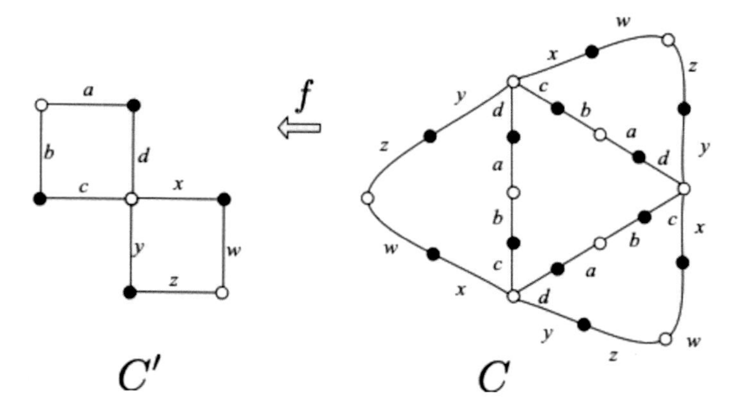

Fig. 1.6 Covering complex and bisimulation

wish to determine what is true in a facet, but also what is true in any simplex. We wish a more *local* semantics. There are two roads towards such local semantics for simplicial complexes. There is another road from a local semantics for simplices to the special case of facets. The purpose of these different perspectives on the semantics is that they give us more modelling flexibility, for example when we reason (and when the intuitions that we wish to formalize are) about faces or vertices instead of facets.

1.5.1 *From Semantics for Facets to Semantics for Simplices*

Multi-Pointed Local Semantics Just as in Kripke semantics we can define the satisfaction relation between multi-pointed simplicial models and formulas, instead of between pointed simplicial complexes and formulas. We then use that for any face of dimension $m < n$, there is a unique set of facets that contain it (we recall that we only consider pure complexes).

A multi-pointed simplicial model $(\mathcal{C}, \boldsymbol{X})$ is a pair consisting of a simplicial model $\mathcal{C} = (C, \chi, \ell)$ and a set $\boldsymbol{X}$ of facets $X \in \mathcal{F}(C)$. The satisfaction relation for such sets of facets is easily defined in the obvious way.

$$\mathcal{C}, \boldsymbol{X} \models \varphi \text{ iff } \mathcal{C}, X \models \varphi \text{ for all } X \in \boldsymbol{X}$$

Consequently, given any simplex $Y \subseteq C$ of $\mathcal{C}$, and the set $\boldsymbol{X} := \{X \in \mathcal{F}(C) \mid Y \subseteq X\}$ of facets containing it as a face, we can define the local truth in Y as follows.

$$\mathcal{C}, Y \models \varphi \text{ iff } \mathcal{C}, X \models \varphi \text{ for all } X \in \mathcal{F}(C) \text{ with } Y \subseteq X$$

Example 16 Consider the simplicial model depicted below, where we recall that for the convenience of the exposition the vertices have been labelled with the colour/agent i and the value of the single local proposition p_i for that agent, where 0 means that p_i is false and 1 means that p_i is true, and where as names of these vertices we use such pairs. The three facets of the complex are $X = \{a0, c1, b1\}$, $Y = \{a1, c1, b1\}$, and $Z = \{a1, c1, b0\}$.

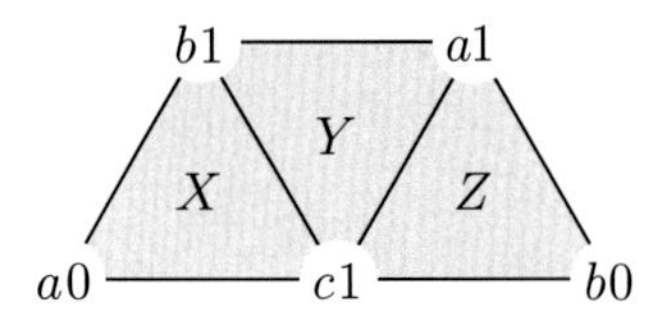

We now have that $\mathcal{C}, X \models K_a \neg p_a$, but $\mathcal{C}, \{b1, c1\} \not\models K_a \neg p_a$. As $\{b1, c1\} = X \cap Y$, $\mathcal{C}, \{b1, c1\} \models \varphi$ is equivalent to ($\mathcal{C}, X \models \varphi$ and $\mathcal{C}, Y \models \varphi$). And from $\mathcal{C}, Y \models p_a$ it follows that $\mathcal{C}, Y \not\models K_a \neg p_a$, and thus $\mathcal{C}, \{b1, c1\} \not\models K_a \neg p_a$. The edge $\{b1, c1\}$ represents that agents b and c are uncertain about what a knows.

Similarly, in vertex $c1$, agent c, who knows there that p_c, is uncertain between valuations $\neg p_a \wedge p_b \wedge p_c$, $p_a \wedge p_b \wedge p_c$, and $p_a \wedge \neg p_b \wedge p_c$. This is described by $\mathcal{C}, c1 \models K_c((\neg p_a \wedge p_b \wedge p_c) \vee (p_a \wedge p_b \wedge p_c) \vee (p_a \wedge \neg p_b \wedge p_c)) \wedge \widehat{K}_c(\neg p_a \wedge p_b \wedge p_c) \wedge \widehat{K}_c(p_a \wedge p_b \wedge p_c) \wedge \widehat{K}_c(p_a \wedge \neg p_b \wedge p_c)$. In particular, from this it also follows that $\mathcal{C}, c1 \models K_c(p_a \vee p_b)$. Although c considers it possible that p_a is false, and also considers it possible that p_b is false, she does not consider it possible that both are false. $\dashv$

Language-Restricted Local Semantics There is also another, slightly more winding, road to a local semantics for simplicial models. This involves restricting the language as well: when interpreting a formula in a simplex, we then only allow knowledge modalities and local variables for the colours of that simplex. Given the language $\mathcal{L}_K(A, P)$, where $P = \bigcup_{a \in A} P_a$, and given a simplicial model $\mathcal{C} = (C, \chi, \ell)$ and a simplex $X \in C$, satisfaction $\mathcal{C}, X \models \varphi$ is only defined for formulas φ in the language restricted to agents in $\chi(X)$ and to local variables for those agents. In other words, we then only formalize what the agents in $\chi(X)$ know and do not know about themselves and about their own local variables, and not what they know about other agents or their local variables, or about what those other agents know. (Similar restrictions abound in game theory and computational social choice; Brandt et al. 2016.) This comes with the following semantics.

The relation $\models$ between a pair $(\mathcal{C}, X)$ and φ is defined for $\varphi \in \mathcal{L}_K(\chi(X), P|\chi(X))$, where $P|\chi(X) := \bigcup_{a \in \chi(X)} P_a$, by induction on the structure of φ. The non-obvious two clauses are as follows. The language restriction implies that, below, the a in p_a and in K_a must be in the set of colours $\chi(X)$.

$$\mathcal{C}, X \models p_a \quad \text{iff } p_a \in \ell(X)$$
$$\mathcal{C}, X \models K_a \varphi \text{ iff } \mathcal{C}, Y \models \varphi \text{ for all } X \in C \text{ with } \chi(X) = \chi(Y) \text{ and } a \in \chi(X \cap Y)$$

Instead of defining this syntactically, we can also define this semantically, as follows.

Let $\mathcal{C} = (C, \chi, \ell)$ be a simplicial model for agents A, and $A' \subseteq A$ a subset of agents. The *restriction* $\mathcal{C}|A'$ *of* $\mathcal{C}$ *to* A' is the simplicial model obtained from $\mathcal{C}$ by keeping only the vertices coloured by A', i.e., $\mathcal{C}|A' = (C', \chi', \ell')$ where $C' = \{X \in C \mid \chi(X) \subseteq A'\}$, where for all $X \in C'$, $\chi'(X) = \chi(X)$, and where for all $p_a \in P_a$ with $a \in A'$ and for all $v \in \mathcal{V}(C')$, $p_a \in \ell'(v)$ iff $p_a \in \ell(v)$. We note that C' is again a pure chromatic simplicial complex and that its dimension is $|A'| - 1$. The language restricted local semantics above is just the regular semantics in $\mathcal{C}|A'$, for formulas in the language $\mathcal{L}_K(A', P|A')$. In other words, for all $\varphi \in \mathcal{L}_K(A', P|A')$:

$$\mathcal{C}, X \models \varphi \quad \text{iff} \quad \mathcal{C}|\chi(X), X \models \varphi$$

Example 17 Reconsider Example 16. We now have that $\mathcal{C}, \{b1, c1\} \models K_a \neg p_a$ and $\mathcal{C}, c1 \models K_c(p_a \vee p_b)$ are undefined, because $a \notin \chi(\{b1, c1\}) = \{b, c\}$ respectively $a \notin \chi(\{c1\})$, i.e., $a \neq c$. However, we have that $\mathcal{C}, \{b1, c1\} \models K_b p_c \wedge \neg K_c p_b$ and, less interestingly, as this is 'all that c knows', that $\mathcal{C}, c1 \models K_c p_c$. ⊣

Let us write $\models_{mp}$ for the satisfaction relation in the 'multi-pointed local semantics' and $\models_{lr}$ for that in the 'language-restricted local semantics'. Then truth with respect to the latter is preserved as truth with respect the former, which is desirable.

Proposition 18 *Let* $\varphi \in \mathcal{L}_K(X)$. *Then* $\mathcal{C}, X \models_{lr} \varphi$ *iff* $\mathcal{C}, X \models_{mp} \varphi$. ⊣

Proof The proof is by induction on φ where the only relevant case is the one for knowledge:

Let us assume that $\mathcal{C}, X \models_{lr} K_a\varphi$, i.e., for all $Y \in C$ with $\chi(X) = \chi(Y)$ and $a \in \chi(X \cap Y)$, $\mathcal{C}, Y \models_{lr} \varphi$. We wish to prove that $\mathcal{C}, X \models_{mp} K_a\varphi$. In order to prove that, using the $\models_{mp}$ semantics, assume that $Z \in \mathcal{F}(C)$ with $X \subseteq Z$, and, using the semantics for knowledge, let $V \in \mathcal{F}(C)$ with $a \in \chi(Z \cap V)$. It then remains to show that $\mathcal{C}, V \models_{mp} \varphi$. Let now $Y \subseteq V$ with $\chi(Y) = \chi(X)$. Note that, as $a \in \chi(X)$, therefore also $a \in \chi(Y)$. From $X \subseteq Z$, $Y \subseteq V$, $a \in \chi(Z \cap V)$, $a \in \chi(X)$, and $a \in \chi(Y)$ it follows that $a \in \chi(X \cap Y)$. From that, the initial assumption $\mathcal{C}, X \models_{lr} K_a\varphi$ and the $\models_{lr}$ semantics it now follows that $\mathcal{C}, Y \models_{lr} \varphi$. By inductive hypothesis it now follows that $\mathcal{C}, Y \models_{mp} \varphi$, and from the $\models_{mp}$ semantics and $Y \subseteq V \in \mathcal{F}(C)$ it then follows that $\mathcal{C}, V \models_{mp} \varphi$, as required.

Conversely, assume that $\mathcal{C}, X \models_{mp} K_a\varphi$. Suppose that $Y \in C$ with $\chi(X) = \chi(Y)$ and $a \in \chi(X \cap Y)$. Let $Z \in \mathcal{F}(C)$ with $X \subseteq Z$, then from the assumption $\mathcal{C}, X \models_{mp} K_a\varphi$ we obtain $\mathcal{C}, Z \models_{mp} K_a\varphi$. Therefore, for all $V \in \mathcal{F}(C)$ with $a \in \chi(Z \cap V)$, $\mathcal{C}, V \models_{mp} \varphi$. Similar to the reasoning in the other direction, for any $Y \subseteq V$ with $\chi(Y) = \chi(X)$ and $a \in \chi(Y)$ we must have that $a \in \chi(X \cap Y)$, such that by definition of $\models_{mp}$ we obtain $\mathcal{C}, Y \models_{mp} \varphi$, and with induction we can then conclude $\mathcal{C}, Y \models_{lp} \varphi$, as required. □

Local Semantics for Distributed Epistemic Models One can just as well consider a similar local semantics for distributed epistemic models. Let $\mathcal{M} = (S, \sim, L)$ be a distributed model for agents A and variables $P = \bigcup_{a \in A} P_a$, so that states $s \in S$ have shape $s = (s_0, \dots, s_n)$. Given such $s \in S$, by $s \subseteq_B s'$ we mean that s is the restriction of s' to the agents $B \subseteq A$. For example, $(s_1, s_2) \subseteq_{\{a_1,a_2\}} (s_0, s_1, s_2)$. We now can, similarly to above, either define, in multi-pointed fashion, that $\mathcal{M}, s|B \models \varphi$ iff (for all $s' \in S$ with $s'|B = s|B$: $\mathcal{M}, s' \models \varphi$), or define, in restricted-language fashion, $\mathcal{M}, s|B \models \varphi$ by induction on $\varphi \in \mathcal{L}(B, P|B)$. Again, truth in the latter is preserved as truth in the former.

1.5.2 A Semantics for Simplices Including Facets

A more rigorous departure from the semantics where the evaluation point is a facet are the following alternative semantics for arbitrary simplices as evaluation points, of which the case for facets is merely an instantiation.

Given are agents A and variables $P = \bigcup_{a \in A} P_a$. Let $\mathcal{C} = (C, \chi, \ell)$ be a simplicial model, $X \in C$, and $\varphi \in \mathcal{L}(A, P)$. Similar to before for facets, we extend the usage of ℓ to include arbitrary simplices as: $p \in \ell(X)$ iff there is a $v \in X$ with $p \in \ell(v)$. Expression $\mathcal{C}, X \models \varphi$ is defined for $\varphi \in \langle(\chi(X), P|\chi(X))$.

$$
\begin{array}{ll}
\mathcal{C}, X \models p_a & \text{iff } p_a \in \ell(X) \\
\mathcal{C}, X \models \varphi \wedge \psi & \text{iff } \mathcal{C}, X \models \varphi \text{ and } \mathcal{C}, X \models \psi \\
\mathcal{C}, X \models \neg\varphi & \text{iff } \mathcal{C}, X \not\models \varphi \\
\mathcal{C}, X \models K_a\varphi & \text{iff } \mathcal{C}, Y \models \varphi \text{ for all } Y \in \mathcal{F}(C) \text{ with } a \in \chi(X \cap Y)
\end{array}
$$

The special case of the knowledge semantics for vertices therefore is:

$$
\mathcal{C}, v \models K_a\varphi \text{ iff } \mathcal{C}, Y \models \varphi \text{ for all } Y \in \mathcal{F}(C) \text{ with } v \in Y
$$

Elementary results corroborating this more general perspective of this semantics are:

Proposition 19 *If $\mathcal{C}, X \models \varphi$ and $Y \in C$ such that $X \subseteq Y$, then $\mathcal{C}, Y \models \varphi$.* $\dashv$

Proposition 20 *If $\mathcal{C}, X \models \varphi$, $Y \in C$ such that $Y \subseteq X$ and $\varphi \in \mathcal{L}(\chi(Y))$, then $\mathcal{C}, Y \models \varphi$.* $\dashv$

Both are easily shown by induction on φ. Proposition 19 says in other words that if φ is true in simplex X then it is true in what is known in combinatorial topology as the *star* of X.

In subsequent sections we will frequently shift between the different semantic perspectives defined in this section, in order to explain the intuitions behind novel concepts.

1.6 Group Epistemic Notions for Simplicial Complexes

1.6.1 Mutual, Common, and Distributed Knowledge

Apart from individual knowledge, group knowledge plays an important role in epistemic logic. There are three well-known kinds of group knowledge. In the first place, there is the notion representing that *everybody knows* a proposition, also known as *mutual knowledge* (Meyer and van der Hoek 1995; Fagin et al. 1995; Osborne and Rubinstein 1994).[3] Then, there is the notion of *common knowledge* of a proposition, which entails not merely that everybody knows it, but also that everybody knows that everybody knows it, and so on (Friedell 1969; Lewis 1969; van Ditmarsch et al. 2009). Finally, the agents have *distributed knowledge* of a proposition if, intuitively (but not technically), they can get to know it by communication (Hayek 1945; Halpern and Moses 1990).[4] This notion is less well studied than the other two. This is somewhat surprising, particularly in the context of distributed systems with communicating agents.

Syntax The notions of mutual, common, and distributed knowledge are formalized by operators E_B, C_B, and D_B, respectively. Mutual knowledge $E_B\varphi$, for $B \subseteq A$, is typically defined by abbreviation as $\bigwedge_{a\in B} K_a\varphi$ and only introduced as a primitive language construct for investigations of succinctness and complexity. Whereas $C_B\varphi$ and $D_B\varphi$ need to be additional inductive constructs in the logical language in order to be able to give a semantics for these notions. Expression $D_B\varphi$ stands for 'the group of agents B have distributed knowledge of φ,' and expression $C_B\varphi$ stands for 'the group of agents B have common knowledge of φ.' Both common knowledge and distributed knowledge are known to enlarge the expressivity of corresponding logics, although in different ways, and to handle distributed knowledge we even need a stronger notion of bisimulation, that may serve us surprisingly well for a generalization of simplicial complexes.

Semantics Given some epistemic model with relations $\sim_a$ for $a \in A$, we define $\sim_B^\cap \;:=\; \bigcap_{a\in B} \sim_a$, and $\sim_B^* \;:=\; (\bigcup_{a\in B} \sim_a)^*$, where $B \subseteq A$ (and where for any relation R, R^* is its transitive and reflexive closure). Both these relations are also equivalence relations. The semantics of common knowledge and of distributed knowledge on epistemic models is now as follows.

---

[3] There is no agreement on terminology here. The notion that everybody knows φ (which is unambiguous) is in different communities called: *shared* knowledge, *mutual* knowledge, *general* knowledge; where in some communities some of these terms mean common knowledge instead, creating further confusion.

[4] The intuition puts one on the wrong foot for distributedly known ignorance: it may be distributed knowledge between a and b that 'p is true and b is ignorant of p', but this cannot be made common knowledge between them. If a were to inform b of this, they would then have common knowledge between them of p, but they would not have common knowledge between them of 'p is true and b does not know p'.

$$\mathcal{M}, s \models D_B\varphi \text{ iff } \mathcal{M}, t \models \varphi \text{ for all } t \text{ with } s \sim_B^{\cap} t$$
$$\mathcal{M}, s \models C_B\varphi \text{ iff } \mathcal{M}, t \models \varphi \text{ for all } t \text{ with } s \sim_B^{*} t$$

It is straightforward to interpret distributed and common knowledge on simplicial models. Let $\mathcal{C} = (C, \chi, \ell)$ be a simplicial model (for A and P), and let $X \in \mathcal{F}(C)$. To interpret common knowledge C_B, define operation $*_B$ for all $B \subseteq A$ inductively as: for all $X, Y, Z \in \mathcal{F}(C)$, $X \in *_B(X)$, and if $Y \in *_B(X)$ and $B \cap \chi(Z \cap Y) \neq \emptyset$, then $Z \in *_B(X)$. Then:

$$\mathcal{C}, X \models D_B\varphi \text{ iff } \mathcal{C}, Y \models \varphi \text{ for all } Y \in \mathcal{F}(C) \text{ such that } B \subseteq \chi(X \cap Y)$$
$$\mathcal{C}, X \models C_B\varphi \text{ iff } \mathcal{C}, Y \models \varphi \text{ for all } Y \in *_B(X)$$

Clearly, for the set of all agents A, $D_A\varphi$ is equivalent to φ given the distributed nature of a simplicial model: all agents together have distributed knowledge of the actual facet, as they colour it. However, when $B \subset A$ then $D_B\varphi$ is typically not equivalent to φ.

Example 21 We recall the simplicial models $\mathcal{C}$ and $\mathcal{C}'$ from Example 10. We now evaluate some propositions involving common and distributed knowledge.

- Let us investigate what a and b commonly know in facet F_1 in $\mathcal{C}'$. To determine common knowledge of a and b in F_1, the commonly known proposition needs to be true in all F_i bordering on edges of the ab-path: $a0$—$b1$—$a1$—$b0$, i.e., in all of F_1, F_2, F_3. In other words, it has to be (in this particular example) a model validity.
  We now have that $\mathcal{C}', F_1 \models C_{ab}(K_c p_c \rightarrow (K_a \neg p_a \vee K_b \neg p_b))$, as the formula bound by C_{ab} is true for all F_1, F_2, F_3. However, $\mathcal{C}, F_1 \not\models C_{ab}(K_c p_c \rightarrow (K_a \neg p_a \vee K_b \neg p_b))$, as $\mathcal{C}, F_4 \not\models K_c p_c \rightarrow (K_a \neg p_a \vee K_b \neg p_b)$. For another example, $\mathcal{C}, F_1 \not\models C_{ab}(K_a \neg p_a \vee K_b \neg p_b)$, namely not in F_2 where c knows $\neg p_c$. And for yet another example we have that $\mathcal{C}, F_2 \models K_c(K_a p_a \wedge K_b p_b)$ whereas $\mathcal{C}, F_2 \not\models C_{ab}(K_a p_a \wedge K_b p_b)$: in F_2 agent c knows more about $K_a p_a \wedge K_b p_b$ than what agents a and b commonly know there.[5]
- In a facet of a simplicial model the agents all together always have distributed knowledge of the valuation in that facet (or of anything else that is true on the facet). For example, $\mathcal{C}, F_4 \models D_{abc}(p_a \wedge p_b \wedge p_c)$. On the other hand, $\mathcal{C}, F_4 \not\models D_{ac}(p_a \wedge p_b \wedge p_c)$, as a and c are both uncertain about the value of b. Still their distributed knowledge in F_4 (of positive formulas) is larger than that of each of them separately: the distributed knowledge of a and c in F_4 is what is true in F_4, F_3; whereas, what a knows in F_4 is what is true in F_2, F_4, F_3, and what c knows in F_4 is what is true in F_1, F_3, F_4. ⊣

[5] Local semantics can also be given for distributed knowledge and for common knowledge. For example, to determine what a and b commonly know on the edge $\{a0, b1\}$ we need to consider facets bordering on the chain $a0|b1|a1|b0$ only. For example, $\mathcal{C}, \{a0, b1\} \models K_a \neg p_a$ whereas $\mathcal{C}, \{a0, b1\} \not\models C_{ab} \neg p_a$.

1.6.2 Common Distributed Knowledge for Simplicial Complexes

Beyond applying the standard notions of group knowledge for epistemic models on simplicial models, we can also consider novel notions of group knowledge for simplicial models, for example, to express dimensional properties of complexes. Consider the following notion in between common knowledge and distributed knowledge.

Let $\mathcal{B} \subseteq \mathcal{P}(A)$. We extend the language with an inductive construct $CB_{\mathcal{B}}\varphi$, for "it is *common distributed knowledge* that φ (with respect to $\mathcal{B}$)." Given epistemic model $\mathcal{M} = (S, \sim, L)$, let $\sim^{*\cap}_{\mathcal{B}} := (\bigcup_{B \in \mathcal{B}} \sim^{\cap}_{B})^*$. This relation interprets operator $CB_{\mathcal{B}}\varphi$ as follows.

$$\mathcal{M}, s \models CD_{\mathcal{B}}\varphi \text{ iff } \mathcal{M}, t \models \varphi \text{ for all } t \in S \text{ with } s \sim^{*\cap}_{\mathcal{B}} t$$

Such a semantics has been proposed in van Wijk (2015), and a complete axiomatization is given in Baltag and Smets (2020). But such 'common distributed knowledge' modalities can be equally well interpreted on simplicial models:

Let $\mathcal{C} = (C, \chi, \ell)$ be a simplicial model. Define operation $*_{\mathcal{B}}$ for all $\mathcal{B} \subseteq \mathcal{P}(A)$ inductively as: for all $X, Y, Z \in \mathcal{F}(C)$, $X \in *_{\mathcal{B}}(X)$, and if $Y \in *_{\mathcal{B}}(X)$ and there is $B \in \mathcal{B}$ with $B \subseteq \chi(Z \cap Y)$, then $Z \in *_{\mathcal{B}}(X)$. Then:

$$\mathcal{C}, X \models CD_{\mathcal{B}}\varphi \text{ iff } \mathcal{C}, Y \models \varphi \text{ for all } Y \in \mathcal{F}(C) \text{ such that } Y \in *_{\mathcal{B}}(X)$$

Here, $\mathcal{B}$ can be any set of subsets of A. Of interest in combinatorial topology may be subsets of agents of the same cardinality, i.e., all subsets of colours/agents of size m, for some $m < n$. Let us therefore define $*_m$ as a special case of the above $*_{\mathcal{B}}$ namely such that: $X \in *_m(X)$, and if $Y \in *_m(X)$ and there is $B \subseteq A$ with $|B| = m + 1$ and $B \subseteq \chi(Z \cap Y)$, then $Z \in *_m(X)$.[6] Geometrically, $Y \in *_m(X)$ means that there is a path from X to Y where each facet shares an m-dimensional face with the next one. We then get:

$$\mathcal{C}, X \models CD_m\varphi \text{ iff } \mathcal{C}, Y \models \varphi \text{ for all } Y \in \mathcal{F}(C) \text{ such that } Y \in *_m(X)$$

Clearly, we have that $CD_0\varphi \leftrightarrow C_A\varphi$. An interesting case to reason about simplicial models of dimension n is $CD_{n-1}\varphi$. This describes truth in a manifold (see Sect. 1.3.2): if $\mathcal{C}, X \models CD_{n-1}\varphi$ and $\mathcal{C}$ is a manifold, then φ should be true on the entire complex. Otherwise, it is true on a certain subcomplex of $\mathcal{C}$, namely on the restriction of $\mathcal{C} = (C, \chi, \ell)$ to the largest set of vertices $v \in \mathcal{V}(C)$ containing X that is a manifold. Clearly, for every $m < n$, $CD_m\varphi$ describes truth on a similar subcomplex of $\mathcal{C}$ but where all facets intersect in dimension m.

[6] The epistemic model equivalent would be: $\sim^{*\cap}_{m} := (\bigcup^{|B|=m+1}_{B \subseteq A} \sim^{\cap}_{B})^*$.

The modalities $CD_m\varphi$ are a mere example. We can also consider modalities describing truth on the boundary of a manifold. We can consider variations of so-called *conditional common knowledge* (van Benthem et al. 2006) targeting features of simplicial complexes. And so on.

Example 22 Reconsider the simplicial model of Example 16, below on the left, call it $\mathcal{C}$, and consider as well a variant $\mathcal{C}'$ extending it, below on the right. We note that $\mathcal{C}$ is a manifold, whereas $\mathcal{C}'$ is not. However, it is tempting to see $\mathcal{C}'$ as two manifolds intersecting in the vertex named $b0$. Simplicial model $\mathcal{C}'$ represents agent b being uncertain between two complexes with only difference the value of c, p_c or $\neg p_c$. With the operator CD_1 we can describe truth in the different regions of $\mathcal{C}'$. We recall that $CD_0 = C_A$. Some typical truths in $\mathcal{C}$ and $\mathcal{C}'$ are:

- $\mathcal{C}, X \models C_{abc} p_c$: this is true as p_c is true in facets X, Y, Z.
- $\mathcal{C}', X \not\models C_{abc} p_c$: this is false as, for example, $\mathcal{C}', V \not\models p_c$, and facets X and V are connected via $c \in \chi(X \cap Z)$ and $b \in \chi(Z \cap V)$.
- $\mathcal{C}', X \models CD_1 p_c$: this is true, as we only need to consider connectivity by a sequence of facets intersecting in dimension 1. We note that $|X \cap Y| = 1$ and $|Y \cap Z| = 1$, and that p_c is true in X, Y and Z. Therefore $\mathcal{C}, X \models CD_1 p_c$.
- $\mathcal{C}', V \models CD_1 \neg p_c$: in the other part of $\mathcal{C}'$, CD_1 describes truth in facets V, U, W.

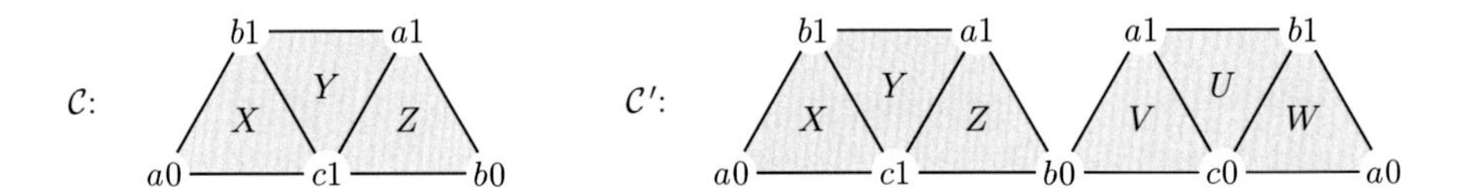

⊣

1.6.3 Bisimulation for Distributed Knowledge

As such, distributed knowledge seems uninteresting on simplicial models, as in any facet of a simplicial model the agents always have distributed knowledge of what is true in that facet (although for proper subsets of all agents distributed knowledge may already be slightly more interesting). Dually, in proper epistemic models, the intersection of all relations is the identity relation (although, also dually, the relation $\sim_B^\cap$ for $B \neq A$ need not be the identity). Somewhat surprisingly, explorations concerning distributed knowledge still seem of large interest to describe the information content of simplicial complexes.

The natural generalization of bisimulation including relations interpreting distributed knowledge not only contains **forth** and **back** clauses for individual agents but now for any group of agents (Roelofsen 2007; Ågotnes and Wáng 2017). Given $\mathfrak{R}ss'$, we recall the **forth** clause:

- **forth**: for all $a \in A$, for all t with $s \sim_a t$, there is a t' with $s' \sim'_a t'$ such that $\mathfrak{R}tt'$.

We now get:

- **forth**: for all $B \subseteq A$, for all t with $s \sim^{\cap}_B t$, there is a t' with $s' \sim'^{\cap}_B t'$ such that $\mathfrak{R}tt'$.

This notion of bisimulation is *stronger* (for a given model it is a more refined relation) than bisimulation for relations $\sim_a$ for individual agents only. See Example 23. This stronger notion may be useful to elicit information from simplicial sets, a generalization of simplicial complexes, as illustrated in Example 24.

Example 23 The typical example where this stronger bisimulation makes a difference is that it distinguishes the following epistemic models.

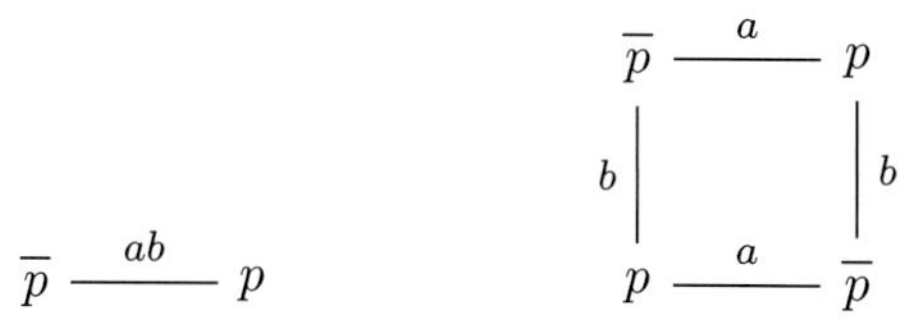

$\overline{p} \overset{ab}{\text{——}} p$

(x)

$\overline{p} \overset{a}{\text{——}} p$, $\overline{p} \overset{b}{|} p$, $p \overset{b}{|} \overline{p}$, $p \overset{a}{\text{——}} \overline{p}$

(y)

In the top-right and bottom-left corner of the square model, agents a and b have distributed knowledge of p; we note that $\sim_a \cap \sim_b$ is the identity relation. The two models are (standardly) bisimilar, but not with the additional clause for the group of agents $\{a, b\}$: in the left one we can then go from a p to a $\neg p$ state by a $\sim^{\cap}_{ab}$ link, but not in the right one, so the **forth** clause fails for $\{a, b\}$. ⊣

Example 24 (Distributed Knowledge for Simplicial Sets) As the models in the previous example are not local, they do not correspond to a simplicial complex. And as the disjunction of the values in indistinguishable states always is the uninformative value 'true', there is no transformation to a local epistemic model (see Sect. 1.5). The only obvious representation is simplicial model (i) below, where the name $a01$ suggestively stands for a vertex coloured with a and valued with the indistinguishable set $\{\neg p, p\}$. A similar quadrangular simplicial model represents that four-state epistemic model. But it is bisimilar to (i).

Distributed knowledge may now be helpful. In model (y), the coalition $\{a, b\}$ (for which we will write ab) can distinguish all four states, we have that $p \rightarrow D_{ab}\, p$ and $\neg p \rightarrow D_{ab} \neg p$ are both true. Now consider adding this coalition as a third agent to model (y). Let us call this model $(y3)$. Then the model is proper, it is bisimilar to a two-state model that is like (x) except that the third agent ab can distinguish the $\neg p$ from the p state, let us call that model $(x3)$, and its corresponding simplicial model is (ii) below. Here, agent ab can 'inform' agents a and b of the value of p.

Clearly, any simplicial model for n colours/agents can similarly be enriched with additional colours for all subsets of agents to thus obtain a simplicial model for $2^n - 1$ agents wherein all distributed knowledge has become explicit. Note that *all* epistemic models thus become proper. It is tempting to model this in combinatorial topology with a generalization of simplicial complexes to so-called *simplicial sets*, as suggested in Ledent (2019, Section 3.7). Just as a complex is a set of subsets of a set of vertices, a simplicial set is a multi-set of its subsets, with the same restriction that it is closed under subsets (see e.g. Hilton and Wylie (1960), where a simplicial set is called a *pseudocomplex*). In (iii) and (iv) we tentatively depict (i) and (ii) as simplicial sets.

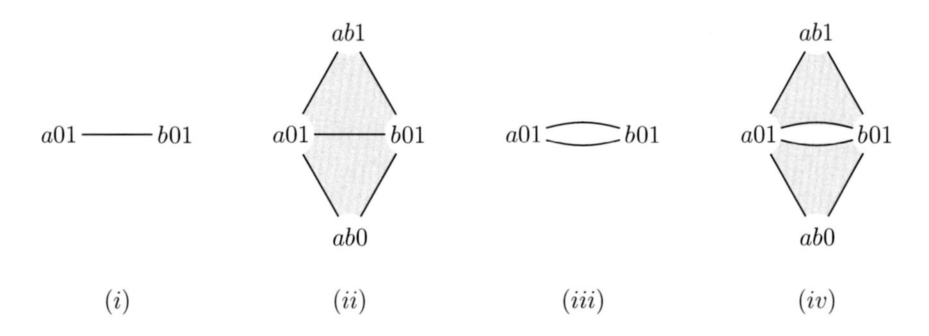

⊣

1.7 Belief for Simplicial Complexes

Apart from knowledge, another common notion in epistemic logics is that of *belief*. Already since the inception of knowledge and belief in the modal framework, the received convention is that, unlike knowledge, beliefs may be false, but that otherwise belief has the properties of knowledge, including the property known as *consistency* (Hintikka 1962; Meyer and van der Hoek 1995; van Ditmarsch et al. 2015). The modality for belief is B_a. That beliefs may be false is embodied in the failure of the validity $B_a\varphi \rightarrow \varphi$. Instead, we have the 'consistency' axiom $B_a\varphi \rightarrow \hat{B}_a\varphi$ that corresponds to seriality of underlying frames (and where $\hat{B}_a\varphi$ is defined by abbreviation as $\neg B_a\neg\varphi$, just as $\hat{K}_a\varphi$ in Sect. 1.3.1 on page 11). Like knowledge, belief should also satisfy the 'introspection' principles $B_a\varphi \rightarrow B_a B_a\varphi$ and $\neg B_a\varphi \rightarrow B_a\neg B_a\varphi$.

More precisely, in the inductive language definition we can replace the inductive clause $K_a\varphi$, for 'agent a knows φ', with the inductive clause $B_a\varphi$, for 'agent a believes φ' (if we were to have both, a more complex proposal is required that will succinctly describe towards the end of this section).

The validity $K_a\varphi \rightarrow \varphi$ for knowledge characterizes the reflexivity of the epistemic models. With the weaker $B_a\varphi \rightarrow \hat{B}_a\varphi$ for belief the relations in epistemic models are no longer equivalence relations. A typical model wherein agent a incorrectly believes p is $\mathcal{M} = (S, R, L)$ with $S = \{s, t\}$, $R = \{(s, t), (t, t)\}$ and $p \in L(t)$, often depicted as $\overline{p} \xrightarrow{a} p$ (this visualization uses the convention for belief that if an arrow points to a state, that state implicitly has a reflexive arrow). However, this visualization is not relevant for modelling belief on simplicial models. See Hintikka (1962), Meyer and van der Hoek (1995), and van Ditmarsch et al. (2015) for more information on belief.

Semantics of Belief We now address the semantics of belief on simplicial models. One should see this as the minimal departure from the notion of knowledge that is needed in order to handle possibly incorrect belief formally. In epistemic models this is realized by replacing the equivalence classes for a given agent by clusters (sets of indistinguishable states, like equivalence classes) together with isolated points/states from which only the states in the cluster are considered possible. Our proposal is the exact realization of that on simplicial models, taking into account that the semantic unit in complexes are not global states but local states.

Let a simplicial model $\mathcal{C} = (C, \chi, \ell)$ be given. For each agent $a \in A$, let f_a be an idempotent function between the a-coloured vertices of C, called *belief function* (a function f is idempotent iff for all x, $f(f(x)) = f(x)$). We emphasize that f_a need not preserve variables: it may well be that $p_a \in \ell(v)$ but $p_a \notin \ell(f_a(v))$, or vice versa.

We recall the semantics of knowledge on simplicial models:

$$\mathcal{C}, X \models K_a\varphi \text{ iff } \mathcal{C}, X' \models \varphi \text{ for all } X' \text{ with } a \in \chi(X \cap X')$$

Somewhat differently but equivalently formulated this is (we recall that X_a is the vertex of facet X coloured a):

$$\mathcal{C}, X \models K_a\varphi \text{ iff } \mathcal{C}, X' \models \varphi \text{ for all } X' \text{ with } X_a \in X'$$

The proposed semantics of belief is now as follows. It is surprisingly straightforward.

$$\mathcal{C}, X \models B_a\varphi \text{ iff } \mathcal{C}, X' \models \varphi \text{ for all } X' \text{ with } f_a(X_a) \in X'$$

Proposition 25 (Properties of Belief) *The following are validities of the logic of belief on simplicial complexes.*

1. $B_a\varphi \rightarrow \hat{B}_a\varphi$ *(consistency)*
2. $B_a\varphi \rightarrow B_a B_a\varphi$ *(positive introspection)*
3. $\hat{B}_a\varphi \rightarrow B_a \hat{B}_a\varphi$ *(negative introspection)* $\dashv$

Proof These properties are easily derived. The requirement that for all $a \in A$ the belief function f_a is idempotent guarantees both introspection properties of belief.

- Belief is consistent, because f_a is a (total) function and the simplicial complex is pure, so that for any a-coloured vertex v, $f_a(v)$ must be contained in at least one facet.
- To show positive introspection we use the contrapositive $\hat{B}_a\hat{B}_a\neg\varphi \rightarrow \hat{B}_a\neg\varphi$. Let $\mathcal{C}, X \models \hat{B}_a\hat{B}_a\neg\varphi$, then there is a facet X' such that $f_a(X_a) \in X'$ and $\mathcal{C}, X' \models \hat{B}_a\neg\varphi$. Therefore, there are X', X'' such that: $f_a(X_a) \in X'$, $f_a(X'_a) \in X''$, and $\mathcal{C}, X'' \models \neg\varphi$. As f_a is idempotent and as X' obviously only contains a single vertex of colour a, $f_a(X'_a) = f_a(f_a(X_a)) = f_a(X_a)$. Therefore, by the semantics of belief, $\mathcal{C}, X \models \hat{B}_a\neg\varphi$.
- Negative introspection is shown very similarly. Let $\mathcal{C}, X \models \hat{B}_a\varphi$ be given, i.e., there is a facet X' such that $f_a(X_a) \in X'$ and $\mathcal{C}, X' \models \varphi$. We need to show that $\mathcal{C}, X \models B_a\hat{B}_a\varphi$. In order to show that, let X'' be such that $f_a(X_a) \in X''$. Again, as $f_a(f_a(X_a)) = f_a(X_a)$ we see that $\mathcal{C}, X'' \models \hat{B}_a\varphi$, thus $\mathcal{C}, X \models B_a\hat{B}_a\varphi$.

□

A special case of belief is when agents correctly believe (and thus know) the value of their local state but may only be incorrect about the beliefs or the variables of other agents. This is guaranteed by requiring that for all $p_a \in P_a$, and all vertices v, $p_a \in \ell(v)$ iff $p_a \in \ell(f_a(v))$. We call such belief functions *locally correct*. It will be obvious that on a simplicial model where belief functions are locally correct, $B_a p_a \rightarrow p_a$ and $B_a\neg p_a \rightarrow \neg p_a$ are valid.

Simplicial Belief on Epistemic Models The belief notion that we proposed on simplicial models results in rather particular models in the Kripke model semantics. Take an epistemic model $(S, \sim, L)$ and let for each agent $a \in A$ belief function f_a be an idempotent function between the equivalence classes $[s]_a$ of a. Define relation $R_a \subseteq S \times S$ as: $R_a ss'$ iff there is a $t \in f([s]_a)$ with $t \sim_a s'$. Then (S, R, L) is a model wherein all R_a are serial, transitive, and Euclidean and thus interpret standard consistent belief (a.k.a. $\mathcal{KD}45$ belief; van Ditmarsch et al. 2015). In other words, the agent's beliefs are uniform: the agent still believes to know her local state, but in all indistinguishable global states she considers it possible that the global states are those of another equivalence class, wherein her local state might be different.

Example 26 Consider again the simplicial model from Example 10 depicted below.

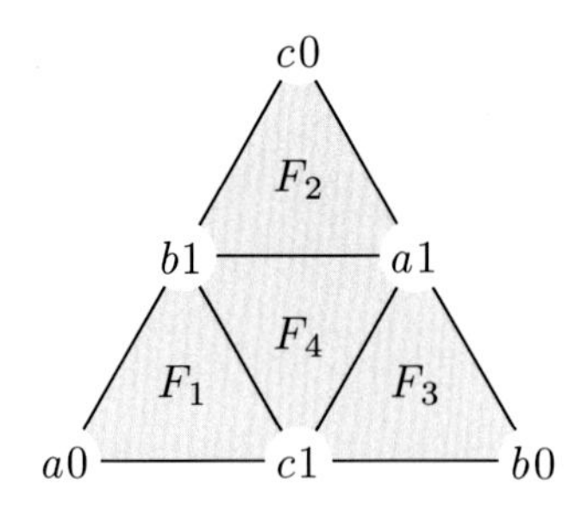

We give three examples of idempotent simplicial functions, and the resulting consequences for the beliefs of the agents:

1. f_a defined as $f_a(a0) = a1$ (and thus $f_a(a1) = a1$).
2. f_c defined as $f_c(c0) = c1$ (and thus $f_c(c1) = c1$).
3. f'_a defined as $f'_a(a1) = a0$ (and thus $f'_a(a0) = a0$), $f'_b(b1) = b0$, and $f'_c(c1) = c0$.

We now can, for example, validate the following statements:

1. $\mathcal{C}, F_1 \models \neg p_a \wedge B_a p_a$ Agent a mistakenly believes that its local state is p_a, whereas in fact it is $\neg p_a$. For the verification of the belief component we note that $\mathcal{C}, F_1 \models B_a p_a$ iff $\mathcal{C}, F' \models p_a$ for all F' containing $f_a(a0) = a1$, which are F_2, F_3, F_4 that all obviously satisfy $a1$.
   $\mathcal{C}, F_1 \models (p_b \wedge p_c) \wedge \hat{B}_a(p_b \wedge p_c) \wedge \neg B_a(p_b \wedge p_c)$
   The first conjunct is obvious, the second is because $a1 \in F_4$, the last is because $a1 \in F_2$ and also because $a1 \in F_3$, neither satisfy $p_b \wedge p_c$.
2. Similarly to the first item, we now have for example that $\mathcal{C}, F_2 \models \neg p_c \wedge B_c p_c$.
3. $\mathcal{C}, F_1 \models B_a(\neg p_a \wedge p_b \wedge p_c) \wedge B_b(p_a \wedge \neg p_b \wedge p_c) \wedge B_c(p_a \wedge p_b \wedge \neg p_c)$
   In F_1, the beliefs of the agents are inconsistent, they all think to know what the actual state of affairs (the designated facet) is, but their beliefs are incompatible. In fact, for any facet, under this f' map, their beliefs about the valuation are mistaken. Also, in any facet, all three agents believe that they are certain (all agents only consider one facet possible).

In this example no belief function is locally correct. If we were to change the value of p_a in vertex $a1$ into false, then f_a and f'_a would be locally correct. ⊣

Combining Knowledge and Belief So far we modelled belief instead of knowledge, while employing the same structures, simplicial models. An alternative modelling of knowledge and belief at the same time is also possible, if we enrich the structures. For Kripke models this means that for each agent each equivalence class of possible states becomes a well-preorder, allowing us to distinguish more and less plausible states among those possible states, while ensuring that there always is a most plausible state. The agent then believes a proposition if it is true in the most plausible states (the well-foundedness of a well-preorder is to guarantee consistency of belief). See for example Baltag and Smets (2008) and van Benthem and Smets (2015). This framework can similarly be employed to combine knowledge and belief for simplicial models, which we illustrate by an example only without formal details.[7]

Example 27 Once more consider the simplicial model $\mathcal{C}$ used in Examples 10 and 26. There are, informally, two 'equivalence classes' for agent a, namely $\{F_1\}$ and $\{F_2, F_3, F_4\}$. In the second, let us assume that F_2 is more plausible than F_3 and that F_2 and F_3 are more plausible than F_4. We then have that $\mathcal{C}, F_1 \models K_a p_c$

[7] Personal communication by Alexandru Baltag.

(a knows that p_c, i.e., a knows that the value of the local state variable p_c of c is 'true', or '1'). Also $\mathcal{C}, F_1 \models B_a p_c$, as the single state of equivalence class $\{F_1\}$ is by definition the most plausible. Then, $\mathcal{C}, F_4 \models p_c$ and $\mathcal{C}, F_4 \models \neg K_a p_c$ but $\mathcal{C}, F_4 \models B_a \neg p_c$, where the last is true because the most plausible state in $\{F_2, F_3, F_4\}$ is F_2 and $\mathcal{C}, F_2 \models \neg p_c$. In facet F_4 the beliefs of a are incorrect. $\dashv$

1.8 Simplicial Action Models

Action model logic is an extension of epistemic logic with operators modelling change of information (Baltag et al. 1998; van Ditmarsch et al. 2008). Although we can, of course, strictly separate syntax and semantics, it is common to see these operators as epistemic models wherein the valuations of atoms have been replaced by formulas (in the logical language), that should be seen as executability preconditions for the states in the epistemic model, that are therefore not called states but actions. Instead of presenting action models (for which we refer to the above references) we will present their simplicial equivalents, the *simplicial action models* introduced in Goubault et al. (2021). We present them with a minor variation, namely with preconditions for vertices instead of facets, and also including so-called *factual change*. We recall that we assume that all simplicial complexes are chromatic pure simplicial complexes. This is also the case for simplicial action models.

Simplicial Action Model A *simplicial action model* $\mathcal{C}$ is a quadruple $(C, \chi, \mathsf{pre}, \mathsf{post})$ where C is a simplicial complex, χ a chromatic map, $\mathsf{pre} : \mathcal{V}(C) \to \mathcal{L}_K(A, P)$ a precondition function assigning to each vertex $v \in \mathcal{V}(C)$ a formula, and $\mathsf{post} : \mathcal{V}(C) \to P \to \mathcal{L}_K(A, P)$ a postcondition function assigning to each vertex and to each local variable for the colour of that vertex a formula.[8]

If $\chi(v) = a$, $\mathsf{pre}(v) = \varphi$, and $\mathsf{post}(v)(p) = \psi$ this should be seen as agent a in v executing the program "If φ then $p := \psi$." Such assignments are simultaneous for all local variables.

Execution of a Simplicial Action Model Let a simplicial model $\mathcal{C} = (C, \chi, \ell)$ and a simplicial action model $\mathcal{C}' = (C', \chi', \mathsf{pre}', \mathsf{post}')$ be given. The *restricted product* of $\mathcal{C}$ and $\mathcal{C}'$ is the simplicial model $\mathcal{C} \otimes \mathcal{C}' = (C'', \chi'', \ell'')$ where:

- C'' is the modal product (Cartesian product) $C \times C'$ restricted to all facets $X \times X'$ such that $\mathcal{C}, X \models \bigwedge_{v' \in X'} \mathsf{pre}'(v')$;[9]
- for all (v, v') in the domain of C'', $\chi''((v, v')) = \chi(v)$ $(= \chi'(v'))$;
- for all (v, v') in the domain of C'' and for all $p \in P$, $p \in \ell''((v, v'))$ iff $p \in \ell(v)$.

[8] In logics that contains modalities for such action models, we need to require that $\mathcal{V}(C)$ is finite and that post is a partial function defined for a finite subset $P' \subseteq P$ only.

[9] Using local semantics, instead of $\mathcal{C}, X \models \bigwedge_{v' \in X'} \mathsf{pre}'(v')$ we can require that $\mathcal{C}, v \models \mathsf{pre}'(v')$ for all $v \in X$ and $v' \in X'$ with $\chi(v) = \chi(v')$. This may be more elegant.

In Goubault et al. (2021), simplicial action models have preconditions associated to facets. The definitions are interchangeable: in one direction, for any facet X, $\mathsf{pre}(X) := \bigwedge_{v \in X} \mathsf{pre}(v)$, as above, and in the other direction, for any vertex v, $\mathsf{pre}(v) := \bigvee_{v \in X} \mathsf{pre}(X)$. Defining preconditions on vertices has however some modelling advantages.

In the first place we can require, as a special case, all preconditions to be *local*, i.e., any $\mathsf{pre}(v)$ for vertex v of colour a has shape $K_a \varphi$ (or, alternatively, any $\mathsf{pre}(v)$ is equivalent to $K_a \mathsf{pre}(v)$ on the simplicial model where the simplicial action model is executed). This would be in accordance with distributed computing methodology, where all actions are sending and receiving actions between agents. It rules out those enacted by the environment feeding new information to the system. Most but not all simplicial action models in Goubault et al. (2021) and Ledent (2019) are local in this sense. An exception are those for binary consensus (Ledent 2019, Ex. 3.48). One could even enforce commonly known preconditions, as in Halpern and Moses (2017).

In the second place we can require, as an even more special case, all preconditions to be in the language restricted to the agent colouring that vertex, i.e., any $\mathsf{pre}(v)$ for vertex v of colour a is a formula $\varphi \in \mathcal{L}_K(\{a\}, P_a)$. As the value of all atoms p_a is known by a, and given the properties of K_a, this implies that preconditions are local. Again, most simplicial action models in Goubault et al. (2021) and Ledent (2019) are even local in that more restricted sense (e.g., for the immediate snapshot).

A more significant difference from Goubault et al. (2021) is that we also incorporate *factual change* (change of the values of local variables) in our simplicial actions models, just as in dynamic epistemic logic (van Benthem et al. 2006; van Ditmarsch and Kooi 2008). We may need factual change to describe snapshot algorithms with write actions, or binary consensus, of which we will give an example.

Example 28 Once more we consider the simplicial model $\mathcal{C}$ from Example 10, in which we execute the action representing that a and b do not know the value of c. We can think of this as a simplicial action model $\mathcal{C}'$ consisting of two facets intersecting in the edge $\{b, a\}$, with preconditions: $\mathsf{pre}(a') = \neg(K_a p_c \vee K_a \neg p_c)$, $\mathsf{pre}(b') = \neg(K_b p_c \vee K_b \neg p_c)$, $\mathsf{pre}(c'_0) = \neg p_c$, $\mathsf{pre}(c'_1) = p_c$. Below we depict that model and its execution.

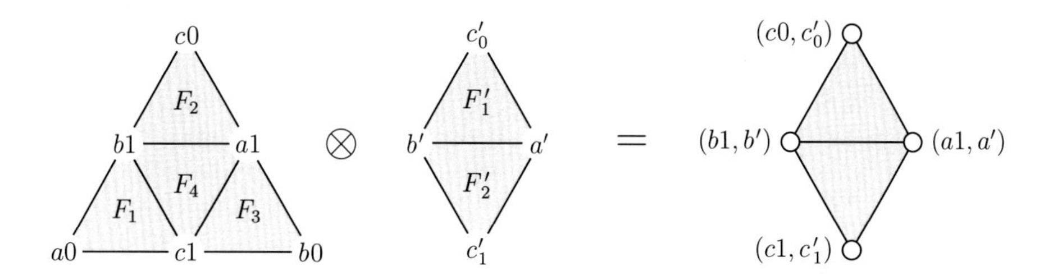

We now have, for example, that after this action b still does not c's value, but knows that c now always knows his value: $\mathcal{C} \otimes \mathcal{C}', (F_4, F_2') \models \neg(K_b p_c \vee K_b \neg p_c) \wedge K_b(K_c p_b \vee K_c \neg p_b)$.[10] We have to choose a perspective of a pair of facets in the updated model in order to be able to evaluate any formula. For example, $\mathcal{C} \otimes \mathcal{C}', (F_4, F_1') \models \neg p_c$ whereas $\mathcal{C} \otimes \mathcal{C}', (F_4, F_2') \models p_c$. ⊣

Example 29 As an example of factual change, consider the following action wherein agent c publicly resets the value of her local variable to true, thus removing the initial uncertainty for a and b about that value. The latter is depicted in $\mathcal{C}$ below.

The simplicial action model $\mathcal{C}'$ consists of a single facet, also depicted, with vertices a, b, c, as usual reusing colours as names, with $\mathsf{pre}(a) = \mathsf{pre}(b) = \mathsf{pre}(c) = \top$ and $\mathsf{post}(c)(p_c) = \top$. We may assume the postconditions for the other variables to be trivial, i.e., $\mathsf{post}(a)(p_a) = p_a$ and $\mathsf{post}(c)(p_b) = p_b$. The action c can be executed in both vertices $c0$ and $c1$, where we note that in the resulting simplicial model $\mathcal{C} \otimes \mathcal{C}'$ the valuation is such that $p_c \in \ell'(c0, c)$ and $p_c \in \ell'(c1, c)$ (so, informally, both are '$c1$' nodes, i.e., nodes coloured c with value 1 or 'true' of local variable p_c). Therefore, the two-faceted updated simplicial model is bisimilar to the single-faceted one depicted to the right of it.

Note that a public *assignment* of p_c to $\top$ is different from a public *announcement* that p_c (is true). This would result in the same updated simplicial model restriction, but as a consequence of a restriction to the facet wherein p_c is true.

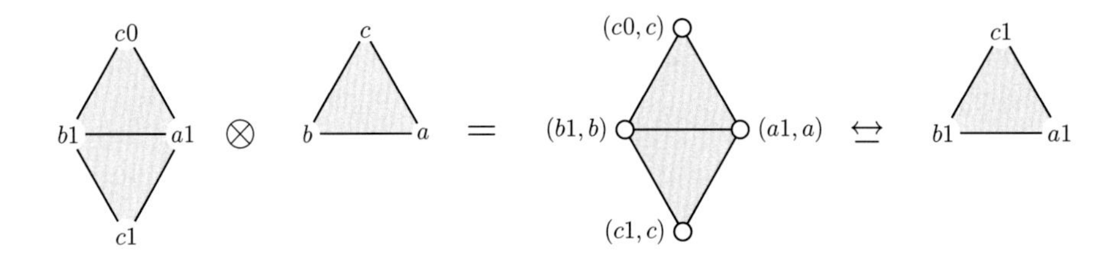

⊣

Example 30 (Simplicial Action Model for Binary Consensus) In binary consensus, agents wish to agree on the value of a binary variable. We may assume that this variable is an atomic proposition, and that each agent has a copy that is its local variable, that they can communicate or manipulate while aiming to obtain consensus. Each agent a thus has a variable 1_a, where $\neg 1_a$ is 0_a, in order to reach consensus on 0 or 1. We assume that all combinations of values are possible, but also that any kind of knowledge about the values of other agents is possible, thus abstracting from the part of the consensus algorithm obtaining such knowledge from

[10] If we were to incorporate modalities for simplicial action models into the logical language, we can also express propositions such as 'before the update c did not know the value of b's variable, but afterwards she knows': $\mathcal{C}, F_4 \models \neg(K_c p_b \vee K_c \neg p_b) \wedge [\mathcal{C}', F_2'](K_c p_b \vee K_c \neg p_b)$.

other agents (e.g., possible knowledge gain modelled as subdivisions). This has the advantage that we can focus on the simplicial action model only.

Let there be two agents a, b with variables $1_a, 1_b$. For both agents there are four different actions, corresponding to four vertices. Consider a. If $K_a(1_a \wedge 1_b)$, then she leaves the value unchanged (aiming for consensus on value 1), if $K_a(0_a \wedge 0_b)$, she also leaves the value unchanged (aiming for consensus on value 0). Otherwise, she considers it possible that her value and that of b might be different. We note that $\neg K_a(1_a \wedge 1_b) \wedge \neg K_a(0_a \wedge 0_b)$ implies $\widehat{K}_a(1_a \vee 1_b) \wedge \widehat{K}_a(0_a \vee 0_b)$. In that case she non-deterministically resets her value to 0 or to 1. This is an assignment $1_a := \top$ or $1_a := \bot$ (i.e., reset to value 0). Agent a can distinguish all these actions, whereas agent b cannot distinguish any of those. The four actions of agent b are analogous, and cannot be distinguished by a.

The simplicial action model is therefore defined as $\mathcal{C} = (C, \chi, \mathsf{pre}, \mathsf{post})$ where $\mathcal{V}(C) = \{u_i \mid i = a, b \text{ and } u = w, x, y, z\}$ and $\mathcal{F}(C) = \{(u_a, v_b) \mid u, v = w, x, y, z\}$, and $\chi(u_i) = i$ for $i = a, b$ and $u = w, x, y, z$. Then, for $i = a, b$:

$$\begin{aligned} \mathsf{pre}(w_i) &= K_i(1_a \wedge 1_b) \\ \mathsf{pre}(x_i) = \mathsf{pre}(y_a) &= \neg K_i(1_a \wedge 1_b) \wedge \neg K_i(0_a \wedge 0_b) \\ \mathsf{pre}(z_i) &= K_i(0_a \wedge 0_b) \end{aligned}$$

and

$$\begin{aligned} \mathsf{post}(w_i)(1_i) = \mathsf{post}(z_i)(1_i) &= 1_i \\ \mathsf{post}(x_i)(1_i) &= \top \\ \mathsf{post}(y_i)(1_i) &= \bot \end{aligned}$$

The assignment $\mathsf{post}(z_i)(1_i) = 1_i$ means that the value of 1_i remains the same: if agent i has a 1, it remains 1, and if she has a 0, it remains 0.

As is to be expected, unless a and b have common knowledge that the value is 1 or common knowledge that the value is 0, they cannot obtain such common knowledge by executing this simplicial action $\mathcal{C}$: binary consensus is impossible. Even when they both know that the values agree, for example when $K_a(1_a \wedge 1_b)$ and $K_b(1_a \wedge 1_b)$, they may still consider it possible that the other does not know that, as when $\widehat{K}_a \neg K_b(1_a \wedge 1_b)$, in which case a considers it possible that b has reset its value to 0, such that consensus would be lost. Although in fact, because $K_a(1_a \wedge 1_b)$ and $K_b(1_a \wedge 1_b)$, they both kept their 1s: there is consensus. And so on, forever.

Validity and agreement are therefore not guaranteed by this simplicial action model. Validity can be obtained for slight variations, for example when agents do not randomly choose a value in case they consider it possible other agents have other values, as here, but choose the value of the majority. Under that policy, then if all agents know all values and there is a majority, validity is guaranteed.

Generalizing the case from 2 agents to n agents is straightforward. Instead of $2 \cdot 4 = 8$ vertices we get $4n$ vertices: *linear* growth. Interestingly, the corresponding action model grows *exponentially*, as each action in the action model has a precondition that is a conjunction of n parts, where each conjunct is one of these 4 options: 4^n actions. The distributed modelling is decidedly more elegant.

Usually, consensus comes with a termination condition, which we did not model here. This is because the action model described here specifies the task without referring to a particular protocol used to solve it. In distributed computing, one usually studies the solvability of this task in a round-based model which is guaranteed to terminate. $\dashv$

1.9 From Global to Local Epistemic Models

Propositions 7 and 8 in Sect. 1.3 reported how epistemic models correspond to simplicial models, however on the assumption that the epistemic models are local. We now investigate when epistemic models that are not local can be transformed into local epistemic models that contain the same information. It is easy to see that changing the domain or the epistemic relations of an epistemic model can result in change of information content, so that the only option is to adjust the valuation by replacing the global variables with local variables.

Given epistemic model $\mathcal{M} = (S, \sim, L)$ for agents A and variables P we therefore wish to construct a local epistemic model $\mathcal{M}' = (S, \sim, L')$ for agents A and variables P' containing the *same information*, i.e., for all $\varphi \in \mathcal{L}_K(A, P)$ there is a $\varphi' \in \mathcal{L}_K(A, P')$ such that $[\![\varphi]\!]_{\mathcal{M}} = [\![\varphi']\!]_{\mathcal{M}'}$, and for all $\varphi' \in \mathcal{L}_K(A, P')$ there is a $\varphi \in \mathcal{L}_K(A, P)$ such that $[\![\varphi]\!]_{\mathcal{M}} = [\![\varphi']\!]_{\mathcal{M}'}$.

In general, as known from Goubault et al. (2021), this quest cannot succeed. A simple counterexample is the *improper* epistemic model

$$\overline{p} \stackrel{a}{\text{———}} p$$

for one agent a and one variable p. Agent a cannot distinguish two states with different valuations. Any local variable q_a of agent a must be true in both states, as they are indistinguishable for a:

$$q_a \stackrel{a}{\text{———}} q_a$$

This model collapses to the singleton

$$q_a$$

We have lost information.

However, on other epistemic models this quest succeeds. For example, consider the following (proper) model for two agents a, b and one variable p.

$$p \stackrel{a}{\text{———}} \overline{p} \stackrel{b}{\text{———}} p$$

We note that p is not local for a and also not local for b. Now consider:

$$q_a q_b \overset{a}{\text{———}} q_a \overline{q}_b \overset{b}{\text{———}} \overline{q}_a \overline{q}_b$$

for variables q_a, q_b. This model is local. As both models are finite and bisimulation minimal, all subsets of the domain are obviously definable in both languages and can thus be made to correspond. We can do this systematically for all formulas by, for example, replacing all occurrences of p by $(q_a \land q_b) \lor (\overline{q}_a \land \overline{q}_b)$ in any formula in the language with p, and in the other direction by simultaneously replacing all occurrences of q_a by $\neg K_a p$ and of q_b by $K_b p$ in the language with q_a, q_b.

In Ledent (2019) the following method is proposed to make arbitrary finite epistemic models local. Given $\mathcal{M} = (S, \sim, L)$ for variables P, consider novel variables $P' = \{p_a^{[s]_a} \mid s \in S, a \in A\}$, i.e., one for each equivalence class $[s]_a$. Then, define $L'(s) = \{p_a^{[s]_a} \in P' \mid a \in A\}$. The model $(S, \sim, L')$ is local, and if $\mathcal{M}$ is proper, then $\bigcap_{a \in A}[s]_a = s$, so that $\bigwedge_{a \in A} p_a^{[s]_a}$ is a distinguishing formula for state s. Therefore, the denotation of p in $\mathcal{M}$ is the same as that of $\bigvee_{p \in L(s)} \bigwedge_{a \in A} p_a^{[s]_a}$ in $\mathcal{M}'$ (see Ledent 2019). Dually, on bisimulation minimal epistemic models, the denotation of any $p_a^{[s]_a}$ in $\mathcal{M}'$ is the same as the of $\bigvee_{t \in [s]_a} \delta_t$ in $\mathcal{M}$, in the original language $\mathcal{L}_K(P)$. This solution is reminiscent of hybrid logic (Blackburn 2000) wherein all states have names, and in particular of the enrichment of models in symbolic model checking (van Benthem et al. 2018; Charrier and Schwarzentruber 2015). This method may result in many more propositional variables. Also, on epistemic models that are not bisimulation minimal, the model with variables $p_a^{[s]_a}$ is not bisimilar to the initial model and then contains more information, as in the following example.

Example 31 Consider the three models below. The leftmost model is improper and the other two are proper. The middle model (with names of states in between parentheses) is proper but not bisimulation minimal. The rightmost model is obtained by having variables correspond to equivalence classes. It is bisimulation minimal. For example, the formula $p_a^{uv} \land p_b^{su}$ is only true in state u, but any formula in the middle model that is true in u is also true in the bisimilar state t.

$$\overline{p} \overset{ab}{\text{———}} p \qquad \begin{array}{ccc} \overline{p}(u) & \overset{a}{\text{—}} & p(v) \\ b \big| & & \big| b \\ p(s) & \overset{a}{\text{—}} & \overline{p}(t) \end{array} \qquad \begin{array}{ccc} p_a^{uv} p_b^{su} & \overset{a}{\text{———}} & p_a^{uv} p_b^{tv} \\ b \big| & & \big| b \\ p_a^{st} p_b^{su} & \overset{a}{\text{———}} & p_a^{st} p_b^{tv} \end{array}$$

⊣

As said, we aim at a transformation that is information preserving in both directions. Therefore, the method from Ledent (2019) is not what we want. We therefore investigated alternative ways to make models local. Before we can report on that, we need to introduce an additional technical tool: the distinguishing formula.

Distinguishing Formulas Given a subset $S' \subseteq S$ of the domain, a *distinguishing formula* for S' is a formula in some logical language typically denoted $\delta_{S'}$, such that: for all $s \in S'$, $\mathcal{M}, s \models \varphi$ whereas for all $s \notin S'$, $\mathcal{M}, s \not\models \varphi$. A distinguishing formula for a singleton $S' = \{s\}$ is a distinguishing formula for s. If $\mathcal{M}$ is finite (a model is finite if its domain is finite) and bisimulation minimal, a distinguishing formula always exists in $\mathcal{L}_K(A, P)$ for any state. We can construct it as follows (van Ditmarsch et al. 2014).

Let a finite epistemic model $\mathcal{M} = (S, \sim, L)$ be given. Let us call a variable $p \in P$ *redundant* on $\mathcal{M}$ if, whenever $\mathcal{M}$ is a not singleton, $p \in L(s)$ for all $s \in S$ or $p \notin L(s)$ for all $s \in S$ (so the extension of p is S or $\emptyset$); and also if there is a $q \in P$ with $p \neq q$ and with $p \in L(s)$ iff $q \in L(s)$ for all $s \in S$. On a finite model, there is only a finite set $P(\mathcal{M}) \subseteq P$ of non-redundant variables (even if P is countably infinite). Given $s \in S$, let τ_s be the *factual description* of s, i.e., $\tau_s := \bigwedge\{p \in P(\mathcal{M}) \mid p \in L(s)\} \wedge \bigwedge\{\neg p \in P(\mathcal{M}) \mid p \notin L(s)\}$, and consider the following procedure:

$$
\begin{array}{ll}
\delta_s^0 & := \tau_s \\
\delta_s^{n+1} & := \tau_s \wedge \bigwedge_{a \in A} \bigwedge_{t \in [s]_a} \widehat{K}_a \delta_t^n \wedge \bigwedge_{a \in A} K_a \bigvee_{t \in [s]_a} \delta_t^n
\end{array}
\tag{1.1}
$$

Consider the relation $\mathfrak{R}$ defined by $\mathfrak{R}st$ if for all $n \in \mathbb{N}$, $\mathcal{M}, t \models \delta_s^n$. If the model is finite, the relation $\mathfrak{R}$ is a bisimulation, and for all $n \geq |S|$, δ_s^{n+1} and δ_s^n have the same denotation. We therefore take $\delta_s^{|S|}$ as distinguishing formula δ_s for state s.

References for distinguishing formulas in epistemic logic are van Benthem (1998) and van Ditmarsch et al. (2014).

Distinguishing Formulas with Local Variables Let us now adapt procedure (1.1) for distinguishing formulas, using a novel set of local propositional variables. Firstly, write $\tau_{[s]_a}$ for $\bigvee_{t \in [s]_a} \tau_t$. For each $\tau_{[s]_a}$ create a novel variable $p_a^{\tau_{[s]_a}}$ with the same denotation on the given model. Note that this variable is local for a: $K_a p_a^{\tau_{[s]_a}}$ is true iff $p_a^{\tau_{[s]_a}}$ is true. Consider the following procedure:

$$
\begin{array}{ll}
\delta_s^0 & := \bigwedge_{a \in A} p_a^{\tau_{[s]_a}} \\
\delta_s^{n+1} & := \bigwedge_{a \in A} p_a^{\tau_{[s]_a}} \wedge \bigwedge_{a \in A} \bigwedge_{t \in [s]_a} \widehat{K}_a \delta_t^n \wedge \bigwedge_{a \in A} K_a \bigvee_{t \in [s]_a} \delta_t^n
\end{array}
\tag{1.2}
$$

The procedure is now in the language with local variables $P' = \{p_a^{\tau_{[s]_a}} \mid s \in S, a \in A\}$. In the sequel, in order to simplify notation, we may write φ_a for any propositional formula φ equivalent to $\tau_{[s]_a}$.

Example 32 Continuing the above example for three states, the following local model is created. As $p \vee \neg p$ is equivalent to $\top$, we write $\top_a$ instead of $p_a^{p \vee \neg p}$, etc., and we only depict the valuations.

$$\top_a p_b \stackrel{a}{——} \top_a \top_b \stackrel{b}{——} p_a \top_b$$

The resulting distinguishing formulas are

$$\begin{aligned}\delta_s^1 &= \top_a \wedge p_b \\ \delta_t^2 &= \top_a \wedge \top_b \\ \delta_u^1 &= p_a \wedge \top_b\end{aligned}$$

Also taking into account that agents know their local variables, an alternative description is:

$$\begin{aligned}\delta_s^1 &= p_b \\ \delta_t^2 &= \widehat{K}_a p_b \wedge \widehat{K}_b p_a \\ \delta_u^1 &= p_a\end{aligned}$$

$\dashv$

Recalling the relation defined as: $\mathfrak{R}st$ if for all $n \in \mathbb{N}$, $\mathcal{M}, t \models \delta_s^n$, it is unclear if on any given model it is a bisimulation with the new local variables if and only if it is a bisimulation with the old global variables. We conjecture that on finite models $\mathfrak{R}$ is again a bisimulation, thus achieving our aim to construct a local epistemic model with the same information content as any given epistemic model. We further conjecture that on any given epistemic model where $\mathfrak{R}$ is a bisimulation but not so with the local variables, no local epistemic model exists with the same information content. Example 33 demonstrates this for an infinite model.

Example 33 Consider the following (image-finite) infinite chain $\mathcal{M}$.

$$\overline{p} \stackrel{a}{——} p \stackrel{b}{——} \overline{p} \stackrel{a}{——} p \stackrel{b}{——} \overline{p} \stackrel{a}{——} p \stackrel{b}{——} \dots.$$

Let us name the states 0, 1, ... from left to right. The model $\mathcal{M}'$ with local variables is:

$$\top_a \overline{p}_b \stackrel{a}{——} \top_a \top_b \stackrel{b}{——} \top_a \top_b \stackrel{a}{——} \top_a \top_b \stackrel{b}{——} \top_a \top_b \stackrel{a}{——} \top_a \top_b \stackrel{b}{——} \dots.$$

State 0 is the unique state where $K_b \neg p$. All finite subsets of the chain can be distinguished by their relation to this endpoint, and therefore in particular all equivalence classes of relations $\sim_a$ and $\sim_b$, i.e., all local states of a and b. As an example, we show how three local b-states and three local a-states are described in the language for the local variables corresponding to booleans of the single variable p. For a local state consisting of states j and k we write jk, etc.

eq. class	distinguishing formula
01	$\widehat{K}_a \overline{p}_b$
23	$\widehat{K}_a \widehat{K}_b \widehat{K}_a \overline{p}_b \wedge \neg \widehat{K}_a \overline{p}_b$
45	$\widehat{K}_a \widehat{K}_b \widehat{K}_a \widehat{K}_b \widehat{K}_a \overline{p}_b \wedge \neg \widehat{K}_a \widehat{K}_b \widehat{K}_a \overline{p}_b$
0	$\overline{p}_b$
12	$\widehat{K}_b \widehat{K}_a \overline{p}_b \wedge \neg \overline{p}_b$
34	$\widehat{K}_b \widehat{K}_a \widehat{K}_b \widehat{K}_a \overline{p}_b \wedge \neg \widehat{K}_b \widehat{K}_a \overline{p}_b$

The variable p is true in all odd states. However, this subset of the domain is not definable in the language with local variables. In other words, there is a $\varphi \in \mathcal{L}_K(\{a, b\}, \{p\})$, namely $\varphi = p$, such that for all $\varphi' \in \mathcal{L}_K(\{a, b\}, \{\overline{p}_b, \top_a, \top_b\})$, $[\![\varphi]\!]_{\mathcal{M}} \neq [\![\varphi']\!]_{\mathcal{M}'}$. The models $\mathcal{M}$ and $\mathcal{M}'$ do not contain the same information. $\dashv$

Above all we hope that the two mentioned conjectures may be answered in due time.

1.10 Conclusion and Further Research

We have presented a survey of epistemic tools and techniques that can be used to model knowledge and change of knowledge on simplicial complexes, which are a representation of high-dimensional discrete topological spaces. Many of these notions bring in novel issues in this topological setting. In future research we wish to apply these tools to model typical distributed computing tasks and algorithms, beyond the situations studied in van Ditmarsch et al. (2021), Goubault et al. (2021) and Ledent (2019), and to explore further epistemic horizons for these and other applications.

An interesting possibility of interaction emerges with algebraic topology, a very deep and mature area of mathematics. For example, the well-known notion of *consensus* depends on the connectivity of an epistemic frame and the related notion of common knowledge (Halpern and Moses 1990); this is a one-dimensional topological property. However, other notions of knowledge appear to be related to higher-dimensional topological properties, and to forms of agreement weaker than consensus (Goubault et al. 2019).

With the exception of Example 5 modelling synchronous computation, we did not investigate impure complexes. Such complexes may well be interpreted in terms of unawareness of local variables as in Fagin and Halpern (1988) (and as unawareness of agents). Special notions of bisimulation would fit impure complexes.

As suggested in Ledent (2019), distributed knowledge may be employed to describe *simplicial sets*, the generalization of simplicial complexes to multi-sets (the *pseudocomplexes* of Hilton and Wylie (1960), with a similar duality to graphs; Bracho and Montejano 1987). Then, we could have two facets that are indistinguishable to *all* agents, aiming at removing the requirement that the epistemic model be proper.

It would also be interesting to consider infinite models. In distributed computing, infinitely many states and infinitely many agents (processes) have been considered in e.g. Aguilera (2004). In modal logic, many tools and techniques apply just as well to the countably infinite situation.

Acknowledgments We thank the reviewers for their very detailed comments. The second author was partially supported by AID project "Validation of Autonomous Drones and Swarms of Drones" and the academic chair "Complex Systems Engineering" of École Polytechnique-ENSTA-Télécom-Thalès-Dassault-Naval Group-DGA-FX-Fondation ParisTech. The fourth author was partially supported by grants UNAM-PAPIIT IN106520 and CONACYT-LASOL.

References

Ågotnes, T., and Y.N. Wáng. 2017. Resolving Distributed Knowledge. *Artificial Intelligence* 252: 1–21.

Aguilera, M.K. 2004. A Pleasant Stroll Through the Land of Infinitely Many Creatures. *SIGACT News*, 35(2): 36–59.

Alur, R., T.A. Henzinger, and O. Kupferman. 2002. Alternating-Time Temporal Logic. *Journal of the ACM* 49: 672–713.

Baier, C., and J.-P. Katoen. 2008. *Principles of Model Checking*. Cambridge: The MIT Press.

Balbiani, P., H. van Ditmarsch, and S. Fernández González. 2019. Asynchronous announcements. *CoRR*, abs/1705.03392.

Baltag, A., L.S. Moss, and S. Solecki. 1998. The logic of public announcements, common knowledge, and private suspicions. In *Proc. of 7th TARK*, 43–56. Burlington: Morgan Kaufmann.

Baltag, A., and S. Smets. 2008. A qualitative theory of dynamic interactive belief revision. In *Proc. of 7th LOFT*, Texts in Logic and Games 3, 13–60. Amsterdam University Press.

Baltag, A., and S. Smets. 2020. Learning what others know. In *Proc. of 23rd LPAR*, eds. E. Albert and L. Kovacs, EPiC Series in Computing, vol. 73, 90–119.

Banerjee, M., and Md.A. Khan. 2007. Propositional Logics from Rough Set Theory. *Transactions on Rough Sets VI* 6: 1–25.

Biran, O., S. Moran, and S. Zaks. 1990. A Combinatorial Characterization of the Distributed 1-Solvable Tasks. *Journal of Algorithms* 11(3): 420–440.

Blackburn, P. 2000. Representation, Reasoning, and Relational Structures: A Hybrid Logic Manifesto. *Logic Journal of the IGPL* 8(3): 339–365.

Blackburn, P., M. de Rijke, and Y. Venema. 2001. *Modal Logic*. Cambridge Tracts in Theoretical Computer Science 53. Cambridge: Cambridge University Press.

Bracho, J., and L. Montejano. 1987. The Combinatorics of Colored Triangulations of Manifolds. *Geometriae Dedicata* 22(3): 303–328.

Brandt, F., V. Conitzer, U. Endriss, J. Lang, and A.D. Procaccia, eds. 2016. *Handbook of Computational Social Choice*. Cambridge: Cambridge University Press.

Charrier, T., and F. Schwarzentruber. 2015. Arbitrary Public Announcement Logic with Mental Programs. In *Proc. of AAMAS*, 1471–1479. New York: ACM.

Chellas, B.F. 1980. *Modal Logic: An Introduction*. Cambridge: Cambridge University Press.

Dabrowski, A., L.S. Moss, and R. Parikh. 1996. Topological Reasoning and the Logic of Knowledge. *Annals of Pure and Applied Logic* 78(1–3): 73–110.

Degremont, C., B. Löwe, and A. Witzel. 2011. The Synchronicity of Dynamic Epistemic Logic. In *Proc. of 13th TARK*, 145–152. New York: ACM.

Dixon, C., C. Nalon, and R. Ramanujam. 2015. Knowledge and Time. In *Handbook of Epistemic Logic*, eds. H. van Ditmarsch, J.Y. Halpern, W. van der Hoek, and B. Kooi, 205–259. London: College Publications.

Edelsbrunner, H., and J. Harer. 2010. *Computational Topology - An Introduction*. Providence: American Mathematical Society.

Fagin, R., and J.Y. Halpern. 1988. Belief, Awareness, and Limited Reasoning. *Artificial Intelligence* 34(1): 39–76.

Fagin, R., J.Y. Halpern, Y. Moses, and M.Y. Vardi. 1995. *Reasoning about Knowledge*. Cambridge: MIT Press.

Fischer, M.J., N.A. Lynch, and M. Paterson. 1985. Impossibility of Distributed Consensus with One Faulty Process. *Journal of the ACM* 32(2): 374–382.

Fisler, K., and M.Y. Vardi. 2002. Bisimulation Minimization and Symbolic Model Checking. *Formal Methods in System Design* 21(1): 39–78.

Fraigniaud, P., S. Rajsbaum, and C. Travers. 2013. Locality and Checkability in Wait-Free Computing. *Distributed Computing* 26(4): 223–242.

Friedell, M. 1969. On the Structure of Shared Awareness. *Behavioral Science* 14: 28–39.

Goubault, E., M. Lazic, J. Ledent, and S. Rajsbaum. 2019. Wait-Free Solvability of Equality Negation Tasks. In *Proc. of 33rd DISC*, 21:1–21:16.

Goubault, E., J. Ledent, and S. Rajsbaum. 2021. A Simplicial Complex Model for Dynamic Epistemic Logic to Study Distributed Task Computability. *Information and Computation* 278: 104597.

Halpern, J.Y., and Y. Moses. 1990. Knowledge and Common Knowledge in a Distributed Environment. *Journal of the ACM* 37(3): 549–587.

Halpern, J.Y., and Y. Moses. 2017. Characterizing Solution Concepts in Terms of Common Knowledge of Rationality. *International Journal of Game Theory* 46(2): 457–473.

Harel, D., D. Kozen, and J. Tiuryn. 2000. *Dynamic Logic*. Foundations of Computing Series. Cambridge, MA: MIT Press.

Hatcher, A. 2002. *Algebraic Topology*. Cambridge: Cambridge Univ. Press.

Hayek, F. 1945. The Use of Knowledge in Society. *American Economic Review* 35: 519–530.

Herlihy, M., D. Kozlov, and S. Rajsbaum. 2013. *Distributed Computing Through Combinatorial Topology*. San Francisco, CA: Morgan Kaufmann

Herlihy, M., and N. Shavit. 1993. The asynchronous computability theorem for t-resilient tasks. In *Proc. 25th Annual ACM Symposium on Theory of Computing*, 111–120.

Herlihy, M., and N. Shavit. 1999. The Topological Structure of Asynchronous Computability. *Journal of the ACM* 46(6): 858–923.

Hilton, P.J., and S. Wylie. 1960. *Homology Theory: An Introduction to Algebraic Topology*. Cambridge: Cambridge University Press.

Hintikka, J. 1962. *Knowledge and Belief. An Introduction to the Logic of the Two Notions*. Ithaca: Cornell University Press.

Knight, S. 2013. *The Epistemic View of Concurrency Theory*. PhD thesis, École Polytechnique, Palaiseau.

Knight, S., B. Maubert, and F. Schwarzentruber. 2019. Reasoning About Knowledge and Messages in Asynchronous Multi-Agent Systems. *Mathematical Structures in Computer Science* 29(1): 127–168.

Kozlov, D. 2008. *Combinatorial Algebraic Topology*. Berlin: Springer.

Lamport, L. 1978. Time, Clocks, and the Ordering of Events in a Distributed System. *Communications of the ACM* 21(7): 558–565.

Ledent, J. 2019. *Geometric Semantics for Asynchronous Computability*. PhD thesis, École Polytechnique, Palaiseau.

Lewis, D.K. 1969. *Convention, a Philosophical Study*. Cambridge: Harvard University Press.

Loui, M.C., and H.H. Abu-Amara. 1987. Memory Requirements for Agreement Among Unreliable Asynchronous Processes. *Advances in Computing Research* 4: 163–183.

Meyer, J.-J.Ch., and W. van der Hoek. 1995. *Epistemic Logic for AI and Computer Science*. Cambridge Tracts in Theoretical Computer Science 41. Cambridge: Cambridge University Press.

Osborne, M.J., and A. Rubinstein. 1994. *A Course in Game Theory*. Cambridge: MIT Press.

Özgün, A. 2017. *Evidence in Epistemic Logic: A Topological Perspective*. PhD thesis, University of Lorraine & University of Amsterdam.

Pacuit, E. 2017. *Neighborhood Semantics for Modal Logic*. Berlin: Springer.

Paige, R., and R.E. Tarjan. 1987. Three Partition Refinement Algorithms. *SIAM Journal on Computing* 16(6): 973–989.

Panangaden, P., and K. Taylor. 1992. Concurrent Common Knowledge: Defining Agreement for Asynchronous Systems. *Distributed Computing* 6: 73–93.

Parikh, R., L.S. Moss, and C. Steinsvold. 2007. Topology and epistemic logic. In *Handbook of Spatial Logics*, eds. M. Aiello, I. Pratt-Hartmann, and J. van Benthem, 299–341. Berlin: Springer.

Peleg, D. 1987. Concurrent Dynamic Logic. *Journal of the ACM* 34(2): 450–479.

Pnueli, A. 1977. The temporal logic of programs. In *Proc. of 18th FOCS*, 46–57. Los Alamitos : IEEE Computer Society.

Porter, T. 2002. Geometric Aspects of Multiagent Systems. *Electronic Notes in Theoretical Computer Science* 81: 73–98.

Roelofsen, F. 2007. Distributed Knowledge. *Journal of Applied Non-Classical Logics* 17(2): 255–273.

Rotman, J. 1973. Covering Complexes with Applications to Algebra. *The Rocky Mountain Journal of Mathematics* 3(4): 641–674.

Sangiorgi, D. 2011. *An Introduction to Bisimulation and Coinduction*. Cambridge: Cambridge University Press.

van Benthem, J. 1998. Dynamic odds and ends. Technical report, University of Amsterdam. ILLC Research Report ML-1998-08.

van Benthem, J. 2010. *Modal Logic for Open Minds*. Stanford, CA: Center for the Study of Language and Information.

van Benthem, J., and S. Smets. 2015. Dynamic logics of belief change. In *Handbook of Epistemic Logic*, eds. H. van Ditmarsch, J.Y. Halpern, W. van der Hoek, and B. Kooi, 313–394. London: College Publications.

van Benthem, J., J. van Eijck, M. Gattinger, and K. Su. 2018. Symbolic Model Checking for Dynamic Epistemic Logic - S5 and Beyond. *Journal of Logic and Computation* 28(2): 367–402.

van Benthem, J., J. van Eijck, and B. Kooi. 2006. Logics of Communication and Change. *Information and Computation* 204(11): 1620–1662.

van Ditmarsch, H., D. Fernández-Duque, and W. van der Hoek. 2014. On the Definability of Simulation and Bisimulation in Epistemic Logic. *Journal of Logic and Computation* 24(6): 1209–1227.

van Ditmarsch, H., and T. French. 2009. Awareness and forgetting of facts and agents. In *Proc. of WI-IAT Workshops 2009*, eds. P. Boldi, G. Vizzari, G. Pasi, and R. Baeza-Yates, 478–483. Hoboken: IEEE Press.

van Ditmarsch, H., E. Goubault, M. Lazic, J. Ledent, and S. Rajsbaum. 2021. A Dynamic Epistemic Logic Analysis of Equality Negation and Other Epistemic Covering Tasks. *Journal of Logical and Algebraic Methods in Programming* 121: 100662.

van Ditmarsch, H., J.Y. Halpern, W. van der Hoek, and B. Kooi, eds. 2015. *Handbook of Epistemic Logic*. London: College Publications.

van Ditmarsch, H., and B. Kooi. 2008. Semantic results for ontic and epistemic change. In *Proc. of 7th LOFT*, Texts in Logic and Games 3, 87–117. Amsterdam: Amsterdam University Press.

van Ditmarsch, H., W. van der Hoek, and B. Kooi. 2008. *Dynamic Epistemic Logic*. Volume 337 of Synthese Library. Berlin: Springer.

van Ditmarsch, H., J. van Eijck, and R. Verbrugge. 2009. Common knowledge and common belief. In *Discourses on Social Software*, J. van Eijck and R. Verbrugge, 99–122. Amsterdam: Amsterdam University Press.
van Glabbeek, R.J. 1990. *Comparative Concurrency Semantics and Refinement of Actions*. PhD thesis, Vrije Universiteit Amsterdam.
van Wijk, S. 2015. Coalitions in epistemic planning. Technical report, University of Amsterdam. ILLC report MoL-2015-26 (MSc thesis).

Chapter 2
Meta-Abduction: Inference to the Probabilistically Best Prediction

Christian J. Feldbacher-Escamilla

Abstract In this paper, a taxonomy of abduction in a wide sense and an exact characterisation of probabilistic selective abduction is given. Two important features of such inferences, namely accuracy and simplicity of explanations and predictions, are described in detail. Afterwards, the epistemic merits of simplicity are discussed. They are used to justify probabilistic selective abduction in terms of an inference to the probabilistically best explanation. By help of the theory of meta-induction and its account of induction, this justification is expanded to abduction in terms of an inference to the probabilistically best prediction. This expansion of the theory of meta-induction to a theory of meta-abduction indicates that the framework fruitfully accounts for important inference methods of science.

2.1 Introduction

The theory of meta-induction allows for justifying inductive inferences based on their past successes. Some have argued that this, finally, solves Hume's problem of induction, at least if one considers it in terms of optimisation (cf. Schurz 2019; and Feldbacher-Escamilla under revision). However, there is also another inference method which is widely-used in science and which is also in need of epistemic justification, namely the method of abduction (cf. Peirce 1994; Harman 1965; and Lipton 2004). The question of justifying abduction will be addressed in the present essay. To be more precise, we are interested in a particular species of abductive reasoning, namely the species of an *inference to the probabilistically best explanation and prediction*. We will show that the meta-inductive framework can be interpreted such that it allows for embedding these inferences and applying the meta-inductive optimality results. By this, meta-induction or, as we will call it: *meta-*

C. J. Feldbacher-Escamilla (✉)
Department of Philosophy, University of Cologne, Germany
e-mail: cj.feldbacher.escamilla@uni-koeln.de

B. Lundgren, N. A. Nuñez Hernández (eds.), *Philosophy of Computing*,
Philosophical Studies Series 143, https://doi.org/10.1007/978-3-030-75267-5_2

abduction, can be not only employed to justify induction, but also to justify a species of abduction.

Our investigation proceeds as follows: In Sect. 2.2, we provide a brief outline of abductive inferences. In Sect. 2.3, we characterise in detail the species of abductive inferences we are interested in here, namely selective abduction in form of an inference to the probabilistically best explanation or prediction, where *best* is understood in terms of accuracy and simplicity. Since the epistemic value of accurate explanations and predictions is clear, but that of simplicity is not, the latter needs to be explored further. This is done in Sect. 2.4, where an information theoretical argument in favour of simplicity is discussed and used for fleshing out the notion of the species of abductive inference we have in mind. As we will see, the epistemic justification of accuracy and simplicity of abduction is guaranteed for an inference to the probabilistically best *explanation*. However, abduction in the sense of an inference to the probabilistically best *prediction* needs further steps of justification. As we will argue, the theory of meta-induction allows to perform these steps. For this purpose, we introduce the framework of meta-induction in Sect. 2.5. There we also outline its vindication of induction. In Sect. 2.6, we expand the epistemic justification of abduction as an inference to the probabilistically best *explanation* to the case of abduction as an inference to the probabilistically best *prediction*. We conclude in Sect. 2.7.

2.2 Abduction

There are three major types of inference used in science and philosophy: deduction, induction, and abduction. Deductive inferences are truth preserving. Inductive inferences are not truth preserving, but have conclusions containing terms that occur already in the premisses. Finally, abduction is formally characterised as a non-deductive inference with a conclusion containing also terms that do not occur already in the premisses. It was already in the premisses. It was Charles S. Peirce who first discussed *abductive inferences* as a topic of philosophy of science and logic in the broad sense. This formal aspect of abductive inferences was described by him as follows: "An Abduction is Originary in respect to being the only kind of argument which starts a new idea" (Peirce 1994, p. 5.145). Clearly, this formal characterisation does not provide much of a restriction. However, it is an important feature that distinguishes it from the other forms, deduction and induction. For this reason, we want to call it a characteristic of *abduction in the wide sense*. To give an example of the different forms of reasoning, one can say that, e.g., $\{\forall x R(x)\} \vdash R(c)$ is a deductive inference, because it is truth-preserving and does not (relevantly) introduce new terms. $\{R(c_1), \ldots, R(c_n)\} \not\mid\!\sim \forall x R(x)$ is an inductive inference because it is ampliative and also does not (relevantly) introduce new terms. And, e.g., the inference from $\exists^n_n R(x)$, $\exists^m_m W(x)$ to $\exists^l_l H_R(x)$, $\exists^k_k H_W(x)$, $\exists^j_j M(x)$ is an abductive one, since H_R, H_W, and M are terms (representing ideas) that do not occur in the premise set ('$\exists_n^n$' stands for 'there are exactly n things such that ...'; in the example below, we will use numerical quantification with sortal variables $x_0, x_1, \ldots$ for expressing absolute frequencies within different generations).

Abductive inferences play a major role in natural science as they are widely used for theory construction. Simplified speaking, by abduction one can infer from empirically accessible data theoretical hypotheses that allow for a more or less simple explanation of the data. Prominent is the example of Gregor Mendel who inferred from phenotypic properties of plants, e.g. colours red R and white W, laws of inheritance, e.g. the inheritance of recessive homozygous white H_W, dominant homozygous red H_R and mixed M traits. He hypothesised about the existence of such an—at his times observationally not accessible—theoretical structure. Based on such a structure, he could explain, e.g., the frequencies of the phenotypic properties in different generations. E.g., that 50% of the plants of the mother generation (let us say 100—Mendel cultivated and tested in the years from 1856–1863 about 5000 pea plants) were white ($\exists_{50}^{50} x_0 W(x_0)$) and the other 50% of the plants were dominantly red ($\exists_{50}^{50} x_0 R(x_0)$), i.e. resulted themselves from breeding only red plants over several generations, was interpreted by Mendel as having only homozygous white and homozygous red plants in the mother generation: $\exists_{50}^{50} x_0 H_W(x_0)$ and $\exists_{50}^{50} x_0 H_R(x_0)$. He further hypothesised that by interbreeding the white with the red plants 100% mixed plants arose in the first daughter generation ($\exists_{100}^{100} x_1 M(x_1)$) which, by the assumption that red was dominant, allowed for explaining the fact that the first daughter generation was completely red ($\exists_{100}^{100} x_1 R(x_1)$). By the same type of reasoning he could argue that interbreeding the mixed plants results in 25% homozygous red ($\exists_{25}^{25} x_2 H_R(x_2)$), 25% homozygous white ($\exists_{25}^{25} x_2 H_W(x_2)$), and 50% mixed ($\exists_{50}^{50} x_2 M(x_2)$) traits, which predicted or explained that in the second daughter generation, due to the dominance of red, 75% of the plants were red ($\exists_{75}^{75} x_2 R(x_2)$) and 25% were white ($\exists_{25}^{25} x_2 W(x_2)$). This also explained how a phenotypical property that disappeared in one generation, namely white, could return in another generation.

More generally, we can distinguish two aspects or kinds of abductive inferences in the wide sense (cf. Douven 2018; Schurz 2008a; Aliseda 2006, p. 46): those generating new hypotheses and those aiming at determining the best hypothesis from a set of available candidates. Abductive inferences of the former kind are sometimes called *creative abductions*, and those of the latter kind *selective abductions* (see, e.g. Magnani 2000; Schurz 2008a; Feldbacher-Escamilla and Gebharter 2019).

The account of Peirce is commonly subordinated to creative abduction. Since Peirce coined the term for this form of inference, we call it also *abduction in the narrow sense* here. Peirce provided the following very general inference schema for it (see Peirce 1994, 5.189):

1. The surprising fact, E, is observed;
2. But if H were true, E would be a matter of course;
3. Hence, there is reason to suspect that H is true.

Since abduction in the narrow sense is about generating new hypotheses and theories, it concerns not only the context of justification of theories, but also their context of discovery. Though most philosophers of science are quite sceptical whether a general approach towards a logic of scientific inquiry can be fruitful, there are accounts that allow for a systematic methodology of abductive hypothesis generation in terms of common cause abduction for generating hypotheses featuring new theoretical concepts on the basis of empirical phenomena (cf. Schurz 2008a; and a generalisation of the approach in Feldbacher-Escamilla and Gebharter 2019; but also Glymour 2018; for a historical case study, cf. Feldbacher-Escamilla 2019, sect. 4).

Selective abduction, on the other hand, is often described as an *inference to the best explanation* and most of the philosophical literature on abduction (in the wide sense) focuses on this form of inference (see, e.g., Harman 1965; Lipton 2004; Niiniluoto 1999; Williamson 2016). It is about selecting among a set of possible explanations and predictions that one with most explanatory and predictive virtues.

The exact relation between creative abduction (or abduction in a narrow sense) and selective abduction is a matter of long-lasting philosophical dispute. A reason for this is that Peirce's criteria and their domain of application is not entirely clear and he himself underwent some development regarding his understanding of abduction (Mohammadian 2019, sect. 2, distinguishes, e.g., between "the early theory from 1859 to 1890 and the later theory from 1890 to 1914"; and Hintikka 1998, p. 511, speaks about "Peirce's early perspective on abduction" and "Peirce's mature view"). Also proponents of the selective abductive camp can be blamed for a similar fault. And a further reason for the long-lasting dispute is that particularly adherents of inference to the best explanation seemed to aim at endowing a long tradition by sloppily equating selective with creative abduction. So, e.g., Harman (1965, p. 88) claimed that "'the inference to the best explanation' corresponds approximately to what others have called 'abduction'" and Lipton (2004, p. 56) claimed that "Inference to the Best Explanation has become extremely popular in philosophical circles, discussed by many and endorsed without discussion by many more (for discussions see, e.g., Peirce (1994))".

In stark contrast to the "identifiers" of abduction with inference to the best explanation, there are strong "separationists" such as Hintikka (1998). According to them, abduction is not at all related to inference to the best explanation. For Hintikka (1998), to explain is to perform a deductive activity, whereas Peircean abduction is an activity of introducing new hypotheses into inquiry, and since deduction and the introduction of new hypotheses are completely different matters, also inference to the best explanation and abduction should be understood as completely different forms of inferences:

> An explainer's job description is [...] twofold: on the one hand to find the auxiliary facts A and on the other hand to deduce the explanandum from them together with the background theory T. (cf. p. 507)
> Since the abductive reasoner does not always have at his or her disposal explanations even of the known data, the abductive inference cannot be a step to the known data to a hypothesis or theory that best explains them. (cf. p. 509)

Roughly, one can also describe the "separationist's" argument as that of abduction being a matter of the context of discovery, explanation being a matter of the context of justification, and since both contexts are separated, one ought also to keep abduction separated from inference to the best explanation.

There are, however, also positions strictly within the spectrum spanning from "identifiers" to "separationists", namely those who agree that abduction and inference to the best explanation are different, but nevertheless share important components. So, e.g., Mohammadian (2019) provides such a middle ground by arguing for the formula "Abduction − The Context of Discovery + Underdetermination = Inference to the Best Explanation". Similarly, Aliseda (2006, p. 46) stresses that the "process side" of abduction consists of both, construction and selection, and highlights the constructive part by help of the formula "Abductive Logic: Inference + Search Strategy" (cf. p. 49).

For the purpose of our investigation, we do not need to take a stance on whether and how exactly creative abduction relates to selective abduction. Rather, we need to characterise the kind of abductive inference (in the wide sense) we are interested in and locate it within the general taxonomy. By this, we want to provide a clear picture with sharp boundaries that mark for which form of abductive inference our justification is fit and for which it is not.

Elsewhere, I have provided an approach to creative abduction in terms of common cause/common origin reasoning (Feldbacher-Escamilla and Gebharter 2019). This essay aims at approaching the problem of how to justify selective abduction. As we have outlined above, the main idea of selective abduction is to infer from some evidence that hypothesis or theory from a set of alternative hypotheses or theories, which explains the evidence best. In practice as well as in theory, pretty much everyone agrees on this. However, different accounts of selective abduction result from different ways of fleshing out what is meant with *best explanation*. It is clear that *best* should be spelled out in terms of explanatory virtues, but what exactly are these virtues? Harman (1965, p. 89), when introducing the notion of an inference to the best explanation, considered that explanation to be the best, which is "simpler, which is more plausible, which explains more, which is less *ad hoc*, and so forth". In a similar line, Lipton (2004) ranks explanations as better, if they are more plausible and contrastive (in the sense of specifying and producing discriminating evidence), and among those that are equally plausible and contrastive, being better becomes a "question of comparative loveliness" (cf. p. 90) which is dependent of "precision, scope, simplicity, fertility or fruitfulness, and fit with background belief" (cf. p. 122).

An important problem of inference to the best explanation Harman/Lipton style is the question of its epistemic justification: What epistemic role do simplicity, scope, etc., and loveliness in general have? Lipton (2004, pp. 61f) stressed the role of loveliness to be that of increasing understanding and, in a scientific realist spirit by linking "the search for truth and the search for understanding" also considered the epistemic role of loveliness to be established. However, such a route of loveliness as a reliable indicator of truth or likeliness is not open to anti-realist philosophers of science.

An extreme way out of this solution is to set as a benchmark for an inference to the best explanation something that expresses an epistemic value *per se*, i.e. truth or likelihood. Such an approach would be an account of an *inference to the most likely explanation*. However, as Lipton (2004) has argued, focussing on likelihood alone would render the account trivial in the sense that we would simply re-state something that we wanted to explain:

> [There are] two more versions of Inference to the Best Explanation to consider: Inference to the Likeliest Potential Explanation and Inference to the Loveliest Potential Explanation. Which should we choose? There is a natural temptation to plump for likeliness. After all, Inference to the Best Explanation is supposed to describe strong inductive arguments, and a strong inductive argument is one where the premises make the conclusion likely. But in fact this connection is too close and, as a consequence, choosing likeliness would push Inference to the Best Explanation towards triviality. We want a model of inductive inference to describe what principles we use to judge one inference more likely than another, so to say that we infer the likeliest explanation is not helpful. (Lipton 2004, p. 60):

So, the problem seems to be that selective abduction Harman/Lipton style is explanatorily potent, but epistemically hard to come by. And selective abduction in the sense of an inference to the most likely explanation is epistemically justified, but explanatorily less (or even im-)potent. However, we think that there is a form of selective abduction that can be epistemically accounted for, i.e. which can be epistemically justified, and which is at the same time explanatorily potent. The idea is to spell out the epistemic value of explanatory virtues such as simplicity in probabilistic terms. Since this account aims at phrasing everything in probabilistic terms, we want to call this form of abduction (in the wide sense) an *inference to the probabilistically best explanation*. It is important to note that this form of selective abduction does take the likelihood into account, however, it is no inference to the most likely explanation, because it also takes other explanatory features into account, though these are also spelled out in probabilistic terms. An approach along these lines is, e.g., outlined in Williamson (2016).

Due to the structural identity of explanations and predictions (we subscribe to the so-called *structural identity thesis*, cf. Hempel 1965, pp. 366–376), we can distinguish this form of abductive reasoning further into inference to the probabilistically best *explanation* and inference to the probabilistically best *prediction*. Figure 2.1 provides an overview of our taxonomy of abduction in the wide sense and indicates where to locate the species of abduction we are concerned with in this essay.

To briefly take stock, we have provided a taxonomy of abductive inferences (in the wide sense). We have seen that in the literature there are accounts that lump the different forms together ("identifiers"), there are accounts that keep them strictly separate from each other ("separationists"), and there are intermediary positions (like the approach of "Abduction − The Context of Discovery + Underdetermination = Inference to the Best Explanation" of Mohammadian 2019). However, since we focus on a particular species of abduction and aim to provide an epistemological justification for it, and only for it, we do not need to take a stance on this. Note, however, whereas Peirce' "early perspective on abduction" seems to support the identifiers' position, his "mature view" rules out such an identification, at least with

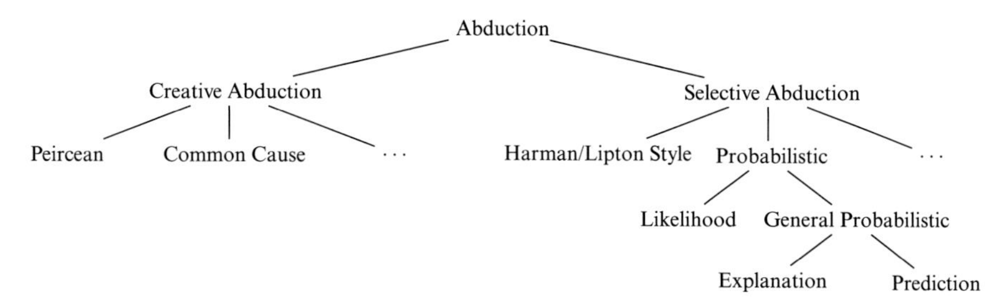

Fig. 2.1 A taxonomy of abduction in the wide sense, where abduction in the narrow sense amounts to Peircean creative abduction; the species we are interested in is abduction in the sense of an inference to the probabilistically best explanation/prediction

respect to the species of probabilistic selective abduction, because in his later view he definitely ruled out probabilistic considerations (thanks to an anonymous referee for pointing this out to me):

> [In my early perspective on abduction] my conceptions of Abduction necessarily confused two different kinds of reasoning. When, after repeated attempts, I finally succeeded in clearing the matter up [i.e. in the mature view], the fact shone out that probability proper had nothing to do with the validity of Abduction[.] (Peirce 1994, 2.102)

In the next Sect. 2.3, we provide more details on abduction in the sense of an inference to the probabilistically best explanation. In the subsequent Sect. 2.4, we will argue for its epistemic justification. Sections 2.5 and 2.6 are devoted to the case of abduction in the sense of an inference to the probabilistically best prediction.

2.3 Inference to the Probabilistically Best Explanation

Now, what are particular characteristics of selective abductive inferences in the sense of an inference to the probabilistically best explanation? Besides the formal constraint of introducing new (theoretical) vocabulary, materially speaking characteristic for this form of abduction is its validation of explanations. So, usually an abductive inference has no single statement as a conclusion, but laws and regularities that can be used in an explanation or that even form a whole theory. In the case of Mendel's abductive inference, given the premise set outlined in our discussion of the example above, the laws and regularities of a validated explanation might consist of assumptions about the traits (the initial traits being dominant or recessive and the daughter traits being dominant, recessive, or mixed) as well as probabilistic reasoning based on assumptions about the average number of descendants of each possible pair per generation.

What are the constraints for validating such an explanation? Abduction in the sense of an *inference to the probabilistically best explanation* (see, e.g., Williamson 2016) is usually supposed to maximise the data's plausibility (in the sense of the likelihood) and the hypotheses' simplicity. Typically, also further features such as scope, fruitfulness or non-ad hocness are put forward, however, for a lack of space we will focus on simplicity as a proxy for such virtues (the probabilistic reconstruction of the epistemic value of other virtues such as scope, fruitfulness, non-ad hocness etc. is in line with that of simplicity; for details cf. Forster and Sober 1994).

Regarding plausibility/likelihood, the parameter consists in the probability of the premise P (also: the *explanandum*) in the light of laws and regularities used in the explanation (the conclusion C or also: the *explanans*). The simplicity constraint is considered to be necessary in order to rule out ad hoc explanations. For, if one takes, e.g., scientists' conditional degrees of belief Pr of the explanandum P in the light of the explanans C as a measure for plausibility/likelihood: $Pr(P|C)$, which is the likelihood of P given C, then it is clear that choosing a C such that $C \vdash P$ maximises the explanandum's plausibility/likelihood in the light of the explanans. In the simplest case one might set ad hoc: $C = P$. However, what we aim at are not ad hoc explanations that might be even trivial, but universal explanations. Since ad hoc explanations usually turn out to become more and more complex with an increased number of data, abductive validation of an explanation hinges not only on $Pr(P|C)$, but also on C's simplicity. If we assume that there is some way of measuring C's complexity via a non-negative function $c(C)$, then we can characterise the validation procedure of an abductive inference as trying to maximise $Pr(P|C)$ on the one side, and minimise $c(C)$ on the other.

Several remarks are in place: First, in order to remain applicable, in this method the aim of maximising $Pr(P|C)$ and minimising $c(C)$ is to be understood not in absolute terms, but in relative ones. We might aim at $Pr(P|C) = 1$ and $c(C) = 0$, but we will almost never achieve this goal. In particular, it is presupposed that we exclude trivial abductive inferences to P itself ($Pr(P|P) = 1$ is maximal). As we have mentioned above, with an increased number of data P to be plausibly (in the sense of likelihood) explained by C usually also the complexity of C increases (this is a case where scope opposes simplicity). And on the other hand, reducing the complexity of C usually leads to generalisations of C that are not in full agreement with P, for which reason $Pr(P|C)$ decreases. Since these two measures are intertwined in many applications, often finding a C such that $Pr(P|C) = 1$ and $c(C) = 0$ is not achievable. This was highlighted, e.g., also by Karl R. Popper, who claimed that the aim of increasing $Pr(P|C)$ "inadvertently but necessarily, implies the unacceptable rule: always use the theory which is the most *ad hoc*, i.e. which transcends the available evidence as little as possible [i.e. which sets $C = P$]" (see Popper 2002, p. 61). However, what is clearly achievable is a comparative task: Assume that the only available *potential explanantia* are $C_1, \ldots, C_n$. Then it holds:

If there is a $i \in \{1, \ldots, n\}$ such that for all $j \in \{1, \ldots, n\} \setminus \{i\}$:

$$c(C_i) \leq c(C_j) \;\&\; Pr(P|C_i) > Pr(P|C_j)$$

$$\text{or} \qquad \text{(Abd)}$$

$$c(C_i) < c(C_j) \;\&\; Pr(P|C_i) \geq Pr(P|C_j),$$

then infer from P by probabilistic selective abduction C_i.

(Abd) demands that in case there is an explanans C_i which plausibilises P better (in the sense of increased likelihood), but not at cost of being more complex, or which is simpler, but still not at cost of less plausibilising P than all the other possible explanantia, that in such a case C_i is to be inferred from P. If one generalises this comparative validation to the set of all potential explanantia *one has thought of* (see Williamson 2016, p. 267), then one might regain an absolute phrasing of selective abductive inferences in the sense of selecting the probabilistically best explanation that is still applicable.

Second, $Pr(P|C)$ and $c(C)$ can be balanced in several ways. One might consider, e.g., a combination of the form $(1 - Pr(P|C)) \cdot c(C)$ which is the product of the inverse of the likelihood and complexity that is to be minimised, but one might also think of maximising $Pr(P|C) - c(C)$. These possibilities of balancing lead to different inferences in at least some applications. However, what is important to note is that they still satisfy (Abd). This is also the minimal constraint we want to put forward for abduction and as long as a non-deductive inference rule introducing new vocabulary satisfies it, we think it is fine to call it an 'abductive' one. In the next Sect. 2.4, we will consider another way of balancing that also satisfies (Abd).

Third, the two parameters $c(C)$ (as a proxy for further explanatory virtues such as scope, fruitfulness, non-ad hocness etc.) and $Pr(P|C)$ are not sufficient for providing a fully adequate account of selective abduction. Usually, also the prior probabilities of the hypotheses used in an explanation are relevant. E.g., if $Pr(C_i)$ is very close to 0 and $Pr(C_j)$ is high, then one will still tend to opt for C_j instead of C_i, although $Pr(P|C_i)$ might be greater than $Pr(P|C_j)$. For simplicity reasons, we restrict the application of (Abd) to cases with close prior probabilities of the alternative hypotheses $C_1, \ldots, C_n$ (other approaches like that of Solomonoff 1964a,b, link prior probabilities to complexity c). Also, as is pointed out in Schurz (2008a), other theoretical virtues of C as, e.g., use-novelty, unification, etc. are typically considered to be relevant for abductive inferences. Again, we restrict the intended application of (Abd) to cases where these theoretical virtues are considered to be satisfied equally well. The reason for this strong restrictions is twofold. First, some of these further parameters might be reducible to the two we are proposing. So, e.g., regarding unification and use-novelty, Forster and Sober (1994) provide reduction strategies which might be cashed out be allowing for complex P and C (conjunctions of descriptions of phenomena and hypotheses). In principle, one might even think of reducing the prior probability of a hypothesis ($Pr(C)$) as relevant parameter via inversely relating it to the complexity measure $c(C)$ (as mentioned, this would be along the lines of Solomonoff 1964a,b). The second reason is that in this essay we are only interested in an exemplary application of meta-induction to abductive inferences. For this purpose, it suffices to show how the theory can be applied in case one scores not only according to accuracy, but also according to some theoretical value like simplicity/complexity. So, we should mention that our aim is to theorise about a simple model of abduction, and we want to stress that, clearly, this model has lots of limitations in comparison to the full repertoire of abduction in the wide sense.

In this model of an abductive inference there are basically two main ingredients: $Pr(P|C)$ and $c(C)$. One might wonder why $c(C)$ is relevant here. It is not hard to provide an epistemic rationale for maximising $Pr(P|C)$ in an inference of C out of P, since it is a central aim of science and philosophy to provide *good* explanations. In the traditional *deductive nomological* model of explanations, the paradigmatic case of a good explanation consists of a deductively valid argument with true laws and auxiliary assumptions as premisses and the claim to be explained as the conclusion of the argument (see Hempel 1965). Now, a high likelihood of P given C *approximates* deduction of P from C for which reason maximising $Pr(P|C)$ also serves for approximating the paradigmatic case of a good explanation (in this respect an inference to the probabilistically best explanation is along the lines of an inference to the most likely explanation). But what about $c(C)$? In how far does decreased complexity or increased simplicity serve the epistemic goals of science and philosophy? Clearly, without taking into account $c(C)$ we would lack a criterion of selecting among a multitude of potential explanations. But if it were just for reducing the number of potential explanations then also a random choice would serve the aim. According to the argument above, not considering $c(C)$ would allow for ad hoc explanations. But what is the epistemic rationale of excluding ad hoc explanations? One argument which is brought forward quite often is that ad hoc explanations *overfit* the data and so in case there is some error in the data, ad hoc explanations also fit errors. So, the argument is that since P might contain false values or statements, validating explanations that perfectly explain erroneous P are themselves defective and their explanantia C wrong. Since decreased complexity $c(C)$ allows for avoiding overfitting, less complex Cs are also less prone to fit errors. As the literature on model selection shows, this can provide a rationale for also taking into account $c(C)$ in choosing among accessible potential explanations.

To sum up, we are interested in abductive inference in the sense of an inference to the probabilistically best accessible potential explanation. 'Best' is understood as balancing two measures of an explanation of P by help of C: the likelihood $Pr(P|C)$ should be high and the complexity $c(C)$ should be low. An epistemic rationale for the former constraint results from approximating traditional models of explanation. Such a rationale for the second constraint might result from considerations of the literature on model selection showing that $c(C)$ influences C's proneness of overfitting, and by this C's proneness of also fitting errors. In the following Sect. 2.4 we are going to make this argument in favour of minimising $c(C)$ explicit.

2.4 Simplicity and the Akaike Information Criterion

One way of arguing for minimising $c(C)$ is to postulate as aim of science and philosophy not only to provide true explanations, but also non-ad hoc, universal, simple explanations. In this way, already by convention about the aim of science and philosophy a demand of minimising $c(C)$ follows. However, there is also the possibility of trying to reduce the epistemic value of minimising $c(C)$ to the epistemic value of providing true explanations. The most famous approach in this direction is an application of an information theoretical framework to the problem of how to epistemically justify simplicity. The main line of argumentation is as follows (see Forster and Sober 1994):

(i) Data P might be noisy and involve *error*.
(ii) An accurate fit of an explanans C to the data P fits also error, it overfits the data.
(iii) Whereas a less accurate fit of C to P may depart from error: Closeness to the truth is different from closeness to the data.
(iv) Fact: The more parameters an explanans C has, the more prone it is to overfit P.
(v) Hence: Simplicity in the sense of having less parameters may account for inaccuracy w.r.t. data P in order to achieve accuracy w.r.t. the truth. So, simplicity is instrumental for truth.

This argument is valid along general lines. But what about the truth of the premisses? Considering applications of the abductive methodology to the natural sciences, premise (i), the assumption of error in the data, is a very natural assumption. But then also premise (ii) and (iii) are straightforward: Assuming that P contains errors one only has a chance of achieving the truth by deviating from P. Intuitively and qualitatively speaking, premise (iv) is also straightforward: If an explanans is complex, it allows for fitting a simple as well as a complex explanandum. If an explanans is simple, it might fit a simple explanandum, but it cannot fit a complex one. But clearly, this is an argument too coarse-grained in order to be convincingly applied for quantitative considerations regarding minimising $c(C)$. However, there is also a much more fine-grained version of premise (iv) stemming from the literature on model selection and curve fitting—here we focus on the latter, since it became a quite influential approach to the epistemic value of simplicity (see Forster and Sober 1994). Note, however, that due to this setup we also focus on a very particular notion of *simplicity/complexity* only, namely parametric simplicity (for other notions of and approaches to simplicity and how to generalise the account presented here, cf. Kelly 2007).

For illustrative purposes, we will make only very simplified considerations here. The idea of model selection is that, given a data set $X = \{x_1, \ldots, x_n\}$, one is looking for a curve $F = \{f_1, \ldots, f_n, \ldots\}$ that adequately fits X. Now, it is assumed that X might contain errors, so X deviates from the truth $T = \{y_1, \ldots, y_n, \ldots\}$ (see premise (i)). Clearly, the perfect choice would be $F = T$, regardless of X, but since only X is available to us, we have to base our choice of F on X. It is also clear that for any data set X with $n = |X|$ elements, choosing as F a polynomial of degree $n - 1$ allows one to perfectly fit X. One can always find parameters $a_{n-1}, \ldots, a_0$ such that for all $x \in X$ there is a $z \in \mathbb{R}$: $\langle z, x \rangle \in F$, given $F(z) = x = a_{n-1} \cdot z^{n-1} + \cdots + a_1 \cdot z^1 + a_0$. So, n parameters $(a_{n-1}, \ldots, a_0)$ allow for defining an F that perfectly fits X. If F has less parameters than n, then it is possible that there are cases where one cannot fit F perfectly to X. So, the number of parameters of F determines possibilities of perfect fitting. However, fitting X perfectly might deviate from the truth T, whereas fitting X imperfectly might allow for achieving the truth T (see premisses (ii) and (iii)).

Clearly, whether an inaccurate fit brings us closer to the truth or not depends on the exact specifics of error, namely the distance between X and T. If there were no error ($X \subseteq T$), then a more accurate and complex model would be better off than a simpler but less accurate model. However, a famous result of Hirotugu Akaike shows that on average (i.e. in estimating) simplicity matters. Forster and Sober (1994) have transformed Akaike's result to the philosophical debate of problems surrounding curve fitting. The result is the following one (see Forster and Sober 1994, p. 10): The estimated predictive accuracy of the family of a model F given some data X, which is also called the *Akaike information measure according to the Akaike information criterion*, is determined by (this criterion serves only as a proxy here and one can draw the same epistemic lesson about the value of simplicity by other information criteria like the Bayes information criterion):

$$AIC(F, X) \propto \log(Pr(X|F)) - c(F) \tag{AIC}$$

Where $c(F)$ is the number of parameters of F (i.e. the degree of the polynomial F plus 1) and F is supposed to be most accurately parameterised regarding X (i.e. it is the/a polynomial of degree $c(F)$ that is closest to X in terms of the sum of squares of the differences).

Note that the idea of the Akaike information criterion is to select an F such that the estimated accuracy regarding the truth T of the family of F given some data X is maximised. Now, as (AIC) tells us, maximising $AIC(F, X)$ is twofold: It consists of maximising the log-likelihood of F given X while at the same time one needs to keep an eye on holding complexity or the number of parameters of F low.

This framework has a wide range of applications (cf. Forster and Sober 1994), however, here we are interested on cashing out (AIC) also for reducing the value of simplicity of the abductive methodology presented before—namely the value of $c(C)$ in (Abd)—to the epistemic value of gaining truths. At least at first glance it is quite straightforward to implement (AIC) into the abductive methodology outlined above: The data set X is to be identified with the premise of the abductive inference P, the explanandum. And the conclusion of the abductive inference C, the explanans, is to be identified with the curve that tries to fit X, i.e. F. The result is an Akaike-motivated characterisation of abductive reasoning: Assume that the only available *potential explanantia* are $C_1, \ldots, C_n$. Then it holds:

C_i can be inferred from P by probabilistic selective abduction iff

$$\text{for all } j \in \{1, \ldots, n\}:$$

$$\log(Pr(P|C_i)) - c(C_i) \geq \log(Pr(P|C_j)) - c(C_j) \qquad \text{(AIC-Abd)}$$

(In case more than one C_i satisfy this constraint one might freely choose among them.)

According to this characterisation, every inference to an explanation is abductively permitted if it manages to get the best balance between likelihood and simplicity. Note that since $Pr(P|C) \in [0, 1]$, $\log(Pr(P|C)) \in (-\infty, 0]$. Furthermore, in principle the complexity of C might have no upper limit (F might be a polynomial of arbitrarily high degree), so $c(C) \in [0, +\infty)$. So, in trying to maximise $Pr(P|C)$ and minimise $c(C)$ one also tries to maximise $\log(Pr(P|C)) - c(C)$.

More generally, (AIC-Abd) also satisfies the constraint (Abd), and since (AIC-Abd) is stronger than (Abd), it is also more specific. What is important for our argumentation is that the implementation of (AIC) in agreement with (Abd) in the criterion (AIC-Abd) seems to do the job of reducing the value of simplicity (low $c(C)$) to the epistemic value of gaining truth.

In this sense, selective abduction as an inference to the probabilistically best *explanation* is epistemically justified. However, selective abduction can not only be used as an inference to the probabilistically best *explanation*, but also as an inference to the probabilistically best *prediction*. And the question is how one can epistemically justify such an abduction, if one is missing future data for figuring out what is the best balance between accuracy and simplicity. In the following, we aim at outlining how the epistemic justification of abduction in the sense of inferring an explanation can be expanded to a justification of abduction in the sense of inferring a prediction. For this purpose, we first need to introduce the framework of the theory of meta-induction.

2.5 Meta-Induction and the Justification of Induction

Meta-induction is a theory which overcomes David Hume's problem of induction. It generalises Hans Reichenbach's *best alternatives approach* (cf. Schurz 2008b, sect. 2). Reichenbach was the first to propose to consider the problem of induction not with respect to the strong requirement of proving that inductive methods are successful, but with respect to the much weaker, but epistemically still highly relevant, requirement of proving that inductive methods are the best methods accessible to us for making predictions. Since our best methods might be still predictively unsuccessful, this requirement is weaker than the one put forward by Hume. Reichenbach argued that if we cannot realise the sufficient conditions of success, we shall at least realise the necessary conditions. In order to spell out this idea, he defined a world to be *predictable*, if it is sufficiently ordered to enable us to construct a series with a limit. Since the principle of induction leads to the limit, if there is a limit, convergence with the principle of induction is a necessary condition to be successful (cf. Reichenbach 1938, pp. 348f; cf. also the explication in Feldbacher-Escamilla 2017, p. 421).

This solution to the problem of induction or this vindication of induction is very simple, but also narrow in the sense that it holds only for ordered worlds. Since we do not know whether our world is ordered in this sense or not, the more challenging task is to vindicate induction also for the case of an un-ordered or—according to Reichenbach's terminology—"unpredictable" world. Exactly this is done within the approach of meta-induction (cf. Schurz 2008b): Here the idea is that if we shift our application of induction from the level of object-induction about the outcomes of an event sequence to the meta level about the success of such methods, one can generalise the Reichenbachean idea also to cases of an unordered world. Reichenbach himself suggested already such a move from the object to the meta level (1938, p. 353), but his reasoning was incomplete in the sense that it remained open how an adequate prediction of the limit of a sequence on the meta level of success can be linked to an adequate prediction of the limit of a sequence of event outcomes on the object level (this criticism was put forward particularly by Skyrms 2000, p. 49). It was not until 70 years after Reichenbach's proposal that this gap could be closed by the approach of meta-induction of Schurz (2008b).

In the following, we outline foundational parts of the framework of meta-induction (for a comprehensive discussion cf. Schurz 2019; and Feldbacher-Escamilla under revision). Its most central part are so-called *prediction games*. A prediction game consists of the following ingredients:

Events, Predictions, and Truth.

- Y_t^s: $Y_1^1, Y_2^1, \ldots; Y_1^2, Y_2^2, \ldots$ are infinite series of events.
- $\mathcal{Y} = \langle\langle y_1^1, y_2^1, \ldots\rangle, \langle y_1^2, y_2^2, \ldots\rangle, \ldots\rangle$ are quantified representations (within the interval $[0, 1]$) of the true (or actual) outcomes (or values) of the events (event variables) to be predicted: $y_t^s \in [0, 1]$.
- F_i: $F_1, \ldots, F_n$ are the prediction or forecasting methods of n predictors or forecasters.
- $\mathcal{F} = \langle\langle\langle f_{i,1}^1, f_{i,2}^1, \ldots\rangle, \langle f_{i,1}^2, f_{i,2}^2, \ldots\rangle, \ldots\rangle : 1 \leq i \leq n\rangle$ are the predictions or forecasts of the single events within the interval $[0, 1]$ of the predictors or forecasters $1 \leq i \leq n$: $f_{i,t}^s \in [0, 1]$

Given these ingredients, we define a prediction game by the following 4-tuple:

Prediction Game.
G is a prediction game (with the true values $\mathcal{Y}$ and the predicted values $\mathcal{F}$) about events of type(s) $I \subseteq \mathbb{N}$ iff

$$\begin{aligned} G = \langle\, & \{\langle s, t, Y_t^s\rangle : t \in \mathbb{N} \;\&\; s \in I\}, \\ & \{\langle s, t, y_t^s\rangle : t \in \mathbb{N} \;\&\; s \in I\}, \\ & \{F_i : 1 \leq i \leq n\}, \\ & \{\langle s, i, t, f_{i,t}^s\rangle : 1 \leq i \leq n \;\&\; t \in \mathbb{N} \;\&\; s \in I\}\,\rangle \end{aligned}$$

Fig. 2.2 Prediction game with event outcomes y and predictions $f_i, \ldots, f_n$ of n predictors

y_1	y_2	y_3	y_4	y_5	$\ldots$
$f_{1,1}$	$f_{1,2}$	$f_{1,3}$	$f_{1,4}$	$f_{1,5}$	$\ldots$
$\vdots$	$\vdots$	$\vdots$	$\vdots$	$\vdots$	$\vdots$
$f_{n,1}$	$f_{n,2}$	$f_{n,3}$	$f_{n,4}$	$f_{n,5}$	$\ldots$

I is a set of indices of the event types in question ($I \subseteq \mathbb{N}$). E.g., a prediction game with $I = \mathbb{N}$ amounts to a task of predicting everything (assuming that the set of all properties is countably infinite); $I = \{3, 4, 5\}$ might filter out, e.g., a prediction game on weather where events of type 3 might be about sunny, events of type 4 might be about rainy, and events of type 5 might be about cloudy days; one might put forward probabilistic constraints for connecting the predicted values $f^3_{i,t}$, $f^4_{i,t}$, $f^5_{i,t}$ as well as for the outcomes y^3_t, y^4_t, y^5_t such that they are non-negative and sum up to 1 and for the ys one typically assumes that they are $\in \{0, 1\}$ (for all $i, t \in \mathbb{N}$); given these constraints, one can interpret such a prediction game also as a probabilistic one. On the other hand, setting $I = \{3\}$ filters out a simple prediction game on all events of type Y^3 (whether it is sunny or not or to which degree it is sunny etc.). If I is a singleton and the specific event type is irrelevant, then one can just omit super-indices. One can speak then also about a 'prediction game' simpliciter (this notion of a prediction game allows also for probabilistic settings and generalises that of Schurz 2008b). In most parts of what follows we only have in mind such simple prediction games with $|I| = 1$.

Relevant for the evaluation of predictions within a prediction game are especially the values of y and f_i (see Fig. 2.2): The closer $f_{i,t}$ is to y_t, the better the prediction of i. And the closer the $f_{i,t}$'s are to the respective y_t's, the better i is a predictor in general.

How good a prediction $f_{i,t}$ is with respect to the true outcome y_t is measured by help of a loss function ℓ, which is a monotonically increasing function with the arguments $f_{i,t}$ and y_t that is convex. The so-called *natural loss* is, e.g., the absolute difference between $f_{i,t}$ and y_t. The *quadratic loss* consists in the squared difference of these two values, etc. Based on such a loss function, we can define the success of a prediction method f_i up to a specific event or round t as the average of the inverse of the loss $(1 - \ell)$ up to round t:

Success.

$$\mathcal{s}ucc(F_i, t) = \frac{\sum_{u=1}^{t} 1 - \ell(f_{i,u}, y_u)}{t}$$

Given a prediction game G with a prediction method F_i and the predictions f_i, $\mathcal{s}ucc(F_i, t)$ expresses how well F_i scored on average until round t in predicting $f_{i,1}, \ldots, f_{i,t}$ given the true event outcomes $y_1, \ldots, y_t$. The higher $\mathcal{s}ucc(F_i, t)$, the better the prediction method is.

Now, in vindicating induction, one aims to show that employing induction is the best thing one can do, because by doing so one scores, at least in the long run, best among all the competitors of such a prediction game. The theory of meta-induction fleshes out this idea by defining a meta-prediction method, which takes for each prediction round t of a prediction game G as input the success of all the predictors $F_1, \ldots, F_n$ up to round $t-1$, and makes a prediction for round t by help of success-based weighting of the predictions $f_{1,t}, \ldots, f_{n,t}$. Such a meta-inductive predictor F_{mi} predicts $f_{mi,t}$ for each round t. And the weights used for this predictions might be, e.g., a normalisation of the difference of the successes in the exponent (many more other meta-inductive prediction methods are studied in Schurz 2019):

Meta-Inductive Weights.

$$w(F_i, t) = \frac{e^{\sqrt{8 \cdot ln(n) \cdot (t-1)} \cdot (succ(F_i, t-1) - succ(F_{mi}, t-1))}}{\sum_{j=1}^{n} e^{\sqrt{8 \cdot ln(n) \cdot (t-1)} \cdot (succ(F_j, t-1) - succ(F_{mi}, t-1))}}$$

Given these weights, the meta-inductive prediction consists simply in linear weighting the predictions $f_{1,t}, \ldots, f_{n,t}$ for round t based on the weights calculated for the respective prediction methods' successes until round $t - 1$:

Meta-Inductive Predictions.

$$f_{mi,t} = \sum_{i=1}^{n} w(F_i, t) \cdot f_{i,t}$$

Since the prediction of F_{mi} for round t is based on the successes of the prediction methods until round $t - 1$, the meta-inductive prediction can be considered to be an *inductive* prediction method. It inductively infers future success from past success. Also, this way of employing induction is not about the values of the event outcomes, but about the successes of other methods, it is a *meta*-method. Hence, it is a *meta-inductive* prediction method.

Now, we cannot go into many details regarding the performance of meta-induction here, but based on theorems of the machine learning literature, one can show that the difference between the successes of the predictors $F_1, \ldots, F_n$ of G and that of the meta-inductive predictor F_{mi} is bounded as follows (cf. Cesa-Bianchi and Lugosi 2006, sect. 2.1f; Schurz 2008b, sect. 7; and the proof in the appendix of Feldbacher-Escamilla 2020; and a slight generalisation in Feldbacher-Escamilla and Schurz 2020):

Theorem about Meta-Inductive Success Bounds.
Given the underlying loss function ℓ (which is used for defining $succ$) is convex, it holds:

$$succ(F_i, t) - succ(F_{mi}, t) \leq \sqrt{const \cdot ln(n)/t} \quad (\forall i \in \{1, \ldots, n\})$$

This bound (*const* is just a particular constant number) is of particular interest, because it shows that the difference between the success rates grows at most sublinearly with t for the meta-inductivist. This means that in the long run, i.e. in the limit, the meta-inductive method cannot be outperformed by any other predictor of the prediction game G. In other words, meta-induction is long run optimal in comparison to any prediction method of G:

Meta-Inductive Optimality.

$$\lim_{t\to\infty} \mathit{succ}(F_i, t) - \mathit{succ}(F_{mi}, t) \leq 0 \quad (\forall i \in \{1, \ldots, n\})$$

It is important to note that this is an analytic result about how we defined F_{mi} and that this holds for any sequence $y_1, y_2, \ldots$ of event outcomes. So, even if the event series in question is not predictable in the sense of Reichenbach, meta-inductive predictions are still long run optimal and in this sense performing meta-inductive predictions is the best one can do. So, we have some form of a priori or deductive justification of meta-induction.

Now, given the meta-inductive optimality result, one can provide also a justification for classical (object-)induction. The general form of reasoning is as follows: As the above-cited result shows, meta-induction for a weighted selection of predictions of any accessible method is proven to be optimal in the long run. If we take for granted the past success of classical inductive methods—something that is clearly a contingent and a posteriori matter of fact and also not scrutinised by Hume himself (cf. Howson 2003, p. 4)—it follows that a meta-inductive selection of such classical inductive methods for predictions of future events is guaranteed to be long run optimal. This holds at least as long as there are no alternative methods in G that outperform classical inductive methods. And since, at least up to now, classical inductive methods of science outperformed any other prediction method, our use of them is a posteriori justified.

Having outlined the meta-inductive framework and how it is employed to justify induction, we want to show how this approach can be also employed to justify abductive reasoning in the sense of an inference to the probabilistically best prediction in next Sect. 2.6.

2.6 Meta-Abduction: Inference to the Probabilistically Best Prediction

In this section, we apply the theory of meta-induction in order to show how not only induction, but also a species of abduction can be justified. For this purpose, we need to show how abduction in the sense of an inference to the probabilistically best prediction can be embedded into the meta-inductive framework—resulting in a theory of meta-abduction—and then outline how the optimality result of meta-induction can be also employed to argue for the optimality of this kind of abduction.

Given the epistemic relevance of simplicity as outlined above, how should we select among hypotheses, explanations, theories? According to probabilistic selective abduction (AIC-Abd), we should try to maximise the information theoretical balance between accuracy ($Pr(P|C)$) and simplicity ($c(C)$). By choosing that hypothesis, explanation or theory which has the *best* balance, we will be closest to the truth, which might be different from being closest to the data P (see Forster and Sober 1994, p. 6). So, given the epistemic aim of *being close to the truth*, (AIC-Abd) seems to be an optimal means to achieve this end. However, this is with respect to explanation. What about predictions? What about choosing the *best* balanced hypothesis or theory for prediction?

The theory of meta-induction can be applied for optimising predictions in any respect, as long as the formal conditions of the framework are satisfied. In our application to Hume's problem of induction we interpreted the framework plainly epistemically: Given a prediction game G with $\mathcal{Y}$ and $\mathcal{F}$, we interpreted $\mathcal{Y}$ as the truth and $\mathcal{F}$ as prediction methods or hypotheses about the truth. However, we can also take in a more pragmatic standpoint and interpret $\mathcal{Y}$ as past, present, and future data, and $\mathcal{F}$ as prediction methods or hypotheses about which data will be gathered in the future. Since data typically contains error and noise, it easily falls apart from the truth, hence this interpretation does not coincide with the former. And in this sense it seems to be perfectly fine that also the criteria for success fall apart: Epistemically speaking, we still aim at predictions that are as closest to the truth as possible. However, given our noisy data, we know that we need to aim at predictions that are best balanced between accuracy (fitting) and complexity (overfitting). Success consists not in minimising the distance from the data, but making a prediction which is *best* balanced between these two parameters. So, in order to achieve this goal in the long run, the idea is to use a normalisation of (AIC). If r is the highest polynomial we are going to consider and Pr is ϵ-regular (i.e. only $Pr(\bot) = 0$ and all other probabilities are $> \epsilon > 0$), then $AIC(C, P) \in [\log(\epsilon) - r, -r]$. Hence, we can normalise $AIC(C, P)$ to $[0, 1]$ by taking

$$\frac{AIC(C, P) - (\log(\epsilon) - r)}{-\log(\epsilon)}$$

which is in $[0, 1]$.

Now, let us consider a prediction game G with $\mathcal{Y}, \mathcal{F}$. Let G be about predicting the *best* balancing for making explanations or predictions. Assume that $\mathcal{Y}$ is a series of objectively *best* balancing. This will still fall apart from providing a most accurate prediction, i.e. a true prediction, because the degree of an extension of a polynomial predicting up to round $t - 1$ might be increased by 1 if one predicts for round t the true value with probability 1, whereas it might not be increased at all by predicting the true value with close to 1 probability and hence deviation from the truth might have a higher AIC (the AIC is higher, if the deviation allows for no change in the degree of the polynomial and the basis of the logarithm in equation (AIC) is $> 1/(1 - Pr(P|C)))$. Given such an "objective" best balancing, we can interpret $\mathcal{F}$ as a set of theories or hypotheses which provide predictions of what will be the best balance once new data enters the game (i.e. once one moves forward to the next round). We take the predictions in $\mathcal{F}$ to be the actual AICs of the same methods in predicting some event in another prediction game, let us say G'. So, if for $f_i' \in \mathcal{F}'$, $AIC(f_i', \{Y_1, \ldots, Y_{t-1}\}) = a$, then the respective $f_i \in \mathcal{F}$ predicts for round t as best balance the normalisation of a, i.e. $\frac{a-(\log(\epsilon)-r)}{-\log(\epsilon)}$. G is, so to say, a meta game where any prediction method of the ordinary game G' predicts that it has the right balance for future predictions. In other words, playing G' and making predictions comes with the commitment of claiming also that one's prediction is right in the sense of best balanced—that is a claim in G.

Now, again by success-based mixing of the forecasts about the best balance to be expected, a meta-inductive learner achieves long run optimality in predicting the best balance in G. If we assume, e.g., that in science creative abductive methods as hinted at in the introduction with high unificatory power had the best balance in the past, then using such creative abduction for inferring theoretical frameworks is epistemically justified, since using them is, at the current state of science, the best thing to do: Following the meta-inductive (or better: meta-abductive) selection allows for optimality in predicting the best balance in G (and actually having the best balance in one's events predictions in G').

Note that given this assumption, anti-abduction fails—which would be to get the worst possible balance of accuracy and simplicity—to be justified: Disunification and theoretically laden hypothesis invention fared suboptimal in past (in G') and hence its predictions of the best balance in G were also wrong. Hence, meta-inductive selection ignores these methods and this is the best thing to do, at least given their past performance. Given this assumption, anti-abduction is by far no optimal means to achieve the epistemic end of being best balanced in G.

Another note is in place: In the argument above, we implicitly made the assumption that success in the meta game G and success in G' are synchronous: Whenever one was relatively successful in choosing the right balance for theory and hypothesis invention (G), one was also likewise successful in predicting events (G'). The problem with this assumption is that in principle nothing hinders an adversary in letting fall things apart from each other. So, in principle one could allow for a method performing good in G', but at the same time failing in G. However, we can argue for our assumption of a parallel development of G and G' by hinting at or assuming a past correlation between abductive theory building and predictive success, and then employ induction as we justified by help of meta-induction in the preceding Sect. 2.5. By this we can inductively transfer this correlation of the past also to the future. And we are epistemically justified in doing so.

Finally, we should also mention that such an approach of meta-abduction as outlined here can be considered more generally as introducing cognitive costs in prediction games. General features of introducing cognitive costs are studied, e.g., in Schurz (2019, sect. 7.6).

2.7 Conclusion

To briefly sum up: In this essay we have provided a taxonomy of abductive inferences (in the wide sense) and an exact characterisation of a species of selective abduction, which aims at inferring the probabilistically best hypothesis, explanation or theory on the basis of data; the two main relevant factors in doing so are likelihood of the data given the inferred hypotheses and simplicity of the hypotheses, where simplicity was used as a proxy for many other explanatory virtues such as scope, non-ad hocness, unification; we have provided an argument for the epistemic value of simplicity and have shown how inferences of *explanations* based on these factors are justified. Finally, we have introduced the framework of meta-induction, outlined its justification of induction, and have also sketched how abductive inferences regarding *predictions* can be justified by employing the framework of meta-induction.

Acknowledgments This research was supported by the *German Research Foundation* (DFG), research unit *Inductive Metaphysics* FOR 2495. I would like to thank two anonymous referees for very helpful comments on an earlier version of this paper. For valuable discussion of this topic, I would like to thank Atocha Aliseda and the audience of *IACAP 2019* (annual meeting of the *International Association for Computing and Philosophy*, June 5–7, 2019, Mexico City). Thanks a lot also to Nancy Abigail Nuñez Hernandez and Björn Lundgren for organising the proceedings to this conference.

References

Aliseda, A. 2006. *Abductive Reasoning. Logical Investigations into Discovery and Explanation.* Dordrecht: Springer. https://doi.org/10.1007/1-4020-3907-7.

Cesa-Bianchi, N., and G. Lugosi. 2006. *Prediction, Learning, and Games.* Cambridge: Cambridge University Press. https://doi.org/10.1017/CBO9780511546921.

Douven, I. 2018. Abduction. In *The Stanford Encyclopedia of Philosophy (Summer 2018 Edition)*, ed. Zalta, Edward N.

Feldbacher-Escamilla, C.J. (under revision). *Epistemic Engineering. Uncovering the Logic of Deceivability and Meta-Induction.* Book manuscript.

——— (2017). Optimisation in a Synchronised Prediction Setting. *Journal for General Philosophy of Science* 48(3): 419–437. https://doi.org/10.1007/s10838-017-9379-7.

——— (2019). Newtons Methodologie: Eine Kritik an Duhem, Feyerabend und Lakatos. *Archiv für Geschichte der Philosophie* 101(4): 584–615. https://doi.org/10.1515/agph-2019-4004.

——— (2020). An Optimality-Argument for Equal Weighting. *Synthese* 197(4): 1543–1563. https://doi.org/10.1007/s11229-018-02028-1.

Feldbacher-Escamilla, C.J., and A. Gebharter. 2019. Modeling Creative Abduction Bayesian Style. *European Journal for Philosophy of Science* 9(1): 1–15. https://doi.org/10.1007/s13194-018-0234-4.

Feldbacher-Escamilla, C.J., and G. Schurz. 2020. Optimal Probability Aggregation Based on Generalized Brier Scoring. *Annals of Mathematics and Artificial Intelligence* 88 (7): 717–734. https://doi.org/10.1007/s10472-019-09648-4.

Forster, M.R., and E. Sober. 1994. How to Tell When Simpler, More Unified, or Less Ad Hoc Theories Will Provide More Accurate Predictions. *The British Journal for the Philosophy of Science* 45(1): 1–35. https://doi.org/10.1093/bjps/45.1.1.

Glymour, C. 2018. Creative Abduction, Factor Analysis, and the Causes of Liberal Democracy. *Kriterion – Journal of Philosophy*. http://www.kriterion-journal-of-philosophy.org/kriterion/issues/Permanent/Kriterion-glymour-01.pdf.

Harman, G.H. 1965. The Inference to the Best Explanation. *The Philosophical Review* 74(1): 88–95. https://doi.org/10.2307/2183532.

Hempel, C.G. 1965. *Aspects of Scientific Explanation and Other Essays in the Philosophy of Science*. New York: Free Press.

Hintikka, J. 1998. What Is Abduction? The Fundamental Problem of Contemporary Epistemology. *Transactions of the Charles S. Peirce Society* 34(3): 503–533. https://doi.org/10.2307/40320712.

Howson, C. 2003. *Hume's Problem*. Oxford: Clarendon Press.

Hume, D. ed. 1748/2007. *An Enquiry Concerning Human Understanding*, ed. Millican, Peter. Oxford World's Classics. Oxford: Oxford University Press.

Kelly, K.T. 2007. How Simplicity Helps You Find the Truth without Pointing at it. In *Induction, Algorithmic Learning Theory, and Philosophy*, ed. Friend, Michèle, Goethe, Norma B., and Harizanov, Valentina S, 111–143. Dordrecht: Springer. https://doi.org/10.1007/978-1-4020-6127-1_4.

Lipton, P. 2004. *Inference to the Best Explanation*. 2nd ed. London: Routledge.

Magnani, L. 2000. *Abduction, Reason and Science. Processes of Discovery and Explanation*. New York: Kluwer Academic.

Mohammadian, M. (2019). Abduction - The Context of Discovery + Underdetermination = Inference to the Best Explanation. *Synthese*. https://doi.org/10.1007/s11229-019-02337-z.

Niiniluoto, I. (1999). Defending Abduction. *Philosophy of Science* 66: S436–S451. https://doi.org/10.1086/392744.

Peirce, C.S. (1994). *Collected Papers of Charles Sanders Peirce*, eds. C. Hatshorne, P. Weiss, and A.W. Burks, Harvard: Harvard University Press.

Popper, K.R. 2002. *Conjectures and Refutation*. New York: Basic Books.

Reichenbach, H. 1938. *Experience and Prediction. An Analysis of the Foundations and the Structure of Knowledge*. Chicago: University of Chicago Press.

Schurz, G. 2008a. Patterns of Abduction. English. *Synthese* 164(2): 201–234. https://doi.org/10.1007/s11229-007-9223-4.

———. 2008b. The Meta-Inductivist's Winning Strategy in the Prediction Game: A New Approach to Hume's Problem. *Philosophy of Science* 75(3): 278–305. https://doi.org/10.1086/592550.

———. 2019. *Hume's Problem Solved. The Optimality of Meta-Induction*. Cambridge, MA: The MIT Press.

Skyrms, B. 2000. *Choice and Chance. An Introduction to Inductive Logic*. 4th ed. Stamford: Wadsworth.

Solomonoff, R.J. 1964a. A Formal Theory of Inductive Inference. Part I. *Information and Control* 7(1): 1–22. https://doi.org/10.1016/S0019-9958(64)90223-2.

———. 1964b. A Formal Theory of Inductive Inference. Part II. *Information and Control* 7(2): 224–254. https://doi.org/10.1016/S0019-9958(64)90131-7.

Williamson, T. 2016. Abductive Philosophy. *The Philosophical Forum* 47(3–4): 263–280. https://doi.org/10.1111/phil.12122.

Chapter 3
On the Conciliation of Traditional and Computer-Assisted Proofs

Favio E. Miranda-Perea and Lourdes del Carmen González Huesca

Abstract A proof of a mathematical proposition or a program specification obtained by a formal verification process, using an interactive theorem prover, can be questioned as a true demonstration or as having the same purposes of a traditional pencil-and-paper proof. However, in our opinion the verification process of a software component exhibits the same construction phases as a purely mathematical one. A correspondence between both kinds of proofs enables us to give a proposal of what we call *transitional proofs*, a concept that outlines a conciliation between traditional paper-and-pencil and computer-assisted proofs, which can be useful in philosophical problems surrounding formalized mathematics and program verification with proof-assistants.

3.1 Introduction

Since the seminal work of Tymoczko (1979) about the Four Color Theorem a lot of work has been done around the question of whether a "proof" involving a computer is a mathematical proof in an essential way. With the rise of modern interactive theorem provers, also called proof-assistants, the computer-assisted proofs (CAPs) involve sophisticated reasoning procedures and not only computations, like the proof of the Feit-Thompson Theorem developed in the Coq proof-assistant (Gonthier et al. 2013). Such kind of achievements arise the question whether the CAPs can be considered equivalent to traditional proofs (TPs), obtained by human collaboration. This is of course an important question for the philosophy of mathematics (Avigad 2010, 2008), but it is also relevant for the philosophy of computer science due to the fact that when employing an interactive proof-assistant, the formal verification of programs is operationally identical to the process of proving mathematical theorems. On the other hand the problem of distinguishing a case study on program verification from one on formalized

F. E. Miranda-Perea (✉) · L. d. C. González Huesca
Departamento de Matemáticas, Facultad de Ciencias, UNAM, Mexico City, Mexico
e-mail: favio@ciencias.unam.mx; luglzhuesca@ciencias.unam.mx

B. Lundgren, N. A. Nuñez Hernández (eds.), *Philosophy of Computing*,
Philosophical Studies Series 143, https://doi.org/10.1007/978-3-030-75267-5_3

mathematics is interesting and not necessarily trivial. For instance, a case about the correctness of an algorithm for efficient data structures can serve from complex mathematical notions. This matter will not be addressed here, but the interested reader can consult (Krantz 2007, Chapter 7).

There are distinct works relating proofs assisted by computer systems with traditional proofs. For instance Lamport (1995, 2012) discuss how to write traditional proofs in a so-called structured proof style that intends to minimize errors. These proofs can be implemented in the TLA^+ language (Lamport 2002), a tool for specifying and reasoning about algorithms and computer systems. On the other direction we mention Ganesalingam and Gowers (2017) that presents a full automatic prover program for problems mainly in metric space theory and whose principal aim is to construct human-style proofs in English prose.

A proof-assistant (PA) is a computer system that consists of a domain-specific language allowing to represent logical objects, as well as definitions and theorems about them, together with a mechanism that allows for the interactive construction and validation of proofs. This proof construction process is driven through a fully interactive goal-oriented proof search, which means that the search is guided by a human agent, who proposes a specific reasoning method, for instance *modus ponens* or transitivity of equality, but also some structural induction principle, and leaves to the machine the task of verifying if such a choice is adequate to solve the ongoing goal. In the positive case the machine answers by presenting the new subgoals whose provability achieves also the original goal.

In this paper with the term *computer-assisted proof* we refer exclusively to a proof of either a software component or of a mathematical theorem constructed with the help of an interactive proof-assistant. In particular we are not considering full automatic proofs involving automated theorem provers, like the original proof of the Robbins Problem (Mccune 1997), or hybrid proofs employing the computer as an auxiliary device for performing mechanical tasks, like checking the reducible configurations in the original proof of the Four Color Theorem (Appel and Haken 1976) or the more recent proof of the Kepler's conjecture (Hales 2006).

A CAP is built by a so-called proof-script which is a piece of text, that is a program, consisting of a sequence of definitions, statements and commands indicating the proof steps to be performed by the proof-assistant. According to the particular facilities of the PA these scripts can be written either in a procedural style or in a declarative language, akin of the usual mathematical writing. See Harrison (1996) for an overview of proof styles.

The naive way to conciliate traditional and computer-assisted proofs would be to define a general transition methodology from one to the other proof category. In the direction from the human to the machine, the process would consist in the implementation of a paper-and-pencil proof in the proof-assistant language, a task similar to the translation of an algorithm or pseudocode to an actual program in a particular programming language. This exercise could be quite cumbersome, due to the human knowledge base, which includes not only background information but also subtle details like the implicit use of quantifiers or substitutions in equational reasoning. Thus, in the general case such task becomes unfeasible. Although in practice some particular translation cases can be alleviated by the use of libraries or the assumption of needed properties (stated as axioms). Moreover, even if we succeed in transliterating a given TP, the resulting CAP might be quite artificial, for we might compare it with the attempt to translate an idiomatic phrase, say from german to spanish, word by word using literal dictionary definitions. The converse process, from the machine to the human proof possess similar problems. Let us note that both traditional and computer-assisted proofs may contain idioms, the former since they consist of a mix between natural language, symbols[1] and perhaps diagrams; and the latter since their structure highly depends on the particular implementation of the mathematical definitions (see the Theorem 6 and the discussion below). This might announce the impossibility of a general translation process. But even if somehow we succeed in this enterprise, perhaps focusing in a specific area like in Ganesalingam and Gowers (2017), the approach seems to be sterile for our purposes, since the same arguments can be applied in favor or against for either kind of proof. For instance, one can say that even if a CAP of a mathematical theorem is validated by a proof-assistant, in practice the corroboration by the mathematical community is required to some extent. These sociological (Asperti et al. 2009; Richard et al. 1979; MacKenzie 2001; Krantz 2007) and even cognitive issues (Robinson 2000) about and around what is a proof will not be discussed here.

From the above said we think that the translation process between both kind of proofs is unworkable in a general way. Thus, to be able to discuss the problem of the equivalence between TPs and CAPs we need a proof approach that can be accepted from both sides. A notion of *transitional proof* that provides us with a framework for philosophical discussions about what is a proof in the age of modern proof-assistants. Ideally, this approach might also provide us with a methodology that enables a translation process in particular cases.

The aim of this paper is to propose such notion and to review some discussions and technicalities about and around CAPs in an introductory way accessible to a multidisciplinary audience, namely mathematicians, philosophers and computer scientists interested in these topics and their relevance in their areas of expertise. The approach is tutorial and intends to be useful to close an important gap between different scientific communities. An issue that is gaining importance due to the increasing use of proof-assistants by very dissimilar actors. That said, it should be clear that the ideas here discussed are far from being novel, see for example (Bundy et al. 2005). Thus we do not claim originality beyond the choosen train-of-thought which includes some technical details considered folklore until now to the best of our knowledge, like the discussion of formal backward proof given in Sect. 3.4. It is important to remark that although in our presentation we refer to an specific proof-assistant, namely Coq (The Coq Development Team 2020), coming out of our experience as users, most of the general discussion concerns any modern interactive theorem prover.

[1] An example is the symbol for "such that" outside a set definition, which may be a colon :, but also a kind of inverted epsilon ∍, and their usage is certainly non standard.

The proposed notion of transitional proof can be useful in some philosophical discussions around computer-assisted correctness proofs.[2] Let us emphasize that we are in the realm of formal methods, that is, the proof procedures we are considering are those suitable to be formalized in an implemented logical framework. The kind of discussion topics we are thinking of are related, for instance, to the challenges proposed by Turner (2018, Chapter 25) around the question of whether software correctness proofs deliver mathematical knowledge. Let us summarize below some aspects of these complaints as presented by Turner:

- *Mathematical challenge*: Mathematical proofs are conceptually interesting and compelling whereas correctness proofs are long and shallow. The latter are carried out in a formal system dominated by numerous cases and shallow steps. They seldom provide originality. We need patience rather than inspiration to construct them.
- *Mechanical challenge*: Proofs of correctness are rarely carried out by hand due to their size and the number of checking cases required. They are constructed with the aid of theorem provers or models checkers. In the case of theorem provers the underlying mechanisms are enhanced formal logical systems serving as a framework to construct the proof, and their specifications are consequence of those systems. This reduces the correctness of the specific problem to that of the theorem prover or model checker but can trigger an infinite regress argument. The use of theorem proving involves physical devices at the end and it is required to assume that these underlying devices are correct. Hence the verification is empirical resulting in an experimental correctness. This can be alleviated if we demand that the proofs are handmade but computer checked.
- *Pragmatic challenge*: Correctness proofs only deliver empirical knowledge. Real software is tested and verified but rarely formally. The relationship between program and specification (proof and proposition) is empirical and uncovered by testing rather than by a formal proof. Testing introduces an empirical observation about the program (the physical program at the end) relating it with the specification bypassing the symbolic program. A correct physical program implies correctness of the symbolic program, if we assume a correct implementation. This is an indirect correctness that relies on the implementation. A regress problem surges.
- *Scientific challenge*: Are the methods of testing and verification aimed at the same goals of mainstream science? Software development involves simulation and experiments, therefore the knowledge of programs is scientific knowledge. Many of these experiments, designed for exploring the functioning of an artifact, apparently introduce a new style of experiment. The purpose of program verification is not to verify a model or a theory in the usual sense, when the testing

[2] We consider here both the proof of a mathematical proposition and that of an algorithm or program fulfilling its specification.

fails we do not change the specification but the program or device.[3] Falsifying tests lead to revision of the artifact not of the hypotheses as in other sciences. This provides a methodological distinction.

A deep study of the interaction process between machine and human is mandatory to tackle the above interrelated challenges. A starting point in this inquiry is provided by De Mol (2014, 2015), which, although mostly related to computer-assisted experimental mathematics, provides us with an important characteristic of the man and machine interaction, which is relevant for CAPs, namely the internalization of mathematical knowledge and tools into the computer. Internalization refers to the phenomenon of implementing mathematical knowledge and methods by the computer in such a way that it can be hidden, at some extent, from the final user. This circumstance is nowadays increasing due to the evolution of computer systems (speed, memory, interfaces, programming paradigms, etc.). In the case of modern proof-assistants this means that we have more and more native libraries available, implemented by several independent users, which are experts on distinct areas. This collaborative work constitutes a robust knowledge base that enables the final users to focus on the particular aspects of their case studies. For instance, in a paper about the mechanized proof of the Feit–Thompson Theorem (Gonthier et al. 2013) we can read:

> In short, the success of such a large-scale formalization demands a careful choice of representations that are left implicit in the paper description. Taking advantage of COQ's type mechanisms and computational behavior allows us to organize the code in successive layers and interfaces. The lower-level libraries implement constructions of basic objects, constrained by the specifics of the constructive framework. Presented with these interfaces, the users of the higher-level libraries can then ignore these constructions, and they should hopefully be able to describe the proofs to be checked by COQ with the same level of comfort as when writing a detailed page in LaTeX.

This feature of interactive theorem provers deeply modifies the interaction between man and machine and is the fundamental idea behind the notion of transitional proof here introduced.

3.2 An Overview of Computer-Assisted Proofs: The Case of COQ

Verification tools for software or hardware systems are now widely used for many purposes depending on the level of safety required to be guaranteed. For instance, basic safety can be proved using automatic tools like static analysers or model

[3] Let us emphasize that we are giving a summary of Turner's ideas, and as pointed by two anonymous referees this claim is most likely to be false, for is common for automated proofs to reveal faulty specifications that must be changed.

checking systems. For full correctness, verification is conducted with the help of interactive systems such as theorem provers.[4]

Computer-assisted proofs as the product of the interaction between a human agent and a proof-assistant depend on various notions, concepts and languages used in and related to the PA. In this work we take the case of the Coq proof-assistant (The Coq Development Team 2020) to give an overview of the CAP construction process. This state-of-the-art theorem prover is widely used for quite dissimilar purposes, for instance the proof of the Four Color Theorem (Gonthier 2008) or the compiler project CompCert (Leroy 2018, 2009).

The discipline of interactive theorem proving has been described as the use of a verification tool to develop formal proofs through the guidance of the proof-search by the user. Examples of these tools or systems are Agda, Coq, Isabelle, Lean, Lego, Matita, NuPRL, Twelf, among others. These systems are based on different higher-order logics and the correspondence between logic and type systems (Geuvers 2009). This overview is for proof-assistants which are based on type systems and the Curry-Howard correspondence also known as "formulas-as-types" and "proofs-as-programs" paradigm.

An interactive proof-development system is an environment which allows us to develop proofs interacting with a human agent. The construction of a proof and its verification are done independently, this means that the system is a cooperation of (1) a system to develop proofs and (2) a proof checker; both based on type theory. The part of the system which helps the user to construct proofs is called the *proof engine* and ensures that a proof is well conducted. The proof is realized by applying commands and tactics available in the assistant. A *tactic* is a function to transform goals or statements into simpler goals. The use of tactics in proof-assistants is influenced by the work of Robin Milner in LCF (Milner 1972). The sequences or the composition of tactics and commands in a proof is called a *proof-script*, which records the interaction of the user and the assistant during a proof. The use of tactics defines what is called a procedural proof style (Harrison 1996) allowing forward and backward styles of reasoning, the former works with elements in the proof context whereas the latter is focused on the active goal (Gonthier and Mahboubi 2010). The proof engine mantains the congruence of the application of these functions to obtain a proof-term at the end of a proof. Finally, the proof checker must validate the proof-term by means of a type check between the proof-term and the original statement or goal. This is known as the De Bruijn criterion (Geuvers 2009).

Coq, a system developed for almost 40 years which is recently defined as a formal proof management system,[5] is an interactive proof-assistant where the user has control of the proof-search, which is managed by two main components: the proof engine and the kernel. The Coq proof engine is an interactive machine to construct proofs by applying tactics corresponding to inference rules and usual

[4] *Desperately seeking software perfection*, Xavier Leroy, Colloquium d'informatique, UPMC, Paris France, October 2015.

[5] https://coq.inria.fr/.

mathematical (backward) reasoning. Some tactics are intended to perform automation of the proof-search, including backtracking. There are also many formalised theories available as modules of the Standard Library, for instance Peano arithmetic, real numbers, polymorphic lists, etc. These modules include definitions, functions, facts, theorems and lemmas to provide the user with operations, properties or specifications already verified. The logical foundation of COQ was first given by the Calculus of Constructions (Coquand and Huet 1988) and then extended with inductive definitions (Pfenning and Paulin-Mohring 1989). The Calculus of inductive Constructions (CiC) is the higher-order logic or the underlying type system of COQ which is implemented in the kernel for type-checking or proof checking. The tactic collection intends to mimic the usual mathematical proof reasoning and is defined through a set of basic tactics which can be composed. Also, the user can define (or program) more elaborated tactics using the dedicated language $\mathcal{L}_{tac}$ which is akin to the OCaml programming language used for implementing COQ. The tactic language $\mathcal{L}_{tac}$ is an intermediate language which is functional with a call-by-value evaluation strategy, non-typed with side effects and implemented specifically for writing tactics (Delahaye 2000). After completing a proof, a proof-object, that is the proof-term interactively constructed, is then checked by the kernel of COQ. This verification must be performed at the very end to ensure that the term meets the original goal. The verification is a type-checking between the proof-term and a type or formula of the proved statement.

The workflow in the COQ proof-assistant (see Fig. 3.1) can be described as a cycle of interactions where the user first inputs a formal statement, using a particular formalisation or theory and possibly some already proven statements available in modules of the standard library. The proof-assistant considers such statement as a goal to be proved. The interaction is done in the proof phase where the application of tactics dissects a specific goal into ideally simpler subgoals to be proven. The subgoals can be solved in any order and the proof engine handles the correct application of tactics proposed by the user maintaining the consistency of variables through unification. This process continues until all the goals are proved and the proof-object obtained is then checked by the kernel. The original statement, as a higher-order formula, is the type of the proof-term.

COQ is also a functional programming language where programs are total and terminating, and it can be used to construct certified programs as it is an environment to carry out programs and to provide proofs of the program correctness (Chlipala 2013).

3.3 Transitional Proofs

A usual paper-and-pencil proof consists not only of its actual proof argumentations but also of several pieces of implicit knowledge related to the underlying theory or to previous work. However, mechanized reasoning by a proof-assistant requires the explicit development and verification of all the implicit knowledge taken for granted

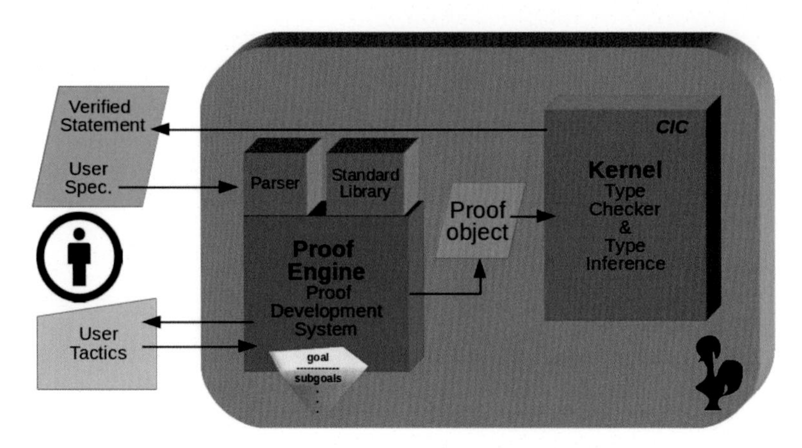

Fig. 3.1 Coq workflow from descriptions in The Coq Development Team (2020) and Geuvers (2009)

and therefore ruled out in a paper-and-pencil proof. This knowledge includes the general results of background theories, usually implemented in independent libraries, which are separately developed and made available perhaps for other human agents; but also some specific methodological details which are obviated in the traditional proof but are mandatory for the mechanization process. For instance the explicit handling of quantifiers and rewriting rules.

We can classify the statements and instructions within a CAP in two levels: those matching the actual traditional proof argumentation, called here high-level results, and the low-level statements and commands corresponding to the details that concern exclusively to the mechanization. For instance, if we define permutations by means of multiset equality and we choose to implement multisets as partial functions, then all results about these functions are considered low-level, whereas permutation properties like symmetry or transitivity are considered high-level.

To construct a CAP we need, ideally, a collaboration between experts on the specific subject and experts on the particular proof-assistant. A proof is build up by fully interacting with the proof-assistant but without abandoning the use of paper or blackboard. This way, the proof is constructed in two directions from the paper to the machine and the other way around. The result of this process is a sketch of a traditional proof together with a raw computer-assisted proof whose validity is unquestionable from the point of view of the proof-assistant but that arguably possess the same problems as a traditional proof with respect to the community validation (Richard et al. 1979). At the end this is a social process, the only difference is that the actors have changed from the experts on the subject to the experts on the proof-assistant.

To generate what we call a *transitional proof* we first separate the statements of the original raw computer-assisted proof obtained from the above described process, according to the mentioned two levels. Then we internalize, ideally all, low-level statements, mostly by hand, in order to automatize its use by means of the proof-assistant mechanisms, that is, either using available facilities or implementing new dedicated libraries and proof tactics. By doing this, such statements become almost invisible to the final user, thus producing an enhanced CAP of the same results. Next, we survey the implementation of the high-level results to generate, together with the original proof sketch, a traditional paper-and-pencil proof. The final outcome is a CAP together with a TP which are surveyable by a wider audience of specialists in the corresponding subjects.

As final desideratum we mention that the CAP and the TP composing a transitional proof should correspond *mutatis mutandis*, the difference being, apart of syntax details, the explicit presence of low-level properties and commands that must be stated in the CAP scripts. This strong correspondence implies that in the process of running the proof-scripts, the interactive proof session that the reader develops, correlates straightforward to the paper-and-pencil proof, thus turning the proof-script of a high-level statement into a WYSIWYG[6] implemented version of the traditional proof. This is an important feature in Mathematician-Computer interactions (De Mol 2014, section 2.1). In this manner, the proposed transitional proof construction would allow to ease the interaction of any standard user (mathematician, computer scientist, philosopher, etc., not necessarily expert in interactive theorem proving) with a proof-assistant while formalizing any given property.

We just described a process to construct a transitional proof from scratch in the sense that we are not assuming the existence neither of a TP nor of a CAP but constructing a traditional proof together with a raw CAP, as mentioned before. However, there are two other possibilities, going from a previously constructed CAP to a TP and vice versa. The first case is very unlikely, for the possibility of detaching the interaction with paper-and-pencil from the proof construction process seems impossible at this time. This way, a CAP is usually derived by mechanizing an already available TP or is the result of the above described process. The remaining direction, from TP to a CAP posses an additional challenge: a paper-and-pencil proof is usually a combination of inference reasoning patterns in both directions: forwards, going from assumptions to conclusions and backwards going from goals to subgoals. However, this bidirectional approach may not be fully supported in some PAs.

Proof-assistants support both forwards and backwards chaining inference processes, to some extent, but our target here are those having backward proof search as their native mechanism. It is worth noting that, since a bidirectional approach is quite necessary, in such systems we still have important efforts to incorporate forward reasoning (Ganesalingam and Gowers 2017; Corbineau 2008; Sacerdoti Coen 2010), thus providing a proof construction mechanism closer to the human procedures, although the backward chaining strategy is still at the heart

[6] "What You See Is What You Get".

of the implementation of such proposals.[7] But also note that there are state-of-the art proof-assistants where the bidirectional approach is better implemented and understood like `Isabelle/Isar` (Wenzel et al. 2008; Wenzel 2002).

That said, if in order to give a correspondence between TPs and CAPs, we are to understand and abstract the native mechanisms of (backward) proof-assistants at a mathematical level, we cannot directly resort to formal logic since by definition its derivations are constructed in a forward fashion. This is an important obstacle to fulfill the above mentioned desideratum. To alleviate this problem we propose in Sect. 3.4 an internalization in the opposite direction, namely from the machine to the human agent, by means of a formal notion of backward proof.

The notion of transitional proof we will propose here applies to general case studies, like the one presented in Sect. 3.5, although to introduce it we simplify the discussion to the case of a proof consisting of a single theorem and its corresponding proof script. Nevertheless, in the computer-assisted proofs community, mechanized proofs concerning software verification or formalized mathematics go from single self-contained scripts about one single property or theorem, perhaps including auxiliary lemmas, to huge projects consisting of several libraries, modules and structures whose development follows software engineering methodologies. Prominent examples of such enterprises are described in Gonthier et al. (2013) and Leroy (2009). It is worth noting that a well-structured proof development,[8] by means of an optimal use of modules and libraries does not automatically yields a transitional proof. Such development of course documents the proof, provides readability and simplifies the code management and understanding to any agent not involved in the original development, but under the assumption that such agent is familiar with the mechanisms and the particular implemented definitions of the PA (a small example of this situation is given by Theorem 6 below). In contrast transitional proofs, introduced next, should provide a more fine-grained link making them easier to be understood by a non-expert on proof-assistants. This way, transitional proofs are not aimed at PA experts, for they can be verbose and poorly structured for such community.

In order to introduce the definition of transitional proof, let us analyze a toy example involving an elementary theorem, namely the non-existence of a natural number between 0 and 1. The demonstrations below exemplify the usual proof construction methodologies and are the result of an experiment consisting of mechanizing the following elementary theorem.

Theorem 1 *There does not exist a natural number n such that* $0 < n < 1$.

Proof This is an immediate consequence of the well-ordering principle. □

---

[7] The direct way of implementing a forward reasoning step is by the backward reading of the cut rule, see page 93.

[8] Either of a single self-contained proposition or of a more ambitious theorem requiring the development of several auxiliary results.

The above can be considered as a traditional "proof" completely correct and acceptable from an advanced point of view. However, it does not provide enough elements to be mechanized or written as an actual rigorous proof. With the purpose of introducing transitional proofs we developed several traditional proofs and mechanizations of this theorem and present next those that best serve our further aims. The CAP counterparts of theorems 3, 4, 5 and 6 are available in the Appendix.

Let us first present a textbook traditional proof.

Theorem 2 (A Textbook Proof) *There does not exist a natural number n such that* $0 < n < 1$.

Proof Let $S = \{n \in \mathbb{N} \mid 0 < n < 1\}$ and assume $S \neq \varnothing$. Take $r \in S$ to be the least element in S. Therefore $0 < r < 1$ which implies $0 < r^2 < r < 1$. Thus $r^2 \in S$, which contradicts the minimality of r. Therefore S must be empty. □

This traditional textbook proof, though rigorous, is not suitable to be directly formalized in a PA, for apart of the implicit background knowledge required, it does not make explicit a train of thought useful to guide the mechanization. Our experiment consisted of mechanizing the above theorem according to different common strategies of reasoning. First we use forward reasoning, which means going exclusively from the hypotheses towards the conclusion.

Theorem 3 (Forward Proof) *There does not exist a natural number n such that* $0 < n < 1$.

Proof We assume that $\exists n \in \mathbb{N}$ such that $0 < n < 1$. The well-ordering principle allows us to take a minimal $r \in \mathbb{N}$ such that $0 < r < 1$. Since $0 < r$ we get $0 = 0 \cdot r < r \cdot r$ and as $r < 1$ we obtain $r \cdot r < 1 \cdot r = r$. Thus $0 < r^2 < r$. From this, as $r < 1$, we get $0 < r^2 < 1$. The minimality of r implies now that $r \leq r^2$, which leads us to $r^2 \not< r$, yielding a contradiction. □

Let us observe that in this proof there is no reference to the set S used in the proof of Theorem 2, for its use would complicate the mechanization in an unnecesary way. The proof looks similar to the previous one but it has an explicit forward structure and it provides us with more explanations. The proof script was constructed having Theorem 2 as a guideline using forward reasoning exclusively. This means we avoid the native backward mechanisms of the PA, which resulted in an awkward CAP.

Next we explore the opposite approach by constructing a proof using exclusively the backward strategy. The result is Theorem 4 and was obtained discarding the previous proofs as a guideline, using only the same auxiliary results, namely the transitivity of the order relation, its compatibility with respect to the product and the relationship between the order relations (Lemmas 1, 2, and 3 below). Since the example is quite elementary, this proof was constructed directly in the PA without resorting to pencil and paper.

Theorem 4 (Backward Proof) *There does not exist a natural number n such that* $0 < n < 1$.

Proof We assume that $\exists n \in \mathbb{N}$ such that $0 < n < 1$. Thus, the well-ordering principle allows us to take a minimal $r \in \mathbb{N}$ such that $0 < r < 1$. We will arrive to a contradiction by showing that $r \cdot r \not< r$ and $r \cdot r < r$. To show $r \cdot r \not< r$ it suffices to show $r \leq r \cdot r$. To prove this, we use the minimality of r, and are required to show $0 < r \cdot r < 1$. We separately prove the two inequalities: the first one $(0 < r \cdot r)$ is a consequence of $0 \cdot r < r \cdot r$, which is solved by the known fact $0 < r$. For the second inequality it is enough to prove $r \cdot r < r < 1$. Again, it suffices to prove the two inequalities separately. The first one is a consequence of $r \cdot r < 1 \cdot r$, which in turn is entailed by the known fact $0 < r < 1$. The second inequality $r < 1$ is already known. This finishes the proof of $r \cdot r \not< r$. The remaining inequality $r \cdot r < r$ was already proven within the previous case. □

Theorem 4 presents a verbose proof which, in comparison with the usual mathematical proof-writing style, has a cumbersome structure due to the exclusive use of backward reasoning. On the other hand, its mechanization is concise and more readable than the previous one. Let us also observe that the mechanization necessarily contains some mandatory low-level parts, like the need to repeat the proof (that is, the sequence of tactics) corresponding to show $r \cdot r < r$, which is undesirable from a programmer point of view. This and some other low-level issues can be partially prevented[9] but others like the explicit term rewriting of term r with term $1 \cdot r$ are unavoidable.

Although both theorems 3 and 4 have mechanized counterparts neither is a good candidate for what we mean to be a transitional proof. In the first case the mechanized version is awkward, for the forward strategy obliges us to use the same tactic all the time in a backward-oriented PA (in the case of COQ, the `assert` tactic). In the second case the traditional proof contains an informal detail in order to avoid the repetition of a subproof, namely that of $r \cdot r < r$. This part does not have an adequate counterpart in the succint mechanized proof, for instead of appealing to a result already proven in a previous subcase, we have to repeat the whole command sequence or to invoke an auxiliary lemma twice. We also observe that in this case the traditional proof is cumbersome in comparison with that in theorem 3.

Consider now the following proof, which was obtained by using Theorem 2 as a guideline but also mantaining a full interaction with the PA.

Theorem 5 (Bidirectional Proof) *There does not exist a natural number n such that* $0 < n < 1$.

Proof We assume that $\exists n \in \mathbb{N}$ such that $0 < n < 1$ and arrive to a contradiction. The well-ordering principle allows us to take a minimal $r \in \mathbb{N}$ such that $0 < r < 1$. We first claim that $r \cdot r < r$. Since $0 < r < 1$, we get $r \cdot r < 1 \cdot r$, that is $r \cdot r < r$. Next, we explicitly contradict the previous claim by showing $r \cdot r \not< r$. To this purpose it suffices to show $r \leq r \cdot r$. From $0 < r$ we get $0 = 0 \cdot r < r \cdot r$. Further,

[9] For instance by first proving $r \cdot r < r$ as an auxiliary lemma or at the beginning of the proof as in Theorem 5.

$r \cdot r < r$ and $r < 1$ yield $r \cdot r < 1$. By the minimality of r, to show the required $r \leq r \cdot r$, it suffices to show $0 < r \cdot r < 1$, but this has already been proven. □

As expected this proof has the best of both worlds, namely a well organized text together with a concise proof script. This structure was obtained by a parallel construction of both proofs, one that adequately combines forward and backward reasoning. The full interaction with the PA helped to enhance the argumentation, in particular this proof lacks repeated subproofs, like the one for $r \cdot r < r$ in the proof of Theorem 4. Thus, Theorem 5 is the best candidate to be a transitional proof. Further, we think that our small experiment already provides some evidence about the mentioned ineffectiveness of a translation process. Let us discuss next the level classification of proof components in detail.

An important question in the course of constructing a transitional proof is to classify the propositions and instructions of a CAP in low and high levels. This issue is deeply related to the question of deciding to what extent we should provide background or technical details in a paper or presentation. This question also depends on the intended audience.[10]

Thus, we lack a precise definition of high and low level propositions. However, according to the experiment just discussed and to our experience with formal verification we can give the following rough classification of low-level items:

- Reasoning patterns: corresponding to some applications of native inference rules, like equational reasoning (term rewriting, symmetry or transitivity) but also to some specific inferences related to the specific subject of the current mechanization case.
- Background knowledge: this agrees with the underlying theories or results needed for the particular case study. For instance, the arithmetic operations and definitions as well as the order properties of natural numbers in our toy example. This kind of results are usually implemented as independent libraries or modules.
- Very low-level operational tasks: like unfolding definitions, the rewriting of specific term occurrences, the explicit need for generalizing or specializing some statements manually, or the need for copying important information that might be transformed or deleted by some specific tactic application.

On the other hand, we can say that the high-level propositions and commands are those that do not fit in the above categories. These include all explicit statements of the TP but also the calls to the tactics and previous results that constitute the desired train-of-thought, for instance the call to a case analysis or a specific induction principle, as well as the application of a background theorem. The decision about what is a low-level proposition is not categorical and depends always on the specific proof, for instance, the substitution (rewriting) of t with s, when $t = s$ is known, can be high-level if this specific substitution is made explicit on the TP. But it also can be low-level if we are dealing with some basic background knowledge like with the

[10] A typical example of this scenario arises in Group Theory where the simplicity of the alternating group A_5 can be proved either by a direct argument or as an application of the Sylow Theorems.

equation $r = 1 \cdot r$. Moreover, sometimes a tactic call corresponding to a reasoning pattern should be classified as a very low-level task like in the case of the transitivity of equality, an inference that is seldom explicity mentioned in a TP.

The question now is how and to what extent we can internalize the low-level components of a CAP. The reasoning patterns are either part of the native implementation or can be defined by the human agent using the tactical facilities of the PA. The same happens with the background knowledge whose internalization takes place with the help of libraries, modules and tactics. The third group of low-level items is mostly unavoidable, for it corresponds to very sensitive details like the application of commutativity or associativity in a specific occurrence of an expression or, as mentioned before, the rewriting of some particular term occurrence, say r with $1 \cdot r$ in the third occurrence of r in the inequality $r \cdot r < r$. This kind of details permeate traditional proofs but they are usually implicit and not even mentioned. In contrast they must be carried out explicitly in the CAP.

We are now in position to propose a definition of transitional proof.

Definition 1 (Transitional Proof) Let C be a proof-assistant. A transitional proof of a statement A with respect to C is a pair $\mathbb{P} = \langle \mathcal{T}, \mathcal{S} \rangle$ where

- $\mathcal{T}$ is a traditional pencil-and-paper proof of A.
- $\mathcal{S}$ is a valid proof script in C proving A.
- The script $\mathcal{S}$ employs different proof facilities of C in a *balanced* way.
- There is a WYSIWYG mapping from $\mathcal{T}$ and $\mathcal{S}$, that is: for any argumentation step T in $\mathcal{T}$ there is a succession of instructions $\widehat{T}$ in $\mathcal{S}$ implementing T. Moreover, the train-of-thought (order of the steps) in $\mathcal{T}$ must be preserved in $\mathcal{S}$.
- The only instructions on $\mathcal{S}$ not corresponding to a proof step in $\mathcal{T}$ are very low-level operational tasks.

With a WYSIWYG mapping we mean that any step in the traditional proof, that is any natural language phrase corresponding to a specific and independent inference step T, corresponds to a sequence of C valid instructions or commands $\widehat{T}$ capturing exactly the same reasoning pattern. The balanced use of proof facilities is also subject of particular interpretation according to the specific human agent and proof assistant. This requirement allows us to discard cases like Theorem 3 as candidates for transitional proofs. Thus, the proposed definition should only be considered as a useful guideline that intends to capture the essential elements of the transitional proof construction process, a task that ideally is carried out by a deep interaction between the human-agent and the PA. In such cases both proofs $\mathcal{T}$ and $\mathcal{S}$ are built by a collaboration involving, pencil, paper, blackboard and the proof-assistant C. In this process the mechanisms of C dictate in some cases the inferences and writing of $\mathcal{T}$, and vice versa (see the Theorem 6 below).

Hence, Definition 1 should only be considered as a proposal that intends to capture the essential requirements of a transitional proof. We think that a strong definition in the mathematical sense is out of reach due to the fact that both traditional and computer-assisted proofs have other relevant aspects not treated here, like the very fact of deciding what is a valid traditional proof and other related sociological (Richard et al. 1979), (MacKenzie 2001, Chapter 6) and cognitive issues (Robinson 2000).

We summarize the results of our simple mechanization experiment, in the light of Definition 1, as follows:

- Theorem 3 does not belong to a transitional proof for the corresponding script $\mathcal{S}$ violates the condition on the balanced use of the COQ tactics.
- Theorem 4 does not belong to a transitional proof for there can be no WYSIWYG mapping for the informal argument about the already proven inequality $r \cdot r < r$, given in the traditional proof.
- Theorem 5 together with its corresponding proof script constitute a transitional proof.

Let us discuss next, using the same elementary theorem, an important question pointed by one anonymous referee regarding the role played by definitions in CAPs. This variation on the same theme also allows us to present an aspect of proof development showing that a simple CAP does not necessarily corresponds to a simple TP and vice versa. This fact entails possible difficulties on the proof acceptance process by an opposite community. An issue that reinforces the need for transitional proofs.

Theorem 6 *There does not exist a natural number n such that* $0 < n < 1$.

Proof Assuming that there is a natural n such that $0 < n < 1$ we will arrive to a contradiction. We have $0 < n$ and $n < 1$. According to the definition of $n < 1$ we have $S\,n \leq 1$, which by definition yields two cases

- $S\,n = 1$. Then we have $n = 0$ and therefore $0 < 1$ and $0 < 0$, which yields the contradiction.
- $S\,n \leq 0$. This a direct contradiction. □

The above proof is gained directly from a proof script whose motivation is to show a simpler mechanization than those already presented. One that does not use library theorems and which is certainly more succint than the previous ones (see the script in the Appendix). Its simplicity lies on the use of the definitions of $<$ and $\leq$ given in the core library (automatically loaded when starting COQ). The order $\leq$ is *inductively* defined by the following rules:

$$\frac{}{n \leq n}\ \text{LE\_N} \qquad\qquad \frac{n \leq m}{n \leq S\,m}\ \text{LE\_S}$$

whereas $n < m$ is just defined as $S\,n \leq m$.

The structure of the above proof depends on these definitions which, as expected simplify the proof-script, but that in a detailed traditional proof need to be explicitly stated, usually as auxiliary lemmas since an inductive definition of order is not common. Although this proof is clearly correct, it relies on a case analysis on the relation $n < 1$, which perhaps does not correspond to a routinary train-of-thought in mathematics, one that first argues with numbers and not directly with the order relation. This example shows the relevance of definitions in the CAP construction process, an issue that also impacts the construction of transitional proofs. On the other hand, the proof also departs from a transitional proof in that some natural operational choices in the CAP can become clumsy in the TP. For instance, in the first case of the proof of Theorem 6 the inequality $0 < 1$ is an unnecesary information which, in the corresponding CAP, is a consequence of the general substitution tactic `subst`. This phenomenon is common and gives evidence in the direction that a well-structured typical CAP may be quite far from a routine mathematical reasoning, thus departing from a transitional proof. This distancing can grow very easily, even in elementary proofs. For instance, to prove a simple equality on natural numbers, say $n + (n + 0) + 2 = S\,(n + (n + 0) + 1)$, a typical approach is to call a tactic like `lia`, which implements sophisticated algorithms on linear integer arithmetic. However, no traditional proof will appeal to such theory for an almost trivial task that only requires an elementary arithmetical reasoning. This way a simple CAP not necessarily corresponds to a simple TP and vice versa.

To close this section it is important to mention that there are other methodologies of proof construction related to our proposal. For instance Lamport's structured proofs (Lamport 1995, 2012) whose principal aims are to avoid proving false statements in mathematics and making proof verification simpler. Lamport studies different levels of structure within a proof, from these he proposes a formal language for structured proofs named TLA^+ for modeling software and hardware (Lamport 2002). This language is also designed to present proofs using hypertext, to enable the user to read a proof to the extent she wants or in the level of detail desired in order to understand the proof. For instance, an hypertext proof can hide, add or skip details of some proof steps. From our point of view this kind of proofs can be seen as a pseudocode that serves as a bridge between a TP and a CAP, for they can be used as a protocol in order to construct transitional proofs. Let us next present our pet example as a structured proof, using the bidirectional proof of Theorem 5 as TP. First we state some auxiliary lemmas.

Lemma 1 (Compatibility) *For all $n, m, p \in \mathbb{N}$, if $n < m$ and $0 < p$ then $n \cdot p < m \cdot p$.*

Lemma 2 (Transitivity) *For all $n, m, p \in \mathbb{N}$, if $n < m$ and $m < p$ then $n < p$.*

Lemma 3 (Relationship Between Less Equal and Less Than) *For $n, m \in \mathbb{N}$, if $n \leq m$ then $m \not< n$.*

These properties and their proofs are considered low-level background results.

Theorem 7 (Structured Bidirectional Proof) *There does not exist a natural number n such that $0 < n < 1$.*

1. Assume $\exists n \in \mathbb{N}$ such that $0 < n < 1$.
2. Assume $r \in \mathbb{N}$ to be the least element such that $0 < r < 1$. Thus

   2.1 $0 < r < 1$
   2.2 $\forall y(0 < y < 1 \rightarrow r \leq y)$

3. $r \cdot r < r$
   Proof: By compatibility using assumptions 2.1
4. $0 < r \cdot r < r$
   Proof: It suffices

   4.1 $0 < r \cdot r$
   Proof: By compatibility using 2.1
   4.2 $r \cdot r < 1$
   Proof: By transitivity using 3. and 2.1

5. $r \cdot r \not< r$
   Proof: By Lemma 3 it suffices $r \leq r \cdot r$

   5.1 $r \leq r \cdot r$
   Proof: By minimality (assumption 2.2) and 4.1, 4.2.

6. FALSE
   Proof: contradiction of 3 and 5.

3.4 A Notion of Formal Backward Proof

In this section we present a notion of formal backward proof that serves two purposes. First, it intends to fill a gap in the literature on formal logic where, to the best of our knowledge, this concept is not formalized at a high abstraction level, for instance a related robust treatment of goal-oriented proofs à la PROLOG (Filliâtre and Paskevich 2014) presents derivations simply as search trees. Second, it provides an abstraction that gives an accessible technical overview of the backward proof-search mechanisms actually implemented in modern proof-assistants. In our honest opinion the proposed notion, though well-understood and elementary for the PA community, will ease the construction of transitional proofs from other perspectives, that of the mathematician used to informal but rigorous pencil-and paper proofs combining several strategies, and that of the computer scientist used to construct structured and well documented programs or in this case proof-scripts. Moreover, the philosopher can serve from this concept to break the technical computational and logical entry barriers in order to enhance her discussion topics and questions.

Within mathematical and computational logic, a proof is represented by a binary relation between a collection[11] of hypotheses Γ and a formula A, denoted by $\Gamma \vdash A$.

[11] A finite collection for our purposes.

This expression is called a sequent and represents the existence of a proof of A entailed by the information in Γ, which is constructed according to some fixed inference rules. As already mentioned and exemplified, to construct a proof, that is to derive a sequent $\Gamma \vdash A$, we have two fundamental reasoning strategies:

- Forward reasoning: we manipulate the assumptions, that is the elements of Γ, using the inference rules to obtain new information in order to construct the conclusion A. This new knowledge only modifies Γ in a direct way, while the conclusion A remains untouched. Such methodology reflects the use of what is given, namely the starting hypotheses but also previous knowledge, including the underlying theories of the particular problem specified by A.
- Backward reasoning: We manipulate the conclusion A by means of the backward reading of inference rules with the purpose of generating some subgoals that entail the original conclusion A. In such process the context Γ can be modified although not directly but only as a consequence of the subgoal generation. This methodology captures the process of analyzing the result we want to prove and reason about what other intermediate results, hopefully simpler to prove, would imply it.

Since mathematical reasoning is usually identified with deductive inference and the definition of proof in formal logic corresponds to a forward construction from the axioms and hypotheses to the conclusion, it can be said that forward reasoning corresponds to the usual mathematical practice. On the other hand, backward reasoning is usually considered only as an intuitive heuristic aid to proof writing (Solow 2013; Eccles 2012; Velleman 2006). This strategy corresponds to the procedural interpretation of logic and it is deeply related to the binary resolution foundations of automatic theorem proving and to the actual implementations of logic in some proof-assistants like Coq. Backward reasoning is not considered in standard logic books like Mendelson (2009), Manin (2010), and Huth and Ryan (2004) and even in advanced studies on binary resolution like Leitsch (1997) the notion of formal proof is forward. The same happens with the classical reference on Edinburgh LCF (Paulson 1990, section 2.13) which provides us with a sequent calculus for backward reasoning but not with a definition of backward proof. A similar situation happens in Harrison (2009, Chapter 6).

In our opinion, an important obstacle to conciliate TPs and CAPs is the lack of a precise high-level definition of tactic and backward proof. For instance, whereas the foundational logics for proof-assistants, those implemented in the kernel, are deeply investigated in the literature (Norell 2009; Barras and Werner 1997; Sozeau et al. 2019; Kumar et al. 2016; Harrison 2006; Wenzel et al. 2008; Ebner et al. 2017; Anand and Rahli 2014), the notion of tactic à la Edinburgh LCF (Gordon et al. 1979), which permeates modern proof-assistants is only defined at the implementation level, usually in a functional language, of the backward application of a specific inference rule. A notable exception figures in the pioneering works of the philosopher Kanger (1957, 1970, 1963) (a more accessible reference is Holmström-Hintikka 2001) where we find the following definition:

Definition 2 (Kanger's Derivation) A proof or derivation of a judgment $\mathcal{J}$ in a deductive system $\mathcal{D}$ is a finite sequence of judgments $\Pi = \langle \mathcal{J}_1, \ldots, \mathcal{J}_k \rangle$ such that $\mathcal{J}_1 = \mathcal{J}$ and for each $\mathcal{J}_i \in \Pi$, $1 \leq i \leq k$, one of the following conditions holds:

- $\mathcal{J}_i$ is an instance of an axiom or starting rule.
- $\mathcal{J}_i$ is the conclusion of an instance of an inference rule whose premises are $\mathcal{J}_{i_1}, \ldots, \mathcal{J}_{i_m} \in \Pi$ with $i_1, \ldots, i_m > i$.

Kanger's definition starts with the judgment sought after, whereas the usual notion of formal proof ends with it. Also, the justifications of any step appear after it. This gives an idea of backward proof. Regrettably, even when this intuitive notion permeates Kanger's works, the above definition is unexploited. Perhaps because this is a merely declarative very high-level definition. In particular, proofs there are written forwards and just read from the bottom upwards. In conclusion, a formal proof according to Definition 2 departs from the backward style of CAPs due to the absence of the operational details dictated by the tactics.

The purpose of this section is to give a notion of formal proof that captures backward reasoning as in Kanger's definition but in an operational way. This grants the internalization of the proof-search mechanisms of a PA into the formal logic, thus easing the construction of transitional proofs especially in regard to breaking the entry barriers of the human agent to the mechanization process. To this purpose we need a high-level definition of tactic, that is a definition that does not refer to actual implementations. It is worth noting that this task represents an internalization process in the converse direction as discussed in the introduction, namely from the machine to the human agent.

Consider a rule in a deductive system[12] $\mathcal{D}$:

$$\frac{\Gamma_1 \vdash A_1 \quad \ldots \quad \Gamma_n \vdash A_n}{\Gamma \vdash B}$$

The backward reading of this rule is the following: *to prove* $\Gamma \vdash B$ *it suffices to show each of* $\Gamma_1 \vdash A_1 \ \ldots \ \Gamma_n \vdash A_n$. This property is formalized by means of a transition rule:

$$\Gamma \vdash B \ \ \triangleright \ \ \Gamma_1 \vdash A_1 \ ; \ \ldots \ ; \ \Gamma_n \vdash A_n$$

where the symbol $\triangleright$ is read as "it suffices to show". The goal $\Gamma \vdash B$ is solved by applying this specific transition rule, and its solution is supplied by the solutions of all the corresponding sequents on the right of the $\triangleright$ symbol. This means that one goal is transformed into n (expectantly simpler) subgoals. To continue the process one of these subgoals must be choosen. This can be done in a non-deterministic way. The backward reading of the inference rules as transition rules is what we call here

[12] Please note that this is a general train of thought and does not refer to the specific deductive system in Sect. 3.5.

a *tactic* following the tradition of Edinburgh LCF (Gordon et al. 1979), although the notion here presented is simplified in order to keep our introductory approach. Let us just mention that one of the important tactic features not treated here is the use of unification variables in goals, a powerful and helpful device which corresponds to unknown parts (subformulae) of required specifications.

Of course, the operational reading of inference rules just presented is well-known from logic programming and other refutation based proof systems like tableaux. But this does not mean that the notion introduced below is attached to refutation procedures. An important diference is that, as we will see below in Theorem 8, for adequate deductive systems, our notion of backward proof results equivalent to formal deduction in the usual sense. A property that usually does not hold for refutation procedures.

The process of backward proof construction of a given sequent S consists of searching for an inference rule $\mathcal{R}$ in system $\mathcal{D}$ whose conclusion matches S and then continue the proof-search with the subgoals (premises) $S_1, S_2, \ldots, S_n$ of $\mathcal{R}$, this is formalized by the tactic corresponding to $\mathcal{R}$. Thus a backward proof will be a particular coherent sequence of tactics whose application ends when there is no subgoal left to prove. It is important to remark that, although the choice can be non-deterministic, the order in which the subgoals are solved is quite sensitive in regard to its effects on the final proof (the proof-script not necessarily the internal representation or proof-term). We cannot abstract from the order for our interest lies in the shape of proofs and their possible correspondence with human made proofs. This issue supports that, at the desired abstraction level, backward proofs are not just forward proofs in reverse. This remark does not mean that in an actual PA, the subgoal solving process must mantain a specific order. As a matter of fact any native order strategy can be easily overridden by focusing on any specific subgoal.

Let us formalize the above ideas by means of a transition system.

Definition 3 A goal $\mathcal{G}$ is any sequent $\Gamma \vdash A$. The set of finite sequences of goals $GSeq$ is recursively defined as follows:

$$\mathcal{S} ::= [\cdot] \mid (\mathcal{G} :: \mathcal{S})$$

where $[\cdot]$ denotes the empty goal sequence. Moreover, if $\mathcal{S}_1, \mathcal{S}_2 \in GSeq$ then $\mathcal{S}_1; \mathcal{S}_2$ denotes the concatenation of $\mathcal{S}_1$ followed by $\mathcal{S}_2$.

We define now a transition system of tactics corresponding to an interactive backward proof-search.

Definition 4 The transition system of tactics for a deductive system $\mathcal{D}$ is defined as follows:

- The non-empty set of states is the set of goal sequences $GSeq$.
- An initial state is a singleton sequence.
- The empty sequence $[\cdot]$ is the unique terminal state.

- The transition relation $\rhd \subseteq GSeq \times GSeq$ is inductively defined by the inference rules in $\mathcal{D}$, where a transition $\mathcal{S}_1 \rhd \mathcal{S}_2$ can be read as "to prove the sequents in $\mathcal{S}_1$ it suffices to prove the sequents in $\mathcal{S}_2$".

A basic transition transforms a singleton goal sequence into a, perhaps empty, sequence of subgoals dictated by the backwards reading of an inference rule of $\mathcal{D}$.

As a specific tactic example consider the cut or substitution inference rule, which provides us with a useful reasoning method to make use of an auxiliary intermediate result or lemma:

$$\frac{\Gamma \vdash A \quad \Gamma, A \vdash B}{\Gamma \vdash B} \text{ (CUT)}$$

It generates two operationally different tactics, named assert and enough:

Assert A: $\Gamma \vdash B \rhd \Gamma \vdash A\ ;\ \Gamma, A \vdash B$
Enough A: $\Gamma \vdash B \rhd \Gamma, A \vdash B\ ;\ \Gamma \vdash A$

In both cases, the human agent needs to propose the concrete lemma A. However, the difference between them is only operational: either the user first asserts and proves the lemma A and then uses it to prove B or, vice versa, we add first a premise A to the knowlege basis Γ and show that this results in enough knowledge to prove the conclusion B. This kind of tactics emphazise the necessity of the interaction between the human agent and the proof-assistant, it is clear that full automation[13] would be quite difficult here. The cut rule is prominent both in proof theory and theorem proving. Its backward reading is at the heart of the implementation of the declarative style of theorem proving, which provides forward reasoning facilities to proof-assistants, see Wenzel (2002), Corbineau (2008), and Sacerdoti Coen (2010).

With the aim of simplifying the presentation we fix next a choice strategy that determines the order in which goals are solved, namely from the first (most left) goal in the current sequence of pending subgoals. This strategy is captured by the following sequencing rule (SEQ)

$$\frac{\mathcal{S}_1 \rhd \mathcal{S}_2}{\mathcal{S}_1; \mathcal{S} \rhd \mathcal{S}_2; \mathcal{S}} \text{ SEQ}$$

Finally, we give the promised definition of a backward proof.

Definition 5 A backward proof of $\Gamma \vdash A$ is a finite sequence of states $\mathcal{S}_1, \ldots, \mathcal{S}_k$ such that

- $\mathcal{S}_1$ is $\Gamma \vdash A$
- For every $1 \leq i < k$, $\mathcal{S}_i \rhd \mathcal{S}_{i+1}$
- $\mathcal{S}_k$ is $[\cdot]$

[13] Of course there are good heuristics for specific deductive systems but this departs from the approach of this paper.

Therefore a backward proof of a sequent $\Gamma \vdash A$ is a finite sequence of tactics that ends in the empty sequence of goals $[\cdot]$, meaning that the original goal $\Gamma \vdash A$ has no pending subgoals left to prove. This transition sequence of tactics corresponds to the bottom-up construction of a derivation tree by means of a left-to-right depth-first proof search. This captures a linear process corresponding to the fact that, although the power of non-determinism is not absent, the cognitive construction process requires, at any time, a unique choice of the next goal to solve. Let us also note that, according to Definition 5, a backward proof is a chaining sequence of particular instances of the relation $\rhd$. Therefore a backward proof of a sequent $\Gamma \vdash A$ corresponds to an instance of the transitive closure of the $\rhd$ relation, defined next.

Definition 6 The transitive closure of the relation $\rhd$, denoted $\rhd^+$, is inductively defined by the following rules:

$$\frac{\mathcal{S} \rhd \mathcal{S}'}{\mathcal{S} \rhd^+ \mathcal{S}'}\ \mathrm{TC1} \qquad\qquad \frac{\mathcal{S} \rhd \mathcal{S}' \quad \mathcal{S}' \rhd^+ \mathcal{S}''}{\mathcal{S} \rhd^+ \mathcal{S}''}\ \mathrm{TC2}$$

With the notion of backward formal proof at hand, we can discern if a particular deductive system is adequate to conciliate formal logic proofs (those checked by the kernel of a PA) with CAPs (those constructed by the PA). This is the case if both notions of proof coincide in the following sense:

Theorem 8 (Equivalence of $\vdash$ and $\rhd^+$) *Let $\Gamma \vdash A$ be any sequent and $\rhd^+$ be the transitive closure of $\rhd$. The following conditions are equivalent:*

(i) $\Gamma \vdash A \ \ \rhd^+ \ \ \Box$
(ii) $\Gamma \vdash A$ *is derivable.*

A direct consequence of this theorem is that there is no need to pursue *(ii)* once we have *(i)*. That is, the backward proof suffices, in the spirit of the formal verification tasks using a PA.

Our notion of formal backward proof, properly justifies most of the proof strategies usually discussed informally; fills a gap between the theory and practice of constructing formal proofs and, in our experience, has advantages while teaching constructive logic to undergraduate computer science students (Miranda-Perea et al. 2015). It is important to remark that the backward proofs captured here are done by a construction that goes only in the backward direction and it is not a combination of forward and backward steps, which constitutes an important difference with related work (Sieg and Byrnes 1998; Sieg and Cittadini 2005; Ferrari and Fiorentini 2015). This is so, for we are not interested in a fully automated proof-search where the proof sought after is constructed by the software without user interaction.

3.5 A Case Study on Modal Logic

In this section we give an example of a more elaborated transitional proof regarding a case study on modal logic, namely the mechanization of an axiomatic system for the constructive Modal Logic S4 of necessity, a case previously developed by us as part of a bigger development presented in González-Huesca et al. (2019). The focus is on admissible rules in particular the deduction theorem, whose validity was a subject of controversy resolved by Hakli and Negri (2012) in the case of classical modal logic. We follow our original presentation whose argumentation follows the lines of transitional proofs. The presentation gives a general guideline for any case study mechanization by considering four implementation stages: data, operations, properties and proofs. In the following we display either traditional or computer-assisted proofs[14] as we consider they fulfill our Definition 1 of transitional proof.

The data. The goal is to implement a deductive system which handles sequents of the form $\Gamma \vdash A$ where A is a modal formula and Γ is a finite collection of hypotheses. Hence the data is represented by data structures (inductive types) for formulas, contexts (lists) and the derivability relation. These are implemented by inductive types in COQ from the paper-and-pencil definition given by BNF[15] grammars. Modal formulae and contexts are generated by the following:

$$A, B ::= p_n \mid A \to B \mid \Box A \qquad\qquad \Gamma, \Delta ::= \cdot \mid \Gamma, A$$

In formulae, p_n denotes an element taken from an infinite supply of propositional variables, indexed by a natural number. Observe that implication is the only propositional connective. For contexts, a list implementation is given with two constructors: $\cdot$ for the empty list and a comma , for the operation that generates a new list from a given one by adding a new element to its right-end, usually called *snoc* constructor. This choice of implementation for contexts, discarding (multi)sets or mappings, is in favor of a high-level reasoning and to take advantage of tools such as structural induction.

The set of axioms $\mathcal{A}$ consists of the following schemes for S4:

A1 $A \to (B \to A)$
A2 $(A \to (A \to B)) \to (A \to B)$
A3 $(A \to B \to C) \to (B \to A \to C)$
A4 $(B \to C) \to ((A \to B) \to (A \to C))$

$\mathbb{K}$ $\Box(A \to B) \to (\Box A \to \Box B)$
$\mathbb{T}$ $\Box A \to A$
4 $\Box A \to \Box\Box A$

[14] The full COQ development of our article González-Huesca et al. (2019) is available in https://bitbucket.org/luglzhuesca/mlogic-formalverif/src/master/S4.

[15] Backus-Naur Form.

The primitive inference rules are the following

$$\frac{A \in \Gamma}{\Gamma \vdash A}\ (\textsc{Hyp}) \qquad \frac{A \in \mathcal{A}}{\Gamma \vdash A}\ (\textsc{Ax})$$

$$\frac{\Gamma \vdash A \qquad \Gamma' \vdash A \to B}{\Gamma'; \Gamma \vdash B}\ (\text{MP}) \qquad \frac{\cdot \vdash A}{\Gamma \vdash \Box A}\ (\textsc{Nec})$$

This sequent-style deductive system avoids the problems with the necessitation rule that led to the invalidity of the deduction theorem for it captures a notion of derivation with a clear distinction between axioms and hypotheses. Moreover, the rule of necessitation respects the notion of derivability of modal formulas as it allows to derive a boxed formula $\Box A$ only in the case that A is a theorem, that is, A can be derived without hypotheses. Let us observe that the system captures the common way of constructing derivations in axiomatic systems reason why the *modus ponens* is presented in the so-called multiplicative style (independent contexts), a feature that posses a challenge to the mechanization. Also the multiplicative style facilitates a translation between axiomatic and natural deduction systems, objective pursued by the formalization.[16]

The operations. The definition of some basic operations on data is required next, for example concatenation denoted by a semicolon $\Gamma_1; \Gamma_2$, context membership $\in$ and the application of the box operator to all the formulae in a context. We show the implementation of these functions.

```
Fixpoint conc (G G': ctx) : ctx :=
  match G' with
  | empty => G
  | snoc D q => snoc (conc G D) q
  end.

Fixpoint elem (a: Formula) (G: ctx) : Prop :=
  match G with
  | empty => False
  | G',b => b = a ∨ elem a G'
  end.

Fixpoint boxed (c:ctx) : ctx :=
  match c with
  | empty => empty
  | snoc G' b  => snoc (boxed G') (Box b)
  end.
```

[16] In González-Huesca et al. (2019) we proved the full equivalence between axiomatic (with the multiplicative style) and natural deduction systems for constructive S4. For a deep analysis of multiplicative and additive systems, the reader may consult (Plato 2014).

The properties and their proofs. We identify some important properties of contexts (these are low-level) and others which are required by the proof-assistant to complete the formal verification process of the desired high-level statements. Their particular form is designed for an adequate automatization.

Lemma 4 (Decidability) *The following properties hold:*

- *For all A and B formulas, either $A = B$ or $A \neq B$.*
- *For all Γ and Γ', either $\Gamma = \Gamma'$ or $\Gamma \neq \Gamma'$.*

Lemma 5 (Reasoning About Context Membership) *The following properties hold:*

- *For all A, Γ, if $A \in \Gamma$ then there are Γ_1, Γ_2 such that $\Gamma = \Gamma_1, A; \Gamma_2$.*
- *For all A, Γ and Γ', if $A \in (\Gamma; \Gamma')$ then $A \in \Gamma$ or $A \in \Gamma'$.*
- *For all A and Γ, if $A \in \Gamma$ then for any Γ', $A \in \Gamma; \Gamma'$.*
- *For all A and Γ, if $A \in \Gamma'$ then for any Γ, $A \in \Gamma; \Gamma'$.*

Finally a more complex low-level property about context/list decomposition, providing a case analysis proof method involving both, concatenation and snoc patterns, which is preferred over a simple case analysis about the (non) emptiness of a context.

Lemma 6 (Context Decomposition) *Let Γ, Γ' and Δ be contexts and A be a formula. If $\Gamma; \Gamma' = \Delta, A$ then either Γ' is the empty context and $\Gamma = \Delta, A$ or there exists a context Γ'' such that $\Gamma' = \Gamma'', A$.*

The properties displayed in the above lemmas have to be included in some layer of the development, where we can also automatize their use by defining some particular tactics. The question of how to carry out this internalization is not simple. For instance, it is clear that Lemmas 5 and 6 are quite general and reusable and therefore can be put in an independent library and considered general background. On the other hand, the first property in Lemma 4 depends on the particular type of formulae and thus belongs to a more specific background category. Thus, the internalization process is related to a classification of the background information in distinct levels and categories (mathematical, computational, etc.) a task that also depends on the specific implementation of the mathematical definitions. This important issue needs to be discussed in detail elsewhere.
The main purpose in the formal verification for our deductive system is to prove the following high-level properties. We emphazise that this development provides an essential aid when developing a transitional proof contributing to the desired WYSIWYG approach, for it keeps the CAP closer to the paper-and-pencil proof.

Theorem 9 (Deduction) *For any context Γ and A, B propositions, if $\Gamma, A \vdash B$ then $\Gamma \vdash A \to B$.*

Proof Induction on $\Gamma, A \vdash B$.
The case of (HYP) follows by analysis of the hypothesis $B \in \Gamma, A$. Two cases are given: either $B = A$ or $B \in \Gamma$. They are proved using the theorem $B \to B$ and an instance of axiom $\mathsf{A1}$, respectively.

For the case of axioms the proof succeeds in any subcase by *modus ponens* and an adequate instance of axiom A1.
For the case of rule (MP), we have $\Gamma_1 \vdash C$ (**1**) and $\Gamma_2 \vdash C \rightarrow B$ (**2**) where $\Gamma, A = \Gamma_2; \Gamma_1$. From this last equation we consider two cases according to Lemma 6 of context decomposition: either Γ_1 is empty or its last element is A:

1. If $\Gamma_1 = \cdot$, then $\Gamma, A = \Gamma_2$ and hypothesis (1) becomes $\cdot \vdash C$. From (2), the induction hypothesis yields $\Gamma \vdash A \rightarrow C \rightarrow B$. This sequent together with an instance of axiom A3 produces, by (MP), the sequent $\Gamma \vdash C \rightarrow A \rightarrow B$. Finally, $\Gamma \vdash A \rightarrow B$ is obtained by applying again (MP) to the last derivation and to the hypothesis (1).
2. If A is the last element of Γ_1 then there exists another context Γ' such that $\Gamma_1 = \Gamma', A$. This implies that $\Gamma = \Gamma_2; \Gamma'$. From $\cdot \vdash (C \rightarrow B) \rightarrow (A \rightarrow C) \rightarrow A \rightarrow B$ (axiom A4) and hypothesis (2) we obtain $\Gamma_2 \vdash (A \rightarrow C) \rightarrow A \rightarrow B$. Finally, by (MP) of this sequent with $\Gamma' \vdash A \rightarrow C$ (coming from the induction hypothesis applied to (1)), it follows that $\Gamma_2; \Gamma' \vdash A \rightarrow B$.

Finally, for the (NEC) rule the goal is to prove $\Gamma \vdash A \rightarrow \Box C$, assuming that $\cdot \vdash C$. This is achieved by *modus ponens* from axiom A1, in the form $\cdot \vdash \Box C \rightarrow A \rightarrow \Box C$, and $\Gamma \vdash \Box C$, which comes out of the hypothesis by (NEC). □

It is important to remark that this proof, gained from the bidirectional approach is quite different from the original in (Hakli and Negri 2012). Next we show the corresponding script that together with the above proof constitutes a transitional proof of Theorem 9

```
Theorem DeductionTh:
  forall (G: ctx) (A B: Formula), (G,A) |- B -> G |- (A ==> B).
Proof.
intros G A B H.
dependent induction H; rewrite <- (ctx_empty_conc G).
{ apply elem_inv in H.
  destruct H.
  rewrite H. apply AxI.
  eapply (MP _ _ A0 (A==> A0)); intuition. }
1-7: solve_axioms.
{ rewrite ctx_empty_conc.
  assert (x':=x).
  apply ctx_decomposition in x.
  destruct x; destruct H1.
  - rewrite H1 in H.
    apply IHDeriv2 in H2.
    apply Ax3_dett in H2.
    rewrite <- (ctx_conc_empty G).
    apply (MP _ _ A0 (A ==>B)); intuition.
  - rewrite H1 in x'.
    simpl in x'.
    inversion x'.
    assert (empty |- ((A0 ==> B) ==> (A ==>A0) ==> (A ==> B)));
    intuition.
    apply (MP G' empty (A0 ==> B) ((A ==> A0) ==> A ==> B) H0) in
```

```
    H2.
    apply IHDeriv1 in H1.
    eapply MP.
    exact H1.
    rewrite (ctx_empty_conc G') in H2.
    exact H2. }
{ eapply (MP _ _ (□ A0) _).
    apply Nec. exact H.
    apply Ax1. }
Qed.
```

In this script the case of axioms highlights an important issue in the correspondence of traditional and computer-assisted proofs, namely the question of how to capture in the system a proof sketch that informs about a repeated pattern reasoning. This approach, common in mathematical writing, is simulated here by the following very specific user-defined tactic `solve_axioms`.

```
Ltac solve_axioms :=
match goal with
  | |- (Deriv (conc empty ?G) (Impl ?A ?F)) =>
  eapply (MP _ _ F _); intuition
end.
```

This tactic implementation abstracts the sketched reasoning pattern in order to keep the TP and the CAP closer. On the other hand, a proof-script with separated manual proofs for each axiom would depart not only from the spirit of transitional proofs but also from the standard proof development practices both traditional and computer-assisted. This example also shows, as already mentioned, that apart from the mathematical background, captured by libraries or modules, we also need another backgound category, in this case one that collects implementations of specific pattern reasonings, a task eased by the native tactic definition mechanisms.

Next we present some other relevant admissible rules.

Theorem 10 (Principle of Detachment) *For any context* Γ *and* A, B *propositions, if* $\Gamma \vdash A \rightarrow B$ *then* $\Gamma, A \vdash B$.

Proof By *modus ponens* of the hypothesis and $A \vdash A$. □

The above proof corresponds to the proof-script:

```
Theorem Detachment:
  forall (G: ctx) (A B: Formula), G |- (A ==> B) -> G,A |- B.
Proof.
intros.
assert ((empty,A)|- A); intuition.
change (G,A) with (G; (empty,A)).
eapply MP.
exact H0.
exact H.
Qed.
```

The last proof was quite straightforward due to the multiplicative *modus ponens*. The next proofs are conducted not by the usual induction on the derivation (structural induction on the derivation data type) but by induction on the context. It is mentioned in González-Huesca et al. (2019) that the use of this method is new, which seems to be correct, at least on traditional proofs. This shows how definitions in CAPs also influence on the operational arguments of traditional proofs, sometimes producing non-standard statements of known results, like in the case of the next theorem and lemmas. This is a direct consequence of the transitional proof construction process since the contexts are inductively defined thus making available proof mechanisms like pattern matching and induction principles, which may not be the first choice reasoning methodologies with a different context definition.

Theorem 11 (Generalized Deduction Theorem) *Let* Γ, Γ' *be contexts and* A, B *be propositions. The following rule is admissible:*

$$\frac{\Gamma, A; \Gamma' \vdash B}{\Gamma; \Gamma' \vdash A \to B}\ (GDT)$$

Proof Induction on Γ'. The base case corresponds to the already proven Deduction Theorem 9. For the inductive step with Γ', C let us assume that $\Gamma, A; \Gamma', C \vdash B$. By the simple Deduction Theorem we have $\Gamma, A; \Gamma' \vdash C \to B$ which, by the Inductive Hypothesis yields $\Gamma; \Gamma' \vdash A \to C \to B$. From this by *modus ponens* with the axiom A3 we get $\Gamma; \Gamma' \vdash C \to A \to B$. Finally the principle of Detachment 10 yields the desired $\Gamma; \Gamma', C \vdash A \to B$. □

Since contexts are lists they are sensitive to the order of their elements, hence the Deduction Theorem enables us to discharge only the last formula in a context. Hence, the generalized version is mandatory for discharging any formula in the current context.

Lemma 7 (Context Permutation) *The following rule is admissible for any* Γ, Γ' *contexts and any proposition* A*:*

$$\frac{\Gamma; \Gamma' \vdash A}{\Gamma'; \Gamma \vdash A}\ (\textsc{Ctx-Perm})$$

Proof A straightforward induction on Γ using Theorem 11 and detachment. □

Lemma 8 (Context Contraction) *The following rule is admissible:*

$$\frac{\Gamma; \Gamma \vdash A}{\Gamma \vdash A}\ (\textsc{Ctx-cont})$$

Proof Induction on Γ. The base case is obvious. For the inductive step we assume that $(\Gamma, B); (\Gamma, B) \vdash A$ from which by the deduction theorem and its general version, we obtain $\Gamma; \Gamma \vdash B \to B \to A$. Next, with help of axiom Ax2 and (MP) we get $\Gamma; \Gamma \vdash B \rightarrow A$ and the induction hypothesis yields $\Gamma \vdash B \rightarrow A$. Finally the detachment principle allows us to conclude $\Gamma, B \vdash A$. □

Context permutation allows us to prove the *modus ponens* rule in an additive style (shared context), which is the more traditional formulation. Further, the peculiar shape of the above structural rules for permutation and contraction eases the proof and presentation of further inference rules, for instance the one for generalized necessitation whose specific formulation internalizes the notion of context weakening. We show next a full transitional proof of this useful rule.

Lemma 9 (General Necessitation) *Let* Δ, Γ *be contexts and* B *be a proposition. The following inference rule is admissible*

$$\frac{\Delta^{\Box} \vdash B}{\Delta^{\Box}; \Gamma \vdash \Box B}\ (\textsc{GenNec})$$

Proof Let $\Delta^{\Box} \vdash B$. By the property of context permutation given as an admissible rule, it suffices to show that $\Gamma; \Delta^{\Box} \vdash \Box B$. We proceed now by induction on Δ. If $\Delta = \cdot$ then we are done by the primitive (NEC) rule. For the inductive step let us assume that $(\Delta, A)^{\Box} \vdash B$. From this, the deduction theorem yields $\Delta^{\Box} \vdash \Box A \rightarrow B$ and the induction hypothesis leads us to $\Gamma; \Delta^{\Box} \vdash \Box(\Box A \rightarrow B)$. Next, using axioms $\mathbb{K}$ and *4*, as well as the transitivity of implication, which is a consequence of axioms A4 and A3, we obtain $\Gamma; \Delta^{\Box} \vdash \Box A \rightarrow \Box B$. Finally the principle of detachment produces $(\Gamma; \Delta^{\Box}), \Box A \vdash \Box B$, which is the same as $\Gamma; (\Delta, A)^{\Box} \vdash \Box B$ and the inductive step is finished. □

The corresponding CAP follows:

```
Lemma GenNec:
  forall (D:ctx) (A:Formula),
   boxed D |-A -> forall (G:ctx), boxed D; G |- □ A.
Proof.
intros D A H G.
apply ctx_permutation.
generalize A H G; clear.
induction D; intros.
- simpl in *.
  apply Nec.
  assumption.
- simpl in H.
  apply DeductionTh in H.
  assert(G; boxed D |- □ (□f ==> A)); intuition.
  + apply AxBoxK_dett in H0.
    apply boxtrans in H0.
    apply Detachment in H0.
    assumption.
Qed.
```

This finishes the case study. We close now the paper with a general discussion.

3.6 Discussion

The problem of elucidating if a CAP can be considered as equal to a traditional paper-and-pencil proof is quite far to be solved. For instance, we might believe that the need in a CAP to make explicit some reasoning steps omitted in a traditional proof, for they belong to a base of previous knowledge and methodologies, suffices to conclude that CAPs are closer to usual formal logic proofs and thus it is clear that they are different from paper-and-pencil proofs. Nevertheless, there are certain characteristics of proof-assistants and CAPs, that speak in the contrary direction. Hopefully, our notion of transitional proof would collaborate in this and other discussions around alternative proofs in both computer science and mathematical (Dawson 2015) practice.

The present work is just the beginning of a long term project involving not only the development of more and more complex case studies and experiments regarding transitional proofs but also the collection and generation of user experience reports especially those coming out of independent works, like Picard and Matthes (2012), which include a formal verification and whose presentation is related to our notion of transitional proof. This will allow us to validate and to refine the here proposed notion of transitional proof which, as a matter of fact, came out of our experience in mechanizing the results on modal logic presented in González-Huesca et al. (2019). In that work our starting goal was just to provide a formal verification of the mathematical results therein but at the end and due to the full synergy between pure paper-and-pencil mathematical reasoning and several mechanized proof attempts, employed during the development, we decided to reconcile, in both the text and the proof scripts, the high-level mathematical reasoning with the low-level procedures related to interactive proof-development. This experience led us to propose the concept of transitional proof. A first suitable candidate to start this user experience enterprise is the so-called mutilated checkerboard problem (Black 1946; McCarthy 1995; Huet 1996). This problem has been formally verified several times Kerber and Pollet (2006), going from fully automated to interactive proofs. For instance in Avigad (2019) a CAP of this problem, developed in the LEAN proof assistant, is presented as an answer to a challenge about formalizing the intuitive argument of the solution in a straightforward way. An analysis and comparison of the available proofs will allow to measure the contribution of the current approach in the communication processes between experts of different communities. This task must also include a detailed investigation on the role of the particular definitions used in a CAP and their impact in the corresponding TP.

Let us discuss now some aspects, in relation to CAPs as understood in this paper, of the challenges introduced in Turner (2018) and that we already summarized in Sect. 3.1. With respect to the mathematical challenge we can say that CAPs can be both long and shallow but also interesting and compelling. CAPs are Leibnizian proofs, every step is laid out and is checked line by line in a mechanical way. Moreover, enhanced CAPs which are part of transitional proofs are also Cartesian proofs Hacking (2014), for they are enforced to be explicit and obvious, which makes them graspable. Whether the CAPs deliver new mathematical knowledge, maybe it is not the time to know, but certainly they deliver methodologies that might lead to new knowledge. A particular example of this behavior appeared in the COQ-mechanized proof of the Four Color Theorem (Gonthier 2008), which originated the Mathematical Components COQ library that has been used for another ambitious formalization projects like the Feith-Thompson Theorem. This library is accompanied by a tutorial book (Mahboubi and Tassi 2018). Here is a extract of the introduction:

> Mathematical Components is the name of a library of formalized mathematics for the COQ system. It covers a variety of topics, from the theory of basic data structures (e.g., numbers,lists, finite sets) to advanced results in various flavors of algebra. This library constitutes the infrastructure for the machine-checked proofs of the Four Color Theorem [Gon08] and of the Odd Order Theorem [Gon+13a]. The reason of existence of this book is to break down the barriers to entry. [...] the Mathematical Components library is built in an unconventional way. As a consequence, this book provides a non-standard presentation of COQ, putting upfront the formalization choices and the proofstyle that are the pillars of the library. This books targets two classes of public. On the one hand, newcomers, even the more mathematically inclined ones, find a soft introduction to the programming language of COQ, Gallina, and the SSReflect proof language. On the other hand accustomed COQ users find a substantial account of the formalization style that made the Mathematical Components library possible.

This constitutes an example of a bidirectional communication effort in the spirit of transitional proofs.

CAP construction certainly requires patience but also originality encouraged by creativity. An example of this effect, more accessible than the huge developments of the above mentioned prominent theorems, is the proof method of admissibility of an inference rule by induction on the hypotheses context given in Sect. 3.5. Another interesting example is provided by a case study on coinductive graphs (Picard and Matthes 2011, 2012). These works introduce a peculiar definition of finite lists by means of total functions whose domains are finite sets of a fixed size. The need for this intricate definition lies in the need for "convincing" the proof-assistant to accept a definition that otherwise would be rejected due to a technical requirement.[17] This depicts the relevance of interfaces as intermediate languages suitable for bidirectional communication between human and machine.

Transitional proofs as understood in this paper are constructed by hand together with help of the proof-assistant. Thus they are only checked and not carried out by the computer alone. This speaks against the mechanical challenge. Moreover the prover systems we are considering here have consistent higher-order logics at their cores, in the case of COQ the Calculus of Inductive Constructions (Pfenning and Paulin-Mohring 1989). This sustains the position that we can trust the proof-checkers (Geuvers 2009; Pollack 1997).

The relationship between proof/program and proposition/specification enforced by a proof-assistant is a strong one based on type theories. The proof terms

[17] The guardeness condition of COQ.

correspond to pure functional programs, a paradigm that maximizes the correspondence between physical and symbolic program, due to the fact that the execution mechanisms native of this programming paradigm are also symbolic to a great extent. This situation departs from the pragmatic challenge.

According to the scientific challenge the purpose of program verification is not to verify a model or a theory in the usual sense: when the testing fails we do not change the specification but the program or device. Falsifying tests lead to revision of the artifact not of the hypotheses as in other sciences, except mathematics. This provides a methodological distinction. As already mentioned in this paper, this claim is more likely to be false in the case of CAPs and it seems more related to computer-assisted experimental mathematics. Though in our case it might assert a difference between some kind of program verification and formalized mathematics: in the latter we cannot say that we do not change the proposition but the proof.[18] However, this scenario depends on testing failure, a concept that, while not directly present in CAPs can appear at some extent by means of the intercommunication between proof-assistants and fully automatic yes/no-systems like automated theorem provers, model checkers or SAT/SMT-solvers (Comar et al. 2012; Moy et al. 2013; Ekici et al. 2017). An example is provided by the Why3 platform (Filliâtre and Paskevich 2013). This matter raises a more precise question about the different kinds of CAPs from the fully interactive, motive of this paper, to the fully automated[19] like the original proof of the Robbins Problem (Mccune 1997). A non-internalized CAP is a formal proof belonging to the realm of pure logic. On the other hand a fully internalized CAP resembles the outcome of a yes/no automated theorem prover thus becoming a black box that departs from what we expect for a train of thought to be comparable with a traditional proof. The notion of transitional proof here presented intends to provide a guideline to calibrate the internalization level, thus playing a useful role in the task of CAPs classification. An interesting related topic is presented in Böhne (2019); Böhne and Kreitz (2017) which speak of distinct grades of formality in proofs discussing their usefulness in learning how to prove using Coq.

Another relevant issue for the question whether the methodologies of general computer science depart from those of classical sciences is found in the Artifact Evaluation and Badging Scheme recently adopted by the Association of Computing Machinery ACM (Saucez et al. 2019). The purpose of this policy is to review and recognize those research papers that include some dataset or implementation, by validating it independently with respect to some aspects, which are reminiscent of those required by the experimental sciences like repeatability,

[18] Although sometimes it is easier to get a proof by modifying the proposition in a way that the original idea remains as a corollary of the new statement.

[19] This means that the system that solved the problem, namely EQP, is a fully automated system. However, the full non-interactive proof was developed during 5 weeks but the development project that lead to the implementation of the EQP prover took 10 years. Thus, an interesting question would be to discuss, knowing this long effort from a research team, to what extent the final proof of the Robbins Conjecture can be considered as fully automated.

replicability or reproducibility. Badges are awarded according to the availability, functionality or reusability of the implementation accompanying the paper. If this policy will become a standard, like peer-reviewing, for evaluating computer-assisted research, maybe now it is not the time to know. Anyway, these kind of policies contribute to positions about a modern incarnation of the Leibnizian dream, like the following one[20] related to Voevodsky's Univalent Foundations project (Voevodsky 2010):

> Voevodsky told mathematicians that their lives are about to change. Soon enough, they're going to find themselves doing mathematics at the computer, with the aid of computer proof-assistants. Soon, they won't consider a theorem proven until a computer has verified it.

To close our discussion we briefly point to some other facets of man-computer interaction which, in our opinion, are analogous to some circumstances in the construction of traditional proofs between human agents. Further inspection in the light of transitional proofs is required around these matters.

- *Evolution*: The discovery of new proofs of already proven theorems plays an important role in the evolution of mathematical knowledge. Apparently this does not happen in some cases and approaches to program verification where what matters is the validity of specifications, but the corresponding proofs seem to be disposable (Krantz 2007, Section 7.3). For instance, we can find several CAPs with available scripts which are difficult to examinate due to the fact that they contain some instructions and features that have been deprecated in the new versions of the required proof-assistant. Such developments are often related to published papers. This speaks of disposable CAPs. On the other hand, the evolution of proof-assistants, by updating the native libraries or enhancing the proof-search mechanisms (Chargueraud 2018), together with the interest in mantaining some particular proof developments, for instance those related to actual verification cases in the industry, forces to review the current developments to ensure compatibility.[21]

  If a library $\mathtt{Lib}_A$ is deprecated in favor of library $\mathtt{Lib}_B$ the new developments depending on $\mathtt{Lib}_B$ might need to check and update the old proofs depending on $\mathtt{Lib}_A$ to optimise the current development. Is this matter comparable with the generation of new proofs of the same result in mathematics?
- *Communication*: Each proof-assistant comes with its own exchanging and coding information devices (foundational logic, proof language and format, libraries, etc). Also, non trivial translation issues arise due to the different proof styles, see Wiedijk (2012). This contemporary Babel yields the difficulty of reusing a formal development of one system by another and has motivated research on automated proof translation between proof-assistants (Keller 2013).

[20] http://blogs.scientificamerican.com/guest-blog/2013/10/01/voevodskys-mathematical-revolution/.

[21] See the summary of changes of COQ's versions 8.X after the major revision of the development https://coq.inria.fr/distrib/current/refman/changes.html#recent-changes.

A situation comparable with the communication efforts between human agents that raises questions about the feasibility of translation procedures. Is the scenario of two proofs of the same proposition mechanized in different proof-assistants comparable with the question of alternative proofs in mathematics? Another related issue concerns the question of whether two proofs of the same proposition mechanized in different proof-assistants should be considered equivalent.

- *Beauty and simplicity*: Mathematicians customarily regard a proof as beautiful if it conforms the ideas of brevity and simplicity (McAllister 2005). On the other hand modularity and abstraction are simplicity principles important to Cartesian proofs for they involve comprehension and seek to avoid complexity. Using these principles generates simple programs that should lead to simple correctness proofs (Turner 2018). Are the concept of transitional proof together with an exploration of the role of specific implementations of mathematical definitions in a proof-assistant useful to contribute in a debate about beauty and simplicity on computer-assisted mathematics? Here we perceive a possible connection with Hilbert's 24th problem (Thiele and Wos 2002).
- *Social aspects*: In which sense can a proof-assistant play the role of a collaborator in the proof acceptance process? what is the role of transitional proofs? All matters mentioned in this paper should also be analyzed from the point of view of proof considered as a social process involving scrutiny, acceptance and collaboration (Asperti et al. 2009; MacKenzie 2001; Krantz 2007).

Appendix: CAP Counterparts for No Natural Between 0 and 1

This appendix shows the Coq proofs for the different versions of the non-existence of a natural number between 0 and 1, discussed in Sect. 3.3.

```
Require Import Arith.
Require Import ZArith.
Require Import Classical.

Section No_Nat_Between_0_and_1.

Lemma nat_well_ordered:
  forall (P:nat -> Prop),
     (exists n:nat, P n) -> (exists m:nat, P m ∧ forall k:nat,
     P k ->  m <= k).
Proof.
intros.
assert (exists! x, P x ∧ forall x', P x' -> x <= x') as G.
  apply dec_inh_nat_subset_has_unique_least_element.
  intro n; apply classic.
  assumption.
destruct G as [x G'].
exists x.
```

```
apply G'.
Qed.

Ltac well_ordering_principle H :=
apply nat_well_ordered in H; [destruct H as [r J]; destruct J as
[H minimal_r]].

(* Theorem 3*)
Theorem noNatbtw0and1Forward: ~ exists n:nat, 0 < n < 1.
Proof.
unfold not.
intro H.
well_ordering_principle H.
assert (0 < r*r).
- apply (mult_lt_compat_r 0 r r); intuition.
- assert (r*r < 1*r).
  + apply (mult_lt_compat_r r 1 r); intuition.
  + replace (1*r) with r in H1 by auto with arith.
    assert (0< r*r < 1).
    * assert(r*r<1).
      -- apply (lt_trans (r*r) r 1); intuition.
      -- intuition.
    * assert (r <= r*r).
      -- apply minimal_r; assumption.
      -- assert (~ r*r < r).
         ++ apply le_not_lt; assumption.
         ++ contradiction.
Qed.

(* Theorem 4 *)
Theorem noNatbtw0and1Backward: ~ exists n:nat, 0 < n < 1.
Proof.
unfold not.
intro H.
well_ordering_principle H.
absurd (r*r < r).
- apply le_not_lt.
  apply minimal_r.
  split.
  + replace 0 with (0*r) by auto with arith.
    apply mult_lt_compat_r ; intuition.
  + apply lt_trans with r.
    * replace r with (1*r) at 3 by auto with arith.
      apply mult_lt_compat_r ; intuition.
    * intuition.
- replace r with (1*r) at 3 by auto with arith.
  apply mult_lt_compat_r ; intuition.
Qed.

(* Theorem 5 *)
Theorem noNatbtw0and1Bidirectional: ~ exists n:nat, 0 < n < 1.
Proof.
unfold not.
intro H.
```

```
well_ordering_principle H.
destruct H as [r_gt_zero r_lt_one].
assert (r*r < r).
- assert (rr_lt_r:=mult_lt_compat_r r 1 r r_lt_one r_gt_zero).
  replace (1*r) with r in rr_lt_r by auto with arith.
  assumption.
- assert (rr_lt_r:=H).
  contradict H.
  apply le_not_lt.
  assert (J:=mult_lt_compat_r _ _ _  r_gt_zero r_gt_zero).
  simpl in J.
  assert (rr_lt_1 := (lt_trans (r*r) r 1 rr_lt_r r_lt_one)).
  apply minimal_r.
  intuition.
Qed.

(* Theorem 6: *)
Theorem noNatbtw0and1Native : forall n : nat, ~(0 < n < 1).
Proof.
do 2 intro.
destruct H.
inversion H0.
- subst.
inversion H.
- inversion H2.
Qed.
End No_Nat_Between_0_and_1.
```

Acknowledgments This research has been funded by UNAM DGAPA PAPIIT grant IN119920. The authors would like to thank two anonymous referees as well as the volume editors for many helpful comments to improve the contents of this paper.

References

Anand, A., and V. Rahli (2014). Towards a Formally Verified Proof Assistant. In *Interactive Theorem Proving*, 27–44. Cham: Springer.

Appel, K., and W. Haken (1976). Every Planar Map is Four Colorable. *Bulletin of the American Mathematical Society* 82(5): 711–712.

Asperti, A., H. Geuvers, and R. Natarajan (2009) Social processes, program verification and all that. *Mathematical Structures in Computer Science* 19(5): 877–896.

Avigad, J. (2008). Understanding proofs. *The Philosophy of Mathematical Practice*, 317–353. Oxford: Oxford University Press.

Avigad, J. (2010). Understanding, formal verification, and the philosophy of mathematics. *Journal of the Indian Council of Philosophical Research* 27: 161–197.

Avigad, J. (2019). A Formalization of the Mutilated Chessboard Problem. http://www.andrew.cmu.edu/user/avigad/Papers/mutilated.pdf

Barras, B., and B. Werner (1997). Coq in Coq. Technical Report.

Black, M. (1946). *Critical Thinking an Introduction to Logic and Scientific Method.* Hoboken: Prentice-Hall.

Böhne, S. (2019). *Different Degrees of Formality*. Ph.D. Thesis, University of Potsdam, Faculty of Science.

Böhne, S., and C. Kreitz (2017). Learning How to Prove: From the Coq Proof Assistant to Textbook Style. In *Proceedings 6th International Workshop on Theorem Proving Components for Educational Software, ThEdu@CADE 2017, Gothenburg, Sweden, 6 Aug 2017*, eds. P. Quaresma, and W. Neuper. Electronic Proceedings in Theoretical Computer Science, vol. 267, 1–18.

Bundy, A., M. Atiyah, A. Macintyre, and D. Mackenzie (2005). The Nature of Mathematical Proof. *Philosophical Transactions of the Royal Society A* 363(1835): 2461.

Chargueraud, A. (2018). *Theory and Practice of Automation in Coq Proofs*. Software Foundations Series, vol. 2. Electronic Textbook. Version 5.5. http://www.cis.upenn.edu/~bcpierce/sf

Chlipala, A. (2013). *Certified Programming with Dependent Types - A Pragmatic Introduction to the Coq Proof Assistant*. Cambridge: MIT Press.

Coen, C.S. (2010). Declarative Representation of Proof Terms. *Journal of Automated Reasoning* 44(1–2): 25–52.

Comar, C., J. Kanig, and Y. Moy (2012). Integration von formaler Verifikation und Test. In *Automotive - Safety & Security 2012, Sicherheit und Zuverlässigkeit für Automobile Informationstechnik, 14.-15. November 2012, Karlsruhe, Proceedings*, eds. E. Plödereder, P. Dencker, H. Klenk, H.B. Keller, and S. Spitzer. Lecture Notes in Informatics, vol. P-210, 133–148. GI.

Coquand, T., and G. Huet (1988). The Calculus of Constructions. *Information and Computation* 76(2–3): 95–120.

Corbineau, P. (2008). A Declarative Language for the Coq Proof Assistant. In *Types for Proofs and Programs*, eds. M. Miculan, I. Scagnetto, and F. Honsell, 69–84. Berlin: Springer.

Dawson, J.W. (2015). *Why Prove it Again? Alternative Proofs in Mathematical Practice*. Berlin: Springer.

De Millo, R.A., R.J. Lipton, and A.J. Perlis (1979). Social Processes and Proofs of Theorems and Programs. *Communications of the ACM* 22(5): 271–280.

De Mol, L. (2014). The Proof is in the Process: A Preamble for a Philosophy of Computer-Assisted Mathematics. In *New Directions in the Philosophy of Science*, 15–33. Berlin: Springer.

De Mol, L. (2015). Some Reflections on Mathematics and its Relation to Computer Science. In *Automata, Universality, Computation: Tribute to Maurice Margenstern*, ed. A. Adamatzky, 75–101. Cham: Springer.

Delahaye, D. (2000). A Tactic Language for the System Coq. In *Proceedings of the 7th International Conference on Logic for Programming and Automated Reasoning, LPAR'00*, 85–95. Berlin: Springer.

Ebner, G. S. Ullrich, J. Roesch, J. Avigad, and L. de Moura (2017). A Metaprogramming Framework for Formal Verification. *Proceedings of the ACM on Programming Languages* 1(ICFP): 1–29.

Eccles, P.J. (2012). *An Introduction to Mathematical Reasoning*. Cambridge: Cambridge University Press.

Ekici, B., A. Mebsout, C. Tinelli, C. Keller, G. Katz, A. Reynolds, and C. Barrett. (2017). SMTCoq: A Plug-in for Integrating SMT Solvers into Coq. In *Computer Aided Verification - 29th International Conference*. Heidelberg: Springer.

Ferrari, M., and C. Fiorentini (2015). Proof-Search in Natural Deduction Calculus for Classical Propositional Logic. In *Automated Reasoning with Analytic Tableaux and Related Methods: 24th International Conference, TABLEAUX 2015, Wroclaw, Poland, September 21–24, 2015, Proceedings*, ed. H. De Nivelle, 237–252. Berlin: Springer.

Filliâtre, J.-C., and A. Paskevich (2013). Why3—Where Programs Meet Provers. In *Programming Languages and Systems*, eds. M. Felleisen and P. Gardner, 125–128. Berlin: Springer.

Gabbay, D.M., and N. Olivetti (2014). *Goal-Directed Proof Theory*. Berlin: Springer.

Ganesalingam, M., and Gowers, W. T. (2017). A Fully Automatic Theorem Prover with Human-Style Output. *Journal of Automated Reasoning* 58: 253–291.

Geuvers, H. (2009). Proof Assistants: History, Ideas and Future. *Sadhana* 34: 3–25.

Gonthier, G. (2008). The Four Colour Theorem: Engineering of a Formal Proof. In *Computer Mathematics*, ed. D. Kapur, 333. Berlin: Springer.

Gonthier, G., A. Asperti, J. Avigad, Y. Bertot, C. Cohen, F. Garillot, S. Le Roux, A. Mahboubi, R. O'Connor, S. Ould Biha, I. Pasca, L. Rideau, A. Solovyev, E. Tassi, and L. Théry (2013). A Machine-Checked Proof of the Odd Order Theorem. In *Interactive Theorem Proving*, eds. S. Blazy, C. Paulin-Mohring, and D. Pichardie, 163–179. Berlin: Springer.

Gonthier, G., and A. Mahboubi (2010). An Introduction to Small Scale Reflection in Coq. *Journal of Formalized Reasoning* 3(2): 95–152.

González-Huesca, L.d.C., F.E. Miranda-Perea, and P.S. Linares-Arévalo (2019). Axiomatic and Dual Systems for Constructive Necessity, a Formally Verified Equivalence. Journal of *Applied Non-Classical Logics* 29(3): 255–287.

Gordon, M.J.C., R. Milner, and C.P. Wadsworth (1979). *Edinburgh LCF*. Lecture Notes in Computer Science, vol. 78. Berlin: Springer.

Hacking, I. (2014). *Why Is There Philosophy of Mathematics at All?* Cambridge: Cambridge University Press.

Hakli, R., and S. Negri (2012). Does the Deduction Theorem Fail for Modal Logic? *Synthese* 187(3): 849–867.

Hales, T.C. (2006). Introduction to the Flyspeck Project. In *Mathematics, Algorithms, Proofs*, eds. T. Coquand, H. Lombardi, and M.-F. Roy. Dagstuhl Seminar Proceedings, Dagstuhl, Germany, no. 05021. Internationales Begegnungs- und Forschungszentrum f"ur Informatik (IBFI), Schloss Dagstuhl.

Harrison, J. (1996). Proof Style. In *Types for Proofs and Programs: International Workshop TYPES'96*, eds. E. Giménez and C. Paulin-Mohring. Lecture Notes in Computer Science, vol. 1512, 154–172, Aussois: Springer.

Harrison, J. (2006). Towards Self-Verification of HOL Light. In *Automated Reasoning*, 177–191. Berlin: Springer.

Harrison, J. (2009). *Handbook of Practical Logic and Automated Reasoning*, 1st edn. New York: Cambridge University Press.

Holmström-Hintikka, G., S. Lindström, and R. Sliwinski (eds.) (2001). *Collected Papers of Stig Kanger with Essays on His Life and Work*. Synthese Library, vol. 303, 1st edn. Cham: Springer.

Huet, G. (1996). The Mutilated Checkerboard Problem. https://github.com/coq-contribs/checker.

Huth, M., and M. Ryan (2004). *Logic in Computer Science: Modelling and Reasoning About Systems*. New York: Cambridge University Press.

Kanger, S. (1957). *Provability in Logic*. Acta Universitatis Stockholmiensis. Stockholm Studies in Philosophy, vol. 1. Stockholm: Almqvist & Wiksell.

Kanger, S. (1963). A Simplified Proof Method for Elementary Logic. In *Computer Programming and Formal Systems*, eds. P. Braffort and D. Hirschberg. Studies in Logic and the Foundations of Mathematics, 87–94. Amsterdam: North-Holland.

Kanger, S. (1970). Equational Calculi and Automatic Demonstration. In *Logic and Value: Essays Dedicated to Thorild Dahlquist on His Fiftieth Birthday*, ed. T. Pauli. Filosofiska studier utgivna av Filosofiska fOreningen oeh Filosofiska institutionen vid Uppsala universitet 9, Uppsala, 220–226.

Keller, C. (2013). *A Matter of Trust: Skeptical Communication Between Coq and External Provers. (Question de Confiance: Communication Sceptique Entre Coq et des Prouveurs Externes)*. Ph.D. Thesis, École Polytechnique, Palaiseau.

Kerber, M., and M. Pollet (2006). A Tough Nut for Mathematical Knowledge Management. In *Mathematical Knowledge Management*, ed. M. Kohlhase, 81–95. Berlin: Springer.

Krantz, S.G. (2007). *The Proof is in the Pudding*. Berlin: Springer.

Kumar, R., R. Arthan, M.O. Myreen, and S. Owens. (2016). Self-formalisation of Higher-Order Logic. *Journal of Automated Reasoning*, 56(3): 221–259.

Lamport, L. (1995). How to Write a Proof. *American Mathematical Monthly. Also appeared in Global Analysis in Modern Mathematics, Karen Uhlenbeck, editor. Publish or Perish Press, Houston. Also appeared as SRC Research Report 94.* 102(7): 600–608.

Lamport, L. (2002). *Specifying Systems, The TLA+ Language and Tools for Hardware and Software Engineers*. Boston: Addison-Wesley.
Lamport, L. (2012). How to Write a 21st Century Proof. *Journal of Fixed Point Theory and Applications* 11: 43–63.
Leitsch, A. (1997). *The Resolution Calculus*. Texts in Theoretical Computer Science. An EATCS Series. Berlin: Springer.
Leroy, X. (2009). Formal Verification of a Realistic Compiler. *Communications of the ACM* 52(7): 107–115.
Leroy, X. (2018). Trust in Compilers, Code Generators, and Software Verification Tools. https://xavierleroy.org/talks/ERTS2018.pdf
MacKenzie, D. (2001). *Mechanizing Proof: Computing, Risk, and Trust*. Cambridge: MIT Press.
Mahboubi, A., and E. Tassi (2018). Mathematical Components. https://math-comp.github.io/mcb/
Manin, Y.I. (2010). *A Course in Mathematical Logic for Mathematicians*. Graduate Texts in Mathematics, vol. 53. New York: Springer.
McAllister, J.W. (2005). Mathematical Beauty and the Evolution of the Standards of Mathematical Proof. In *The Visual Mind II*, 15–34. Cambridge: MIT Press.
McCarthy, J. (1995). The Mutilated Checkerboard in Set Theory. http://www-formal.stanford.edu/jmc/checkerboard.html
Mccune, W. (1997). Solution of the Robbins Problem. *Journal of Automated Reasoning* 19: 263–276.
Mendelson, E. (2009). *Introduction to Mathematical Logic*, 5th edn. London: Chapman & Hall/CRC.
Milner, R. (1972). Logic for Computable Functions: Description of a Machine Implementation. Technical Report, Stanford University, Stanford.
Miranda-Perea, F.E., P. Selene Linares-Arévalo, and A. Aliseda-Llera (2015). How to Prove it in Natural Deduction: A Tactical Approach. CoRR, abs/1507.03678.
Moy, Y., E. Ledinot, H. Delseny, V. Wiels, and B. Monate (2013). Testing or Formal Verification: Do-178c Alternatives and Industrial Experience. *IEEE Software*, 30(3): 50–57.
Norell, U. (2009). *Dependently Typed Programming in Agda*, 230–266. Berlin: Springer.
Paulson, L.C. (1990). *Logic and Computation: Interactive Proof with Cambridge LCF*. Cambridge Tracts in Theoretical Computer Science, vol. 2. Cambridge: Cambridge University Press.
Pfenning, F., and C. Paulin-Mohring (1989). Inductively Defined Types in the Calculus of Constructions. In *Proceedings of the 5th International Conference on Mathematical Foundations of Programming Semantics*, 209–228. Berlin: Springer.
Picard, C., and R. Matthes. (2011). Coinductive Graph Representation: The Problem of Embedded Lists. *Electronic Communications of the EASST*, 39.
Picard, C., and R. Matthes. (2012). Permutations in Coinductive Graph Representation. In *Coalgebraic Methods in Computer Science - 11th International Workshop, CMCS 2012, Colocated with ETAPS 2012, Tallinn, Estonia, March 31 - April 1, 2012, Revised Selected Papers*, eds. D. Pattinson and L. Schröder. Lecture Notes in Computer Science, vol. 7399, 218–237. Berlin: Springer.
Pollack, R. (1997). How to Believe a Machine-Checked Proof. In *Twenty Five Years of Constructive Type Theory*, eds. G. Sambin and J. Smith. Oxford: Oxford University Press.
Robinson, J.A. (2000). Proof = Guarantee + Explanation. In *Intellectics and Computational Logic (to Wolfgang Bibel on the Occasion of His 60th Birthday)*, ed. Hölldobler, S., vol. 19. Applied Logic Series, 277–294. Amsterdam: Kluwer.
Saucez, D., L. Iannone, and O. Bonaventure (2019). Evaluating the Artifacts of Sigcomm Papers. *SIGCOMM Computer Communication Review* 49(2): 44–47.
Sieg, W., and J. Byrnes (1998). Normal Natural Deduction Proofs (in Classical Logic). *Studia Logica*, 60(1): 67–106.
Sieg, W., and S. Cittadini (2005). Normal Natural Deduction Proofs (in Non-classical Logics). In *Mechanizing Mathematical Reasoning, Essays in Honor of Jörg H. Siekmann on the Occasion of His 60th Birthday*, 169–191.

Solow, D. (2013). *How to Read and Do Proofs: An Introduction to Mathematical Thought Processes*, 6th edn. Hoboken: Wiley.

Sozeau, M., S. Boulier, Y. Forster, N. Tabareau, and T. Winterhalter (2019). Coq Coq Correct! Verification of Type Checking and Erasure for Coq, in Coq. *Proceedings of the ACM on Programming Languages* 4(POPL): 1–28.

The Coq Development Team (2020). *The Coq Proof Assistant Reference Manual Version 8.11.* https://coq.github.io/doc/v8.11/refman/

Thiele, R., and L. Wos (2002). Hilbert's Twenty-Fourth Problem. *Journal of Automated Reasoning* 29(1): 67–89.

Turner, R. (2018). *Computational Artifacts - Towards a Philosophy of Computer Science*. Berlin: Springer.

Tymoczko, T. (1979). The Four-Color Problem and its Philosophical Significance. *The Journal of Philosophy* 76(2): 57–83.

Velleman, D.J. (2006). *How to Prove it: A Structured Approach*. Cambridge: Cambridge University Press.

Voevodsky, V. (2010). Univalent Foundations Project (A Modified Version of an NSF Grant Application), Unpublished. http://www.math.ias.edu/vladimir/files/univalent_foundations_project.pdf.

von Plato, J. (2014). *Elements of Logical Reasoning*. Cambridge: Cambridge University Press.

Wenzel, M. (2002). *Isabelle, Isar - A Versatile Environment for Human Readable Formal Proof Documents*. Ph.D. Thesis, Technical University Munich.

Wenzel, M., L.C. Paulson, and T. Nipkow (2008). The Isabelle Framework. In *Theorem Proving in Higher Order Logics*, eds. O.A. Mohamed, C. Muñoz, and S. Tahar, 33–38. Berlin: Springer.

Wiedijk, F. (2012). A Synthesis of the Procedural and Declarative Styles of Interactive Theorem Proving. *Logical Methods Computer Science* 8(1): 1–26.

Chapter 4
Is There Anything Special About the Ignorance Involved in Big Data Practices?

María del Rosario Martínez-Ordaz

Abstract Here, I address the question of whether there anything special about the ignorance involved in big data practices. I submit that the ignorance that emerges when using big data in the empirical sciences is *ignorance of theoretical structure with reliable consequences* and I explain how this ignorance relates to different epistemic achievements such as knowledge and understanding. I illustrate this with a case study from observational cosmology.

4.1 Introduction

Cosmology is the branch of astronomy which concerns the studies of the origin and evolution of the universe; some of its objects of enquiry include galaxies, dark matter and dark energy, among others. For a long time, cosmology had been regarded to be very different from other empirical disciplines. Despite its successful predictions and observational discoveries, cosmology was in general perceived as too speculative, having a status even closer to philosophy than to other areas of physics (cf. Massimi and Peacock 2015). Nonetheless, this has changed in the last decades, mostly, thanks to the development of new technological and formal resources that allow scientists to receive, order and integrate enormously large amounts of data. This data is later used in surveys, like Kepler, Gaia and DES, SDSS, DESI, LSST, Euclid and WFIRST, which increase the scope of the cosmologists' predictions, makes their models more accurate and grants cosmologists access to

M. del R. Martínez-Ordaz (✉)
Federal University of Rio de Janeiro, Rio de Janeiro, RJ, Brazil

Department of Logic, Nicolaus Copernicus University in Toruń, Toruń, Poland

B. Lundgren, N. A. Nuñez Hernández (eds.), *Philosophy of Computing*,
Philosophical Studies Series 143, https://doi.org/10.1007/978-3-030-75267-5_4

novel phenomena.[1]But, cosmology is not the only scientific discipline that has been benefited from the emergence of big data and data science; as a matter of fact, the same happened to geology, climatology, biology, and other areas of scientific enquiry that had a long history of working with large datasets.

Unfortunately, and despite its positive outcomes, the incorporation of big data into scientific practice has come with some problems. Associated to the increase in the amount of data there is a significant increase in the scientists' ignorance regarding the ways in which such data hangs together. For example, observational cosmology has made much progress accessing phenomena that were initially considered to be unreachable for us, like galaxies that are million light years away, and that now can be "photographed" by us. However, much of this observational success depends on computational processes that cannot be fully scrutinized, examined and justified by human agents (see Humphreys 2009). This is, we could look at pictures of two galaxies colliding and rely on them as visual representations of the actual phenomena, yet we might not be able to rationally justify such a reliance.[2] This lack of epistemic access to the ways in which the received data holds together when generating certain outputs, such as pictures cosmological phenomena, is called *ignorance of theoretical structure* and it limits the understanding of the inference patterns that hold within a set (or a collection of sets) of data (cf. Martínez-Ordaz 2021).

Here, I address the question of whether there is anything special about the ignorance involved in big data practices. I submit that the ignorance that emerges when using big data in the empirical sciences is *ignorance of theoretical structure with reliable consequences* and I explain how this ignorance relates to different epistemic achievements such as knowledge and understanding. I illustrate this with a case study from observational cosmology.

While philosophy of science has already started discussing the different epistemic challenges and ethical consequences of the use of big data in the scientific endeavor, very little attention has been paid to the individual agents and the ways in which they overcome ignorance and acquire both knowledge and understanding when depending on big data. The novelty of this paper lies in paying attention to the problems that individual epistemic agents face when using big data in the empirical sciences.

The plan for the paper goes as follows. In Sect. 4.2, I discuss the epistemological worries about the use of big data in the empirical sciences. Later on, in Sect. 4.3, I scrutinize the relation between these epistemological worries and ignorance. In Sect. 4.4 I argue that the ignorance that underlies big data practices in the empirical sciences is *ignorance of theoretical structure with reliable consequences*, and in Sect. 4.5, I illustrate this with a case study from observational cosmology.

[1] Thanks to this, much progress is being made in the study of the nature of dark matter and the formation and evolution of galaxies due to the possibility of ordering, integrating and even visualizing the data that different telescopes report. A great example of this are the famous images of the *Bullet Cluster* which integrate optical data, X-ray data, and a reconstructed mass map, and that work as evidence in favor of the existence of dark matter.

[2] These problems in observational cosmology are approached again and in more detail in Sect. 4.5.

Section 4.6 is devoted to drawing some conclusions on the connections between ignorance and big data practices in the empirical sciences.

4.2 Epistemological Worries About Big Data

In this section, I discuss the most distinctive epistemological worries associated with the use of big data in the empirical sciences; I divide these worries into two main categories: the methodological and the understanding-directed. To do so, the section is divided in three main parts: Sect. 4.2.1. introduces some preliminary concepts, Sect. 4.2.2, summarizes the main methodological worries and Sect. 4.2.3. presents two concerns about the scope of these worries. Finally, Sect. 4.2.4. the main concerns related to understanding.

4.2.1 Preliminaries

Big data is the field that concerns ways to work with datasets whose size is beyond the ability of typical database software tools to capture, analyze, store, and manage (cf. Manyika et al. 2011). Note that the name *big data* does not only indicate the amount of data that is managed but, more importantly, the range of computational methods used to work with such data (cf. Arbesman 2013; Boyd and Crawford 2012).

Big data practices are grounded in data science and, due to the human agents' cognitive limitations, make constant use of machine learning algorithms to process, retrieve, analyze and extract information from immensely large and complex datasets. There are five main characteristics of these datasets: *volume*(the amount of data that is being managed, measurable in terabytes, petabytes, and even exabytes), *velocity* (the data generation rate and the processing time requirement), *variety* (the data-type, which can be structured, semi-structured, unstructured, and mixed), *veracity* (how accurate or truthful a dataset or a data source may be) and *value* (the possibility of turning data into something useful).[3]

From the outset I want to be clear about the main purpose of the paper. From now on, I only focus on the epistemic challenges that individual agents face when working with big data in the sciences. My aim in the rest of this section is to show that big data practices have introduced important challenges to the scientific activity—leaving aside philosophical discussions regarding the logical grounds of information and machine learning algorithms, the philosophical approaches to

[3] I am fully aware of the fact that there is an ongoing philosophical debate about the status of the different characterizations of big data, however, I believe this will suffice for the purposes of the paper. If interested in comprehensive philosophical analyses of the inferential mechanisms used when faced with immensely large amounts of data, see: Floridi (2011), and for introductory discussions regarding the epistemology of big data see Floridi (2012) and Leonelli (2014).

computability, the connections between Artificial Intelligence and the human mind, among others.

4.2.2 The Methodological Worries

Traditionally, scientific knowledge has been regarded as hierarchical, explanatory and at least partly unified. First, according "to hierarchical models of science, our scientific knowledge (...) forms a knowledge system that has two properties: (1) it is stratified, and (2) the items of some layer are or should be justified in terms of items of a higher layer" (Batens 1991, 1999). Second, the demand for explanatory power has at least two sources: pragmatic strand in terms of the power to predict and manipulate reality and an epistemic in terms of understanding what reality is like. Pragmatism and manipulability require simplicity and optimization whereas understanding reality through science requires accurate representations. These aims can conflict, but whenever they go together happily our explanatory ambitions tend to be satisfied. Third, the hierarchical spirit of scientific knowledge aided by its explanatory character enable scientists to look for unification, at least, in particular domains. When different theories satisfactorily explain the same system at different levels, the explanations that they provide are compatible, interconnected and mutually reinforcing.[4]

But big data has affected information gathering processes making them more comprehensive and faster than ever before. The incorporation of big data into the empirical sciences has modified the ways in which scientific knowledge was traditionally pursued and scientific methodology followed. The three main methodological worries associated to big data practices that have caught the philosophers' attentions are: the lack of clarity about purposes and uses of data, the reliance on correlations, and the epistemic opacity that surrounds the results of big data (cf. Humphreys 2009, Floridi 2012, Leonelli 2014).

The first two worries come from analyzing the actual novelty of big data practices. First, the increase of data that big data brings to scientific practice must not be conceived as essentially problematic. "Yes, there is an obvious exponential growth of data on an ever-larger number of topics, but complaining about such overabundance would be like complaining about a banquet that offers more than we can ever eat (...) We are becoming data-richer by the day; this cannot be the fundamental problem" (Floridi 2012, 436). As a matter of fact, the novelty of big

[4] However, hierarchical models face important difficulties, like lacking stable justificatory mechanisms that can avoid infinite regress or the absence of (robust) relations that can explain how to increase the order of our knowledge system. This considered, contextual models tend to be more satisfactory in both respects -especially when explaining the ways in which scientists rationally deal with incomplete, incompatible and even inconsistent information in their day-to-day practice (cf. Batens 1991). This has had an important effect on the unificatory character of science, as contextual approaches seem to be the only effective resource to address scientific practice, unification should remain only as a regulatory ideal. Despite the weakening of the hierarchical and unificatory nature of scientific knowledge, it is still expected for science to be mainly an explanation-seeking activity.

data, at least for the epistemology of science, should not lie in the sheer quantity of data involved, but rather in

> (1) the prominence and status acquired by data as commodity and recognized output, both within and outside of the scientific community and (2) the methods, infrastructures, technologies, skills and knowledge developed to handle data. (Leonelli 2014, 2)

The first methodological worry concerns (1) and the need for determining the purposes and uses of data. Nowadays, the key epistemological problem for the use of big data in the sciences is to identify which questions are interesting, or even essential, to answer at a certain moment, as well as the production and selection of the relevant answers (cf. Floridi 2012).

Moreover, regarding (2), the most notorious change when moving to big data driven scientific practices consists of moving from mistrusting *correlations* to ground scientific activity in the search for them. Correlations being “the statistical relationship between two data values, are notoriously useful as heuristic devices within the sciences” (Leonelli 2014, 3). For a long time, they were considered to be confusing and even misleading; as they do not suffice for explanation, it seemed unclear how much correlations could get scientists closer either to truth or to knowledge. Nowadays, correlations are seen as a form of knowledge—even if compared to explanatory knowledge–, “the correlations may not tell us precisely why something is happening, but they alert us *that* it is happening. And in many situations this is good enough” (Mayer-Schönberger and Cukier 2013, 14). While part of our current scientific practices are explanatory, there is another significant part that takes correlations to be keystones for scientific development. Thus, big data practices have shown that the explanatory character of science is not as strong as the traditional view had suggested, and the traditional way in which scientific knowledge was conceived is not enough for capturing and describing the statistical epistemic practices that nowadays ground many scientific disciplines.

The second methodological worry is to determine under which circumstances can scientists rationally trust correlations as legitimate instances of scientific knowledge. This concern is motivated from two sides. First, correlations leave unexplained why and how something is the case, and therefore, it seems hard to trust them as grounds of any epistemic enterprise. Second, given the large amount of information involved in big data practices, it is inevitable that it exceeds our cognitive capacities. Since scientists cannot ever process such quantities of data by themselves, whenever wanting to access and manage this data, they require technological implementation; making any epistemic product linked to those datasets, necessarily constrained by specific technological resources. This is, the rational trust of correlations requires a previous defense of the rational trust on the technological resources involved in their discovery.

The third worry concerns the scientists’ trust on the products of big data despite their lack of epistemic access to the way in which such products are achieved. Due to our physical limitations, the only way in which scientists can approach certain phenomena is through technological implementation. But, once the data is gathered, due to their cognitive constraints, scientists must now rely on computer resources that store, filter, classify and structure the data in the shortest possible time. The combination of these factors causes that the observational reports depend on not necessarily interconnected layers of technological implementation. And despite them being hardly scrutinizable by humans, these technological resources are indispensable for approaching phenomena that were initially inaccessible to us and for structuring data that we would have never been able to compute by ourselves. Such a lack of scrutinizability associated to big data processes has often been explained as a case of *epistemic opacity*.

A process is *epistemically opaque* "to a cognitive agent X at time t just in case X does not know at t all of the epistemically relevant elements of the process" (Humphreys 2009, 618). Many of our epistemic processes are, in different degrees, opaque to us, but what is distinctive of the ones involved in big data is that many of them would be *strongly* opaque to human agents. The third worry captures two main cases of epistemic opacity:

- **Opacity regarding the status of the products:** Nowadays, it is not clear whether the models that are created by computer-based methods are substitutes for empirical experiments in empirically inaccessible contexts or they are closer to theoretical abstractions (cf. Barberousse and Vorms 2014, Morrison 2015, Chap. 7). This opacity has an impact in the way in which these models are and should be endorsed by the scientists and the doxastic commitments that they might have towards them.[5]
- **Opacity regarding the procedures:** In big data practices, "no human can examine and justify every element of the computational processes that produce the output of a computer simulation or other artifacts of computational science" (Humphreys 2009, 618).[6] While the patterns that emerge through these computational processes are often necessary for the scientific enterprise, they are obtained in such unique ways that traditional human modeling techniques would not be able to generate (cf. Bedau 1997). This has as a consequence that some of the procedures that underlie the filtering, the selection and the leading to specific outputs become not-reproducible by, and even inaccessible to, human agents.

  Some of the elements that remain strongly opaque to us include privileged inference patterns that have been produced via machine learning algorithms and

[5] The reader might object that it is not clear if the status of the products of these computer-based methods is epistemically opaque, or if it is simply the case the their philosophical significance is not completely understood. However, while the question of whether a model counts as an experiment or an abstraction seems to be more philosophically oriented—rather than focused on our epistemic access to the world–, the epistemic opacity described here concerns only the status of the outputs of these computer-based methods, and not the methodology in itself. This is, this epistemic opacity concerns the question of whether the measurements, the descriptions and the visualizations of the data should count as observational reports, or only as theoretical expectations within a specific model.

[6] Here I am concerned with processes associated with the realization of algorithms in code as well as to the ways in which programs are actually run in particular instances. If interested in the conditions under which these processes can be made transparent see Creel (2020).

that are now significantly distant from human programmers' initial inputs. The fact that scientists might not be able to scrutinize all the steps through which some outputs were obtained, leaves them lacking inferential explanations for such outputs.

The second and the third worries are interconnected in the following way: while big data practices concern primarily the identification of new correlations, those supposed correlations rest on a multitude of sources—where there is often opacity of the workings of computer systems and reasoning, or research and observational techniques. There is a lot that is not known but that is trusted in big data practices; and therefore, the need for explaining how this ignorance does not affect the scientists' rationality.[7]

4.2.3 But, Why Big Data?

At this point, the reader might wonder whether the second and, specially, the third worries actually address consequences of big data—and not common phenomena in any scientific discipline. Following this intuition, one might consider that phenomena such as reliance on technology are not immediately connected to the use of large datasets, and therefore do not need to be approached by studying the epistemic practices of big data.

This concern is properly grounded: nowadays, different types of computer simulations are key resources in astrophysics similarly as they are in genomics or in fluid dynamics, they are handy when using immensely large datasets but also when working with ordinary-size sets of information (cf. Morrison 2015, Chap. 6 and 7). Even more problematically, some procedures of computer simulations are not strongly opaque to us, because they are only used to make mathematical operations go smoother but not to do work that exceeds human capacities. However, while there are elements that traverse scientific practice regardless of the amount of data that scientists deal with, the main difference—datawise—that exists between big data practices and other scientific practices is that the former constitute limit cases of the amount of data that is processed for scientific purposes.

While many epistemic practices in science might not require to implement methodological networks of high performance computing and deep learning algorithms, it is a fact that any research that aims at working with immensely large datasets would need to do so. And when this happens, at least, some crucial parts of the processes will remain opaque for the scientists. What is characteristic of big data practices is that such an epistemic opacity, necessarily, would surround at least some of the main products and procedures of these practices, and despite this, some of them will have surprisingly novel and seemingly reliable outputs. These outputs would often be of the form of reliable correlations that are susceptible to ground part of the future research.

[7] I am indebted to a referee who helped to give a better phrasing of my ideas on this point.

Furthermore, the number of scientific applications of big data has increased substantially in the last decade, and it is expected to only keep growing in the following years; thus, study of the epistemic successes and disadvantages of big data practices can only shed light on the grounds of our current, and future, science. In addition, if big data practices generate limit cases of epistemic opacity—mostly due to the amount of computational challenges that they deal with–, the ways in which this opacity is sorted out might be illuminating of ways in which similar problems can be tackled within similar ordinary-sized data practices.

4.2.4 The Understanding-Directed Worries

Understanding "consist of knowledge about relations of dependence. When one understands something, one can make all kinds of correct inferences about it" (Ylikoski 2013, 100). Scientific understanding is a fundamental component of any successful scientific enterprise; understanding a theory allows scientists to find new domains of application for it, and understanding an empirical domain makes it possible to build new theoretical approaches to that domain.

There is a common agreement on the fact that the increase of data that the sciences receive, store and manage nowadays should lead scientists to an ever greater understanding of the world. Unfortunately, according to the traditional literature, the more scientists rely in correlations and statistics, while losing grasp of causal explanations, the further away from understanding they are Leonelli (2014).[8] As a matter of fact, for achieving understanding, "the ability to explain why certain behaviour obtains is still very highly valued—arguably over and above the ability to relate two traits to each other" (Leonelli 2014, 6). In what follows, I present three worries that concern the achievement of scientific understanding in big data practices.

The first worry that comes when pursuing understanding in big data practices: understanding requires explanatory knowledge, correlations do not suffice for explanation, and the salient product of big data methodology is the recognition of new correlations. Therefore, understanding and big data methodology might just be going in opposite directions—and more frequent than desirable, one might have to choose between gaining understanding and identifying new patterns.[9]

[8] A group of epistemologists of science characterize understanding as an epistemic achievement that comes only after having obtained explanatory knowledge; this type of understanding has received the name of *explanatory understanding* (cf. Kvanvig 2003; Grimm 2006, 2014; Strevens 2013, Strevens 2017; Kelp 2014; Sliwa 2015; Lawler 2016, 2018). If understanding is essentially explanatory, it would be available only if (1) scientists can provide (causal) explanations for what is being understood, and (2) the content of their beliefs is true.

[9] There is an alternative account for scientific understanding which does nor require the previous acquisition of explanatory knowledge; however, it still requires that the content of the beliefs that will be related and understood is known to be true, which will also conflict with the third

The second worry is that, due to the involvement of epistemic opacity, agents would not be able to identify the relations of dependence between their beliefs. While this worry is clearly close to the ones presented in Sect. 4.2.2, it takes the problem a bit further and consider those cases in which a strong epistemic opacity does not prevent the achievement of knowledge, but conflicts with the pursuit of understanding.

A third worry concerns the quality of the data that inform the agents' beliefs. Scientific data is, and has been, often *defective* (vague, partial, conflicting or even inconsistent). This defective character of information is not only ubiquitous, but inevitable; for this reason, an important part of the scientific activity consists of tolerating the defects of the scientific data while aiming to acquire some scientific success—such as increase of either predictive or explanatory power, accuracy, empirical adequacy, among others. So, it should not come as a surprise that the data that scientists get when working with immense datasets is defective. However, there is a consensus on the factive character of understanding; this is, the content of what will be understood should be true. In the case of defective data, the satisfaction of this factive condition does not seem so straight forward, and therefore, neither does understanding.

I take this section to have shown that, when science incorporates big data to its epistemic practices there are, at least, six important epistemological worries to address. The next section is devoted to explaining how these worries have a common ground: ignorance.

4.3 Types of Ignorance

Here, I take that the study of ignorance can shed light on important peculiarities of big data practices in the empirical sciences; this section is devoted to characterize the different types of ignorance that have been recognized in traditional epistemology.

The section is divided in two main parts: Sect. 4.3.1. explains very briefly how the worries introduced in the previous section indicate different types of ignorance. Section 4.3.2, provides an overview of the different types of ignorance that epistemologists have recognized and they might put forward against the scientific activity.

understanding-directed worry. If interested in this view, see: Pettit 2002; Elgin (2004), 2007, Elgin (2017); De Regt and Dieks (2005), De Regt (2009), De Regt (2015), Khalifa (2013), De Regt and Gijsbers (2017).

4.3.1 A Common Ground

In the previous section, I argued that there are six worries associated with big data epistemic practices. While these worries might seem very different from each other, they have a common ground: ignorance.

Assume for a moment the intuitive characterization of ignorance as *lack of knowledge about something*. Needing to determine the purposes of data reveals that, because big data methodology consists in accumulating as much data as possible without a privileged purpose, when possessing access to immense datasets, scientists often *ignore* the specifics of the domains of application for such data as well as the problems that it can help to solve within the discipline. The trust of correlations when also aspiring to explanation reveals the previous acknowledgment of ignorance of a causal link. The two types of epistemic opacity that I discussed before are clear instances of ignorance, scientists ignore the nature of the products of simulations as well as the procedures through which they were obtained—and most of the time, they cannot perform the inferential procedures that originated such products.

As epistemic opacity indicates ignorance, when it conflicts with understanding, it can be said that because there is still a blank that should be filled—such a blank could be about the status of the models and simulations, or about the mechanisms that generated such models—understanding remains out of reach for the scientists. For the case of the emergence of defective data the role that ignorance plays should not be marginalized. There is the trivial sense in which having incomplete, partial or vague information is only a direct consequence of ignoring important bits of such information—how it connects, how it behaves, how does it relate to other datasets, among other aspects. Yet, there is also a more substantial interpretation of the ignorance behind the use of defective data, which is, even if we know that two mutually conflicting or even inconsistent chunks of information cannot be true at the same time, what scientists ignore is how to determine the truth values of the propositions contained in each chunk, and that uncertainty is what prevents the achievement of understanding.

4.3.2 Ignorance(s)

Traditionally, ignorance has been understood as *lack of knowledge*. In this sense, one can be ignorant via the non-satisfaction of any of the basic conditions for knowledge. This is, by failing at fulfilling a doxastic condition (S believes that p), an alethic condition (p is true), a justificatory condition (S believes that p with justification) or a Gettier-proofing condition (S's justification for believing that p must withstand Gettier-type counterexamples) (cf. Le Morvan and Peels 2016, 18).

Following such characterization, ignorance is often classified in, at least, the following types: (i) *absence of factual knowledge*, (ii) *absence of objectual knowl-*

edge, (iii) *absence of procedural knowledge*, and very recently, another type of ignorance has been added to the list: (iv) *absence of knowledge of theoretical structure* (cf. Martínez-Ordaz 2021). Orthogonally, one can also recognize (v) *erotetic ignorance*—absence of answers to questions.

Let's look at these types of ignorance by paying special attention to corresponding challenges that they (might) impose to the scientific activity:[10]

(i) **Factual ignorance (or absence of factual knowledge):** this ignorance consists in lacking knowledge of either facts or the truth of specific propositions. For instance, let *p* be ⌜The speed of light, in vacuum, is 299,792,458 metres per second⌝. When an agent *S* is factually ignorant of *p* the agent fails at determining the (correct) truth value for the proposition in question. This could happen due to: *S* holding a false belief, *S* struggles at assigning an alethic value to *p* or *S*' cognitive limitations prevent her from knowing a particular fact.
This type of ignorance conflicts with scientific reasoning by limiting the application of certain inferential rules. For example, if *S* fails at assigning an alethic value to *p*, *S* will not be able to detach the consequent of every conditional of the form $p \rightarrow q$.

(ii) **Objectual ignorance:** this ignorance requires absence of knowledge of a particular object. The main characteristic of this ignorance is that one ignores a whole set of properties that an object possesses and that are regarded to be indicative of such an object.
This ignorance conflicts with scientific activity by troubling preventing agents from connecting lists of properties to a particular object. Even if knowing that there is an *x* which has the properties p_1 and p_2, and knowing that there is a *y* that has the properties p_1, p_2 and p_3, one cannot determine whether there is any relation between *x* and *y* until we come to know them. Therefore, the main problem that comes with objectual ignorance is the impossibility of relating lists of properties to a common object, preventing scientists from identifying (new) phenomena and naming them.

(iii) **Procedural ignorance (or absence of procedural knowledge):** this type of ignorance requires agents to not know how to perform a certain task, such as riding a bike, baking a cake, operating a computer, and so on[11] Most of the time, an agent is considered to be procedurally ignorant when she cannot neither explain nor perform a specific task.
This ignorance conflicts with scientific practice especially in experimental contexts. For example, consider a scenario in which all members of a particular

[10] Because there is no clarity regarding its status compared to the other types or whether this ignorance reduces to any of the others, in what follows, I do not focus on this particular type expecting that the characterization of the other four is broad enough to capture the large majority of cases of lack of answers to questions.

[11] According to some epistemologists, this type of ignorance also resembles factual ignorance; Specifically, it can be translated into ignoring lists of causal relations, this is, not knowing what has to be done to obtain certain outcome (cf. Williamson 2001, Snowdon 2004).

scientific community are ignorant of how to reproduce an experiment in order to validate other team's reports; this absence of procedural knowledge becomes an impediment for the other team's results.

(iv) **Ignorance of theoretical structure:** this type of ignorance consists in lacking knowledge of

> the (relevant) inference patterns that scientific theories allow for. When ignoring (the relevant parts of) the theoretical structure of a theory, scientists are not capable of grasping abstract causal connections between the propositions of their theory, they can neither identify the logical consequences of the propositions that they are working with nor can explain under which conditions the truth value of such propositions will be false. (Martínez-Ordaz 2021, 12)

Ignorance of theoretical structure is often the cause of persistent instances of any of the other types of ignorance. Lacking access to a relevant part of the structural conditions of a theory prevents scientists from either inferring the value of certain propositions (causing factual ignorance), identifying whether distinct sequences of properties refer to the same object (causing objectual ignorance) or explaining inferential procedures (causing procedural knowledge).

The partial overcoming of this type of ignorance within a specific set of data, consists in identifying ways to inferentially secure particular regions of the logical space associated to such a set.[12] Determining ways in which reliable outputs are obtained and logical harm is avoided.[13]

Going back to the connection between ignorance and big data, one should still wonder to which extent the analysis of ignorance would be revealing of the epistemic grounds of big data practices. This, in light of the fact that as humans are epistemically limited, they are constantly ignorant of different things at different moments. So, if ignorance is not only common but essential to human agents, why should we worry about it when using big data in the sciences? With this in mind, in the next section I discuss the ways in which ignorance challenges scientific rationality in big data contexts.

[12] *Partial overcoming of ignorance of theoretical structure* means that, when tolerating a contradiction, scientists need not to identify the *ultimate* or the total structure of their theory, but that they can provide a set of inference patterns that allow them to successfully use the theory in question while avoiding logical triviality (cf. Martínez-Ordaz 2021). It is important to remember that, as theoretical structure is a dynamic entity, changing as a theory evolves, new findings can lead to changes in how inferences are made. For example, new findings in methods of approximation greatly affect the inferences made in many sciences.

[13] Faced with $p \rightarrow q$ and $\neg p \rightarrow q$, if S is ignorant of the truth of p (and its negation), many would be happy to detach q; however, when being ignorant of the theoretical structure that relates instances of p, $\neg q$ and q this would not be necessarily possible. Ignoring the theoretical structure that relates a set of data means ignoring how negation works within that particular set, what can be inferred, what is not a consequence of the dataset, among others.

But once this ignorance is partially overcome, propositions will have a specific value in a world like the one described by a specific theory (or model); and this does not necessarily extend to the actual world.

4.4 Big Data, Big Ignorance?

Here I claim that the ignorance that underlies big data practices is, often, ignorance of theoretical structure *with reliable consequences*, and that this ignorance does not prevent scientific understanding from being achieved.

The section is divided in four parts: Sect. 4.4.1. I briefly acknowledge the importance of identifying the ignorance that is involved in big data practices and the ways to overcome it. Section 4.4.2. addresses the type of ignorance that underlies the big data practices and Sect. 4.4.3. sketches the type of understanding that is achievable through these practices.

4.4.1 The Landscape

Big data methodology consists of the recollection of very different types of data (images, redshifts, time series data, and simulation data, among others) that relates to different aspects and facets of the studied phenomena—that is, in the large majority of cases, scientists receive partial information about their object of study. This recollection involves integrating data from various sources and formats which initially might not be fully compatible. Also the data is produced, transmitted and analyzed at an extremely high velocity, which prevents individual agents from keeping a detailed track of how the data changes and relates. In addition, it is well known that the use of defective data comes with the price of different degrees of ignorance (cf. Wimsatt 2007, Norton 2008). But scientific rationality is only met either when the degree of ignorance can be maintained or reduced, or when scientists do not hold any doxastic commitments towards the information that they are working with—in particular, if they do not trust neither the information that they are working with nor their results.[14]

The combination of the above poses the following dilemma against scientific rationality: unless scientists find a efficient way to low the level of ignorance, they are irrational for trusting data that at its best is defective and at its worst might be false; or they are irrational for reasoning under high degrees of ignorance—regardless their doxastic commitments towards the products of using big data. So either we explain how scientists can reliably lower their degrees of ignorance when working with big data or we accept the fact that they are irrational.

[14] While there is not a uniform view on what *scientific rationality* is exactly, there is a common agreement on the fact that reliable indicators of it include: the achievement of knowledge and understanding, instances of scientific success (accurate predictions, the provision of explanations, manipulability via experimentation, among others), reliable mechanisms for constructing, testing, revising, and selecting theories, among others. For the purposes of the paper, I focus only on the relation between scientific rationality, knowledge and understanding.

4.4.2 The Ignorance Behind Big Data

In big data practices, the combination of both physical instruments and formal tools has helped to automate much of the scientists' processes (like pattern recognition and classification) as well as facilitating big tasks (like processing vast amounts of data in hours instead of the months or years it would take for a group of human agents by themselves). This resulted in aiding the identification and scrutiny of newly detected objects. For instance, the fact that the NASA Chandra X-ray Observatory constantly receives, stores, filters, classifies and integrates enormously large amounts of optical data and X-ray data, among others, has enabled the detection, and the later scrutiny, of the so called *Bullet Cluster*, a phenomenon that was expected to occur but which detection would have been impossible without the help of observatories that do not only receive data but also process it (see Sect. 4.5.1.). These practices have allowed scientists to acquire knowledge regarding the objects that were initially inaccessible; this is, to attain objectual knowledge.

However, as the selection of data sometimes depends on epistemically opaque processes, scientists end up ignoring how certain outputs were obtained as well as other possible outputs that were disregarded by the algorithms—meaning that there are going to be some important bits of information about phenomena that will remain ignored by the scientists despite having being initially captured. At this point, it only seems fair to say that what scientists ignore is the way in which particular sets of data hang together in order to entail certain outputs, this is, they ignore the relevant part of the theoretical structure that glues the received data and the products of computational processes.

In addition to reliance on technology, there is another element in data-driven sciences that should also be considered as causing epistemic opacity about processes and products, this element is the *increasing collaboration*.

Big data practices possess a massively cooperative nature which makes the transmission and acquisition of knowledge very opaque as well. Very often scientific communities rely on the quality of the datasets that were initially processed and now shared by other communities, making these practices based on a new type of epistemic trust. What is at stake here is a reliance not on the individuals that integrate the communities, but on their technological choices and the procedures that such choices imply—regardless of how opaque they are for the individuals who have chosen and employ them. The result is that, if what is transmitted and acquired is a type of knowledge, it is not of the kind of knowledge by (expert) testimony (cf. Sullivan 2019).[15]

[15] This, especially when adopting a standpoint similar to the so-called assurance view of testimony, according to which "testimony is restricted to speech acts that come with the speaker's assurance that the statement is true, constituting an invitation for the hearer to trust the speaker. Such views highlight the intention of the speaker and the normative character of testimony where we rebuke the testifier in the instance of false testimony (Tollefsen 2009)" (Sullivan 2019, 21). Because testimony is often perceived as a highly intentional speech act that assumes that the expert can offer explanations to back up her assertions, it is not clear how technological implementations

In big data practices, the source of knowledge is not always an individual that can provide better explanations to support her claims if asked to do so; it is often a combination of methodologies plus machine implementation over inputs that come from very diverse sources in very different formats, and which interconnections are not always clear to us. In the long run, this has the effect of scientists being unable to provide explanations about procedures that might have led to the discovery of novel phenomena. Yet, should this be understood as a case of procedural ignorance? not necessarily. When they cannot provide inferential explanations about why an output obtains, they are not ignoring only a specific recipe, they are ignorant of how the bits of data relate to one another—at least, inferentially; and this is indicative of ignorance of theoretical structure.

Consequently, when working with big data, scientists are trading knowledge of some parts of theoretical structure in exchange for access to inaccessible objects. As a matter of fact, the incorporation of big data to the empirical sciences has created a new epistemic preference: "answers are found through a process of automatic fitting of the data to models that do not carry any structural understanding beyond the actual solution of the problem itself" (Napoletani et al. 2014, 486. in Leonelli 2020).

Nonetheless, if the ignorance that underlies big data practices is ignorance of theoretical structure, one should not overlook the fact that, in the corresponding literature, this ignorance is described as the main source of negative epistemic outputs such as resilient anomalies, mutually conflicting inferential products, among others (cf. Martínez-Ordaz 2021). In light of its negative impact on the pursuit of knowledge and understanding, one should worry that this ignorance causes a larger epistemic harm in big data contexts. In particular, by preventing scientists from determining the inference patterns that govern specific datasets and the selection mechanisms for inferential products from these sets, ignorance of theoretical structure might put in danger the epistemological basis of big data practices.

The challenge seems more complicated when, scientists might not be able to satisfactorily get rid of this ignorance due to the combination of (a) the fact that big data is often used to study phenomena that is hard to verify or intervene without heavy technological implementation, and (b) the strong presence of different types of epistemic opacity and diverse instance of epistemic trust. (a) and (b) might just be enough to undermine the epistemological basis of big data practices in the empirical sciences. This concern can be formulated in the following way:

1. If one cannot explain how certain (heavily) mediated observational reports are generated, one must weaken one's belief of them being evidence of something being the case.
2. In empirical data-driven sciences, certain outputs of big data procedures are expected to count as observational evidence of something occurring in a particular way.

could provide us with something like that; in particular, in cases in which there is a strong epistemic opacity surrounding the outputs of such implementations.

3. But, when present, ignorance of theoretical structure messes up with the scientists' capability to explain how these outputs are generated; without necessarily affecting the scientists' capability to explain why a particular phenomenon would occur in a specific way.

In light of the above, scientists might have to chose between either rejecting these outputs as evidence about a specific phenomenon or, at the risk of being irrational, trusting products that they ignore where they come from and how they were obtained, and use them for testing empirical hypotheses.

The first option might look appealing in cases in which there are alternative methods for gathering evidence about phenomena which do not require the use of big data and opaque computational methods. However, if these phenomena are essentially inaccessible to us without heavy technological implementation, to take the first option would mean to lose our epistemic access to entire empirical domains. This makes the second option more appealing, but also rises the question of how to make outputs of processes that are opaque to us more reliable when using them as evidence in the empirical sciences.

This concern has not been overlooked by the scientists, as a matter of fact, they have constantly sought for methodologies that allow them to preserve and justify the reliability of the data—regardless if they are ignorant of the processes that generated the data. An important instance of how these collaborative work succeeds are

> taxonomic efforts to order and visualise data inform causal reasoning extracted from such data (Leonelli 2016, Sterner and Franz 2017), and can themselves constitute a bottom-up method -grounded in comparative reasoning -for assigning meaning to data models, particularly in situation where a full-blown theory or explanation for the phenomenon under investigation is not available (Sterner 2014). (Leonelli 2020)

The positive outcomes of the joint work of researchers, curators and programmers include accurate predictions, measurements and descriptions. This considered, there is the need to explain the continued trust that scientists posit on big data practices despite the epistemic opacity that surrounds them.

While scientists might be ignorant of the way in which data hangs together in order to generate certain outputs, some of the results that are reached through different computational processes would be extremely novel and accurate. And is in light of such successful results that scientists are justified in trusting the processes that generated them—this justification is of a *reliabilist* nature. Scientists trust big data practices in a similar way than they trust their vision, mainly because they can recognize that the outputs of both big data procedures and processes carried out by the visual system produce more successful than ineffective consequences.

But, in big data contexts in the empirical sciences, what counts as a *successful* consequence? A successful consequence of a computational process would be an output (prediction, description, representation, etc.) that grants access to empirical phenomena—especially if that phenomena that wouldn't be accessible to humans without the aid of big data and computational processes–, and enhances the achievement of objectual knowledge regarding such phenomena. In addition, this output should be:

- novel in its field,
- empirically adequate,[16]
- fruitful—the output seems to be crucial for the development of related research programs, and
- the output holds a possible evidential relation with a model or theory within the discipline.[17]

Note that, when the output concerns an empirical domain that is physically inaccessible to us, the empirical adequacy of such an output might be hard to evaluate observationally; therefore, the success of such a result can be graded considering both its connections with accepted models or theories, and its impact on measuring, predicting and explaining other empirical phenomena. But what is it about the volume, velocity, variety, etc.—what is it about big data's features? While in Sect. 4.2, I emphasised the epistemically negative consequences of these features and their connection with proprietary and opaque software—causing different types of epistemic opacity; here, I want to draw the reader's attention to the benefits of these features for the novelty of the outputs of big data practices.

Gathering largely immense amounts of data of different kinds at an extremely high speed, makes the resulting sets of data, when finally integrated, extremely informative. For instance, the representations that are obtained through big data are not only representations that, computationally and observably, we could not have constructed ourselves alone; but they are also exceptionally comprehensive representations of highly complex phenomena. The fine grained detail that is possible to recognize in big data products constitutes one of the most important parts of their novelty, and this is mostly caused by the fact that the data that is received comes not only in different formats but also refers, in great detail, to the different layers of the studied phenomenon. So, when this data is satisfactorily integrated and structured—even if the integration process is extremely opaque to us–, the result will be a highly detailed map of the phenomenon, in which scientists would be able to zoom in and zoom out, in such a way that will enhance their grasping of the phenomenon at different levels and in different scenarios.

The combination of all the above gives the impression of, while the ignorance that underlies big data practices in the empirical sciences is ignorance of theoretical structure, its main characteristic is that it possesses (significantly) reliable consequences.

[16] Even if ignoring the status of the output; this is, not knowing if it should be accepted as a substitute for empirical experiment or a theoretical abstraction of a specific empirical domain.

[17] I am fully aware of the fact that there are many ongoing debates regarding the sufficient (and necessary) conditions for the evaluation of scientific success in big data practices. And, for that reason, the criteria listed in this section are not expected to be taken as neither universal nor definitive, but only intuitive enough to consider them as indicative of success when satisfied.

4.4.3 Understanding Big Data

According to what has been discussed in previous sections, big data practices have granted scientists access to new phenomena, and have provided them with the opportunity of accurately identifying, measuring and predicting their behavior. In particular, the use of big data has allowed empirical scientists to achieve objectual knowledge of things that, for centuries, were considered to be too complex for the human mind. But this success is not without its downside, it comes with the loss of causal explanations—with respect to, at least, novel phenomena, and therefore, the loss of explanatory understanding with regard to the newly discovered objects. However, is this all we can get from the implementation of big data practices?

In what follows, I argue that there is a way to interpret the epistemic profits of big data practices as a keystone for the achievement of scientific understanding. In particular, I explain how the type of understanding that can be gained through these practices is *modal understanding*.

First, given the multiplicity of approaches undertaken by scientists to extract useful information from big data, it could be said that even in those cases where the data points out the existence of an object and some of is properties, it does not do so in a way that suffices for full blown knowledge of theoretical structure. Crucially, the disjointness in methods, types of information and models used to arrive at the object leaves gaps in the grasping of specific theoretical structures—both in terms of inferences and properties, and in terms of experimental and operational procedures for its use.

Second, it might happen that scientists are able to, from big data analysis, obtain important information regarding an object. If that information is put together with independent theoretical knowledge in the special sciences, it becomes possible for scientists to generate a representation of the object, a possible world or a proper part of one that represents how the object is embedded in a relevant theoretical domain. However, given the distinct sources of knowledge of the object and the lack of unity in methods and conceptual resources scientists cannot be sure that these possible worlds are actual, i.e., that these representations hold some actual empirical domain.

In the corresponding literature, it has been argued that *False Theories can still Yield Genuine Understanding* (see De Regt and Gijsbers 2017). That is, for a given set of propositions, even if the veridicality condition is not satisfied, this would not necessarily prevent scientists from gaining understanding of such a set of data. According to De Regt and Gijsbers, what is needed for understanding is only the satisfaction of an 'effectiveness condition'—where, for this case, 'effectiveness' could be understood as the tendency to produce useful empirical outcomes of certain kinds, such as accurate descriptions and predictions.

One has some modal understanding of some phenomena if and only if one knows how to navigate some of the possibility space associated with the phenomena" (Le Bihan 2017, 112). In the case of big data practices, to achieve modal understanding of the behaviour of novel objects in an established theoretical domain would be to determine the set of possible worlds that correspond to the generic structural features assumed by the theoretical view that such a cluster of data substantiates. An important remark is that the understanding that is obtained is understanding of the relations that hold within (a segment of) a theoretical structure given the presence of a newly discovered object—not to be confused with dependence relations between objects in the actual world.

However, at this point, the reader might wonder whether modal understanding does not assume the previous achievement of knowledge of theoretical structure—which we are supposedly ignorant of in these cases. Such knowledge could be expressed in the form of laws, grounding relations, or any kind of dependence relation within the structure. This considered, it is not clear how big data can aid the scientists to produce such modal understanding, or to even draw out a possibility space without directly positing causal relations and laws, among others. It seems that if there are correlations, even if they be reliable, either no possibility space can be generated solely from them or the possibility space that is obtained is caused by our previous knowledge of the relevant part of the theoretical structure. And therefore, big data practices have no impact on the achievement of modal understanding.

While the reader's concern might sound appealing, it is grounded on a misunderstanding about of the scope of both knowledge of theoretical structure and modal understanding. On the one hand, due to our cognitive limitations, it seems implausible that we can achieve full knowledge of the theoretical structure of a specific set of data—whether it is a theory, a model, or only a set of collected information about a specific phenomenon. But the same happens with ignorance, no scientist working within a specific field will be absolutely ignorant of the theoretical structures of the sets of data that she is working with. Knowing parts of such structure and satisfactorily drawing some inferences when using these sets, is compatible with ignoring other parts of the structure and failing at identifying which inferences are correct within it. So, yes, the scientist that achieves modal understanding can rely on both her previous knowledge of certain segments of the structure and the outputs of big data processes. These two elements can constrain the possibility space, helping the scientist to identify the specific inference patterns that might govern the newly identified objects. On the other hand, modal understanding does not require the possibility space to be constrained by any type of metaphysical assumptions. In particular, while the possibility space that is being understood could reveal dependence relations of the form of grounding relations, in the large majority of cases, the relations that enhance this understanding are only inferential—making the possibility space, logical space.

Big data processes integrate immensely large amounts of data in such a way that they enhance extremely comprehensive representations of new objects; this grants scientists with objectual knowledge of the things that have been newly identified. After the acquisition of such knowledge, and thanks to the comprehensiveness of these representations, scientists are able to incorporate the factuality of these objects into theoretical structures that could explain why they occur the way they do. The comprehensiveness of these representations helps to narrow down the set of alternative structures that scientists can navigate, at least, at an inferential level; gaining understanding of the specific possible worlds that are compatible with what we now know about these new objects—without necessarily recognizing whether any of the alternative structures is isomorphic with a specific chunk of the world. This is, while scientists might ignore the theoretical structure of immensely large datasets and the ways in which certain outputs are produced within the sets, they can still gain understanding of the worlds in which, at least, the most salient outputs of big data processes are true.

In sum, despite the presence of ignorance of theoretical structure in big data practices, there are two main epistemic products that are obtained through them: objectual knowledge regarding objects that were initially unreachable to us, and modal understanding of how such newly identified objects fit in theoretical descriptions of the world. The conquest of modal understanding, allows the scientists to have a clear picture of the set of possible worlds that correspond to the structural connections that are relevant only with respect to some domain of the possibility space associated with the phenomena in question. In the next section, I illustrate this with a case study from cosmology.

4.5 Cosmology and Big Data

This section offers a case study from observational cosmology that illustrates the role that ignorance of theoretical structure plays in big data practices. Furthermore, this case brings the attention to three main phenomena: the successful consequences of computational processes, the acquisition of objectual knowledge and furthermore, the achievement of (some) modal understanding.

4.5.1 The Story

Observational cosmology aims at providing precise agreement between large scale-physical theories, cosmological models and observation. On a daily basis, high-throughput detectors and (land and space based) telescopes generate terabytes of raw data about objects in specific regions of the sky. The data that is gathered comes in heterogeneous formats, once received, it has to go through different isolating, filtering and integrating processes; and only after the data is processed, merely a fraction of it is saved and put to the service of cosmologists.

In light of its observational character, the reduction of data into images constitutes one of the best received outputs of technological implementation in cosmology.

> Much work has also been directed to the automated analysis and classification of objects on images, particularly the discrimination of stars from galaxies on optical band photographic plates and CCD images. Each object is characterized by a number of properties (e.g., moments of its spatial distribution, surface brightness, total brightness, concentration, asymmetry), which are then passed through a supervised classification procedure. Methods include multivariate clustering, Bayesian decision theory, neural networks, k-means

> partitioning, CART (Classification and Regression Trees) and oblique decision trees, mathematical morphology and related multiresolution methods (Bijaoui et al. 1997; White 1997). Such procedures are crucial to the creation of the largest astronomical databases with 1–2 billion objects derived from digitization of all-sky photographic surveys. (Feigelson and Babu 1997, 365)

While the selection of such methods is often described in internal technical memoranda, it commonly goes unnoticed and is almost never subject to public scrutiny. Once this is done, the constructed images should be reduced into catalogues. Some of the most successful results of these processes include the reports of the microwave background from the COBE, WMAP and Planck satellites, the detection of gravitational lensing, the ensemble of surveys such as Kepler, Gaia and DES, SDSS, DESI, LSST, Euclid and WFIRST, and the observation of the *Bullet Cluster*; being the latter one of the most important contributions to the cosmology of this century.

The *Bullet Cluster*, officially named 1E 0657-558, is one of the most energetic known galaxy clusters in the universe (cf. Schramm 2017, 13). The cluster consists of "two merging galaxy clusters, that the hot gas (ordinary visible matter) is slowed by the drag effect of one cluster passing through the other. The mass of the clusters, however, is not affected, indicating that most of the mass consists of dark matter" (Riess 2017). The Bullet Cluster was first discovered in 1998, later on, it was registered by The Chandra x-ray observatory in 2004. And it was only in 2004 when optical images of the Bullet Cluster were integrated by the Magellan telescope and the Chandra x-ray (cf. Markevitch et al. 2004; Clowe et al. 2006).

In the following years, it was possible to provide a picture of the bullet cluster which comprehensively integrated optical data, X-ray data, and a reconstructed mass map, becoming one of the most famous and informative images in all of astronomy (Figs. 4.1 and 4.2).

4.5.2 Evaluating the Case Study

This case study illustrates two main things: (1) the ignorance that underlay the epistemic practices in the discovery of the Bullet Cluster is ignorance of theoretical structure with *reliable consequences*. And, (2) despite their ignorance, scientists were able to achieve objectual ignorance and modal understanding of the world(s) in which the Bullet Cluster could exist.

First of all, as it has been argued in Sect. 4.4.2, due to their methodological basis, many big data practices are underlain by ignorance of theoretical structure, especially the practices in which different instances of epistemic opacity are combined. The epistemic practices associated with the use of big data in observational cosmology are not an exception to this general claim. As a matter of fact, due to the nature of the discipline's main object of study, these practices are the result of the union of reliance on technology and increasing collaboration. The former is clearly linked to the common use of different layers of machine implementation

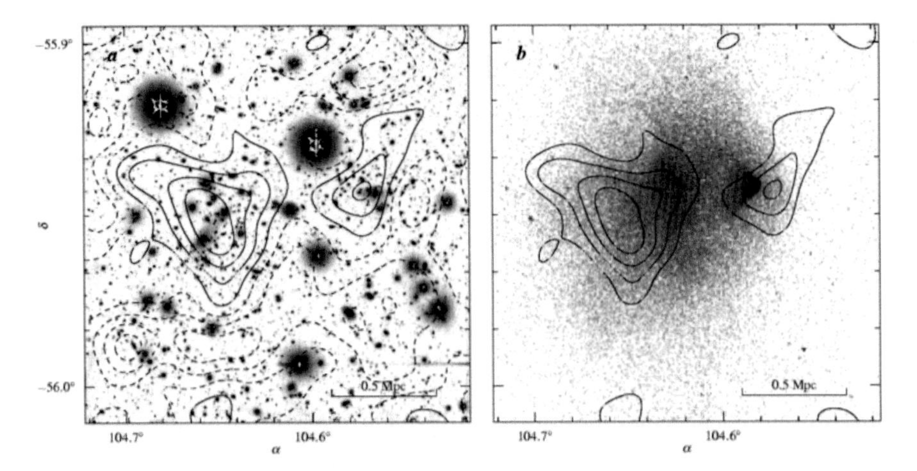

Fig. 4.1 First visual representations of the Bullet Cluster 2004 (gray-scale I-band VLT image). *Left:* Overlay of the weak-lensing mass contours on the optical image of 1E 0657-56. *Right:* Overlay of the mass contours on the X-ray image. From Markevitch et al. (2004), 820. ©AAS. Reproduced with permission

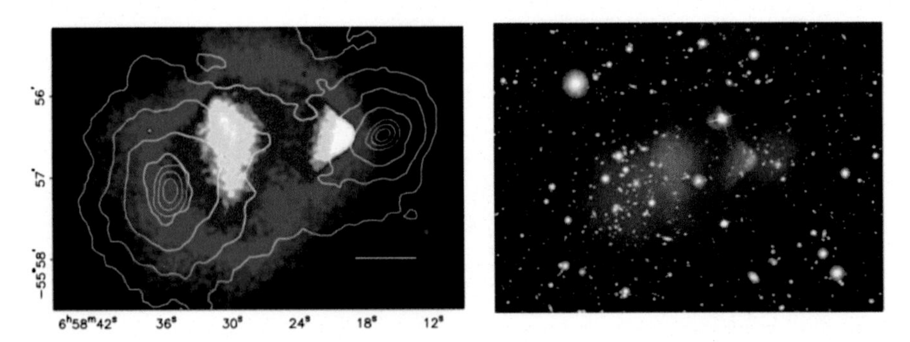

Fig. 4.2 Visual representations of the Bullet Cluster 2006. *Left*: 500 ks Chandra image of the cluster. From Clowe et al. (2006). ©AAS. Reproduced with permission. *Right*: Composite image of the matter of the Bullet Cluster. From NASA/CXC/SAO (2006)

and computational processes—aided by deep learning algorithms—which scope exceeds programmers' original input and human's cognitive capacities. The latter is mostly connected to the way in which raw data is filtered, structured and reduced into catalogues by one community, and later on reused by another community. The choices that the first community made are taken by the second to be at their best correct and at their worst, at least, non-problematic; but, considering that those initial choices were constrained by ignorance makes the foundations of this trust shaky.

Yet, this ignorance does not necessarily undermine the scientists' rationality when relying on big data practices and some of their *successful* outputs; in this particular case, the combined models that enhance (visual) representations of the Bullet Cluster. Intuitively, the criteria that when satisfied, could be indicative of an output being successful includes evidence of the way in which such output grants scientists access to empirical phenomena while enhancing the acquisition of objectual knowledge and modal understanding of such phenomena. In addition, the output should be novel, empirically adequate, fruitful, and there should exist a possible evidential relation between the output and a model or theory within the discipline.

In this respect, the study of the cluster produced by the NASA Chandra X-ray Observatory has provided

> sufficient precision to determine the mass distribution of the underlying galaxies through weak gravitational lensing (...) they made four main observations relating to the mass distributions:
>
> 1. Due to the large distance scales in question, stellar matter was only moderately affected. For the most part, the stars from each galaxy simply passed through the other galaxy without any inelastic interactions. The only visible effect is a velocity reduction through gravitational forces, with the occasional inelastic meeting of stars.
> 2. As usual, the gaseous component of the galaxies is much more spread out. The meeting of two gas clouds results in a significant interaction under the electromagnetic force, due to the shorter length scales (...)
> 3. The centre of total mass of the galaxies, observed through weak gravitational lensing, is offset from the stellar and gaseous matter. This suggests the presence of additional invisible matter.
> 4. The dark matter distributions can be inferred from the total mass contours, and they remain mostly spherical in shape. (Schramm 2017, 13,14)

(1)–(3) indicate (purely) observational discoveries that were achieved only thanks to the observation of the bullet cluster—regardless how heavily meditated this observation was. As a matter of fact, the comprehensiveness, the accuracy and fine granularity of the visual representations of the Bullet Cluster—and the models that ground them–, combined with the impact that they have had in the study of the universe in general, are indicative of the scientists' acquisition of objectual knowledge of, at least, the Bullet Cluster. This reinforces the intuition that, even if cosmologists cannot have sufficient epistemic access to the inferential processes that generated these images, they can consider the outputs of such processes to be evidence of something occurring in a particular way.

Another sign of success is the existence of a possible evidential relation between the results of big data computational processes and a conjecture, a model or a theory within the discipline. (4) can be interpreted as indicative of the fulfillment of this condition. The Bullet Cluster has been taken by many cosmologists as evidence in favor of the existence of dark matter and, transitively, in favor of the model of cosmology- ΛCDM (*Lambda-Cold Dark Matter*) (cf. Lage and Farrar 2015).[18]

[18] This model is a parametrization of the Big Bang cosmological model according to which the universe is composed mainly of three elements: a cosmological constant denoted by Λ and associated with dark energy; the postulated cold dark matter (CDM); and ordinary matter.

But the Bullet Cluster does not only hold a possible evidential relation with ΛCDM, some cosmologists have interpreted the existence of the Bullet Cluster as a challenge to modify alternative models in order to give account of this phenomenon without accepting the existence of dark matter; a good example of this are refined versions of the *Modified Newtonian dynamics* (MOND)—which is a hypothesis that proposes a modification of Newton's laws to account for observed properties of galaxies and constitutes an alternative to the hypothesis of dark matter.

At this point, there is a weak form of underdetermination of theory by data surrounding the Bullet Cluster: As with the mass discrepancies in galactic structures (that were originally explained by the dark matter hypothesis but were eventually explained by MOND), it has been proved that the Bullet Cluster can be explained by both ΛCDM and MOND (cf. Angus et al. 2006).[19] But, while this underdetermination is problematic for explaining the phenomenon in itself, it reinforces the idea of the description and representations of the Bullet Cluster being sufficiently empirically adequate to count as observational evidence that *must* be explained. And at the same time, the fact that nowadays, the discovery and observation of the Bullet Cluster is driving the modifications of ΛCDM and MOND is indicative of its fruitfulness within the discipline.

Finally, the fact that objectual knowledge has been gained with respect to the Bullet Cluster is not enough for neither achieving explanatory knowledge of what causes this phenomenon nor for deciding the truth value of the dark matter hypotheses or the Modified Newtonian dynamics. Yet, what has been gained through the observation of the Bullet Cluster is the opportunity of incorporating its factuality into alternative theoretical structures that could explain why it occurs the way it does and when doing so, exploring the logical space described by such structures.

Thanks to the observation of the Bullet Cluster, cosmologists have been able to develop modal understanding of what the Bullet Cluster might be and how it would behave in, at least, both a ΛCDM-constrained world (cf. Kraljic and Sarkar 2014) and a MOND-constrained world (cf. Angus et al. 2006). While, at least, part of the computational processes associated to the gathering, filtering and structuring of the data might remain opaque to the cosmologists, they have now been equipped with fine grain detailed representations of the Bullet Cluster and its behaviour in different contexts. The value of this structured information is that it provides the scientists with a possibility space constrained by the existence of the Bullet Cluster, which they can navigate in either a ΛCDM-direction or a MOND-direction. This is, while cosmologists might remain ignorant of the theoretical structure that underlies the set of raw data about the Bullet Cluster, they now have access to the inference patterns that the representation of the Bullet Cluster allows for—without necessarily knowing which of these patterns, if any, is satisfied in the actual world.

[19] I am greatly indebted to an anonymous referee for pointing me to this problem.

4.6 Final Remarks

When incorporating big data into the empirical sciences, scientists are able to reach objects that were initially inaccessible to them. However, the outcomes of big data applications often involve high degrees of epistemic opacity about how such outcomes were generated. This leaves scientists having to choose between rejecting these outputs as observational evidence or, at the risk of being irrational, relying on them—even if ignoring where they come from and how they were obtained.

Here I argued that the ignorance associated with the epistemic opacities found in big data practices is ignorance of theoretical structure *with reliable consequences*. Scientists might ignore the structural particularities of how the observational outputs are identified and generated, but, at the same time, they have evidence in favor of the reliability of these outputs—and therefore, of the processes that generated them. Such reliability has made possible that scientists achieve objectual knowledge of initially inaccessible objects as well as modal understanding of how these objects (could) behave and relate to one another, all this while being ignorant of the inference patterns that govern the datasets from which the access to these objects is constructed.

Acknowledgments I am indebted to Moisés Macías-Bustos for his extremely valuable and constant feedback on different versions of this paper. Thanks also to Atocha Aliseda-Llera, Otávio Bueno, Gabrielle Ramos-García, Felipe Rocha, Alexandre Meyer and Carlos César-Jiménez for fruitful discussions on these issues. Thanks to Pawel Pawlowski and Luis Estrada-González for their challenging questions on this project. I owe special thanks to the four anonymous referees for their valuable comments and suggestions. Thanks to the audiences at the *IACAP2019* and the *Colóquio Virtual SELF - Problemas Filosóficos*.

This research was supported by the Programa Nacional de Pós-Doutorado PNPD/CAPES (Brazil) and by the UNAM-PAPIIT Projects IN102219 and IN403719.

References

Angus, G., B. Famaey, and H. Zhao. 2006. Can MOND Take a Bullet? Analytical Comparisons of Three Versions of MOND Beyond Spherical Symmetry. *Monthly Notices of the Royal Astronomical Society* 371(1): 138–146.

Arbesman S. 2013. Five myths about Big Data. *The Washington Post*. Available at: www.washingtonpost.com/opinions/five-myths-about-big-data/2013/08/15/64a0dd0a-e044-11e2-963a-72d740e88c12_story.htm.

Barberousse A., and M. Vorms. 2014. About the Warrants of Computer-Based Empirical Knowledge. *Synthese* 191(15): 3595–3620.

Batens, D. 1991. Do We Need a Hierarchical Model of Science?. In *Inference, Explanation and Other Frustrations. Essays in the Philosophy of Science*, ed. Earman, 199–215. Berkeley-Los Angeles-Oxford: University of California Press.

Bedau, M. 1997. Weak Emergence. *Philosophical Perspectives* 11: 375–399.

Bijaoui, A., F. Rué, and R. Savalle 1997. *Statistical Challenges in Modern Astronomy II*. New York: Springer.

Boyd, D., and K. Crawford. 2012. Critical Questions for Big Data. *Information, Communication & Society* 15(5): 662–679, https://doi.org/10.1080/1369118X.2012.678878.
Clowe, D., M. Bradac, A.H. Gonzalez, M. Markevitch, S.W. Randall, C. Jones, and D. Zaritsky. 2006. A Direct Empirical Proof of the Existence of Dark Matter. *The Astrophysical Journal* 648(2): L109–L113. https://doi.org/10.1086/508162.
Creel, K.A. 2020. Transparency in Complex Computational Systems. *Philosophy of Science* 87(4): 568–589.
De Regt, H.W. 2009. Understanding and Scientic Explanation. In *Scientific Understanding: Philosophical Perspectives*, eds. H.W. de Regt, S. Leonelli, and K Eigner, 21–42. Pittsburgh: University of Pittsburgh Press.
De Regt, H.W. 2015. Scientic Understanding: Truth or Dare?. *Synthese* 192: 3781–97.
De Regt, H.W., and D. Dieks. 2005. A Contextual Approach to Scientific Understanding. *Synthese* 144: 137–70.
De Regt, H.W., and V. Gijsbers. 2017. How False Theories Can Yield Genuine Understanding. In *Explaining Understanding: New Perspectives from Epistemology and Philosophy of Science*, 50–75. Milton Park: Routledge.
Elgin, C.Z. 2004. True Enough. *Philosophical Issues* 14: 113–131.
Elgin, C.Z. 2009. Exemplification, Idealization, and Understanding. In *Fictions in Science: Essays on Idealization and Modeling*, ed. M. Suárez, 77–90. Milton Park: Routledge.
Elgin, C.Z. 2011. Making Manifest: Exemplification in the Sciences and the Arts. *Principia* 15: 399–413.
Elgin, C.Z. 2017. Exemplification in Understanding. In *Explaining Understanding: New Perspectives from Epistemology and Philosophy of Science*, 76–91. Milton Park: Routledge.
Feigelson, E.D., and G.J. Babu. 1997. Stadistical Methodology for Large Astronomical Surveys. In *New Horizons from Multi-Wavelength Sky Surveys*, eds. B. McLean, D.A. Golombek, J.J.E. Hayes, and H.E. Payne, 363–370. Cambridge: Cambridge University Press.
Floridi, L. 2011. *The Philosophy of Information*. Oxford: Oxford University Press.
Floridi, L. 2012. Big Data and Their Epistemological Challenge. *Philosophy & Technology* 25(4): 435–437.
Floridi, L., N. Fresco, G. Primiero. 2015. On Malfunctioning Software. *Synthese* 192(4): 1199–1220.
Fricke, M. 2015. Big Data and Its Epistemology. *Journal of the Association for Information Science and Technology* 66(4): 651–661.
Fulton, B.J., and E.A. Petigura. 2018. The California-Kepler Survey. VII. Precise Planet Radii Leveraging Gaia DR2 Reveal the Stellar Mass Dependence of the Planet Radius Gap. *The Astronomical Journal* 156(6): 1–13.
Garofalo, M., A. Botta, and G. Ventre. 2016. Astrophysics and Big data: Challenges, Methods, and Tools. *Astroinformatics (AstroInfo16) Proceedings IAU Symposium No. 325*: 1–4.
Grimm, S.R. 2006. Is Understanding a Species of Knowledge?. *British Journal for the Philosophy of Science* 57(3): 515–535.
Grimm, S.R. 2014. Understanding as Knowledge of Causes. In *Virtue Epistemology Naturalized*, ed. A. Fairweather, 329–345. New York: Synthese Library.
Humphreys, P. 2009. The Philosophical Novelty of Computer Simulation Methods. *Synthese* 169(3): 615–626.
Kelp, C. 2014. Knowledge, Understanding and Virtue. In *Virtue Epistemology Naturalized*, ed. A. Fairweather, vol. 366, 347–360. New York: Synthese Library.
Khalifa, K. 2013. Is Understanding Explanatory or Objectual?. *Synthese* 190(6): 1153–1171.
Kraljic, D., and S. Sarkar. 2014. How Rare Is the Bullet Cluster (in a ΛCDM universe)?. *Journal of Cosmology and Astroparticle Physics*.
Kvanvig, J. 2003. *The Value of Knowledge and the Pursuit of Understanding*. Cambridge: Cambridge University Press.
Lage, C., and G.R. Farrar. 2015. The Bullet Cluster Is Not a Cosmological Anomaly. *Journal of Cosmology and Astroparticle Physics*, 02 038.

Lawler, I. 2016. Reductionism About Understanding Why. *Proceedings of the Aristotelian Society* 116(2): 229–236.
Lawler, I. 2018. Understanding Why, Knowing Why, and Cognitive Achievements. *Synthese*.
Le Bihan, S. 2017. Enlightening Falsehoods: A Modal View of Scientific Understanding. In *Explaining Understanding: New Perspectives from Epistemology and Philosophy of Science*, 111–136. Milton Park: Routledge.
Le Morvan, P., and R. Peels. 2016. The Nature of Ignorance: Two Views. In *The Epistemic Dimensions of Ignorance*, eds. R. Peels and M. Blaauw, 12–32. Cambridge: Cambridge University Press.
Leonelli, S. 2014. What Difference Does Quantity Make? On the Epistemology of Big Data in Biology. *Big Data & Society* 1(1): 1–11.
Leonelli, S. 2016. *Data-Centric Biology: A Philosophical Study*. Chicago: University of Chicago Press.
Leonelli, S. 2020. Scientific Research and Big Data. In *The Stanford Encyclopedia of Philosophy* ed. E.N. Zalta. https://plato.stanford.edu/archives/sum2020/entries/science-big-data/
Massimi, Michela, and John Peacock. 2015. The Origins of Our Universe. Laws, Testability, and Observability in Cosmology. In *Philosophy and the Sciences for Everyone*, 14–32. Routledge.
Manyika, J., M. Chui, B. Brown, et al. 2011. *Big Data: The Next Frontier for Innovation, Competition, and Productivity*. McKinsey Global Institute.
Markevitch, M., A.H. Gonzalez, D. Clowe, A. Vikhlinin, W. Forman, C. Jones, S. Murray, W. Tucker. 2004. Direct Constraints on the Dark Matter Self-Interaction Cross Section from the Merging Galaxy Cluster 1E 0657-56. *The Astrophysical Journal* 606(2): 819–824. https://doi.org/10.1086/383178
Martínez-Ordaz, M. del R. 2021. The ignorance behind inconsistency toleration, S.I. Knowing the Unknown, *Synthese*. 198: 8665–8686
Mayer-Schönberger, V., and K. Cukier. 2013. *Big Data: A Revolution That Will Transform How We Live, Work, and Think*. ISBN-10: 0544227751, ISBN-13/EAN: 9780544227750.
Morrison M. 2015. *Reconstructing Reality: Models, Mathematics, and Simulation*. New York: Oxford University Press.
Napoletani, D., M. Panza, and D.C. Struppa. 2014. Is Big data Enough? A Reflection on the Changing Role of Mathematics in Applications. *Notices of the American Mathematical Society* 61(5): 485–490.
NASA/CXC/SAO. 2006. X-ray: NASA/CXC/CfA/ M. Markevitch *et al.*; Optical: NASA/ STScI; Magellan/ U.Arizona/ D.Clowe et al.; Lensing Map: NASA/STScI; ESO WFI; Magellan/U.Arizona/D.Clowe et al. https://chandra.harvard.edu/photo/2006/1e0657/
Norton, J. 2008. Ignorance and Indifference. *Philosophy of Science* 75: 45–68.
Riess, A. 2017. Dark Matter. In *Encyclopædia Britannica*, Encyclopædia Britannica. https://www.britannica.com/science/dark-matter
Schramm, S. 2017. *Searching for Dark Matter with the ATLAS Detector*, Springer Theses, Springer.
Sliwa, P. 2015. IV–Understanding and Knowing. *Proceedings of the Aristotelian Society* 115(1pt1): 57–74.
Snowdon, P. 2004. Knowing How and Knowing That: A Distinction Reconsidered. *Proceedings of the Aristotelian Society* 104(1): 1–29.
Stanley, Jason, and Timothy Williamson. 2001. Knowing How. *The Journal of Philosophy* 98.8: 411–444.
Strevens, M. 2013. No Understanding Without Explanation. *Studies in History and Philosophy of Science Part A* 44(3): 510–515.
Sterner, B. 2014. The Practical Value of Biological Information for Research. *Philosophy of Science* 81(2): 175–194.
Strevens, M. 2017. How Idealizations Provide Understanding. In *Explaining Understanding: New Essays in Epistemology and the Philosophy of Science*, eds. S. R. Grimm, C. Baumberger, and S. Ammon. New York: Routledge

Sterner, B., and N.M. Franz. 2017. Taxonomy for Humans or Computers? Cognitive Pragmatics for Big Data. *Biological Theory* 12(2): 99–111.

Sullivan, E. 2019. Beyond Testimony: When Online Information Sharing is not Testifying. *Social Epistemology Review and Reply Collective* 8(10): 20–24.

Wimsatt, W. C. 2007. *Re-Engineering Philosophy for Limited Beings: Piecewise Approximations to Reality*. Harvard University Press.

Ylikoski, P. 2013. The Illusion of Depth of Understanding in Science. In *Scientific Understanding: Philosophical Perspectives*, eds. H. De Regt, S. Leonelli, and K. Eigner, 100–119. Pittsburgh University Press.

Chapter 5
Predictive Fairness

Anders Herlitz

Abstract It has recently been argued that in normal decision circumstances no systematic decision method that predicts the likelihood that individuals possess some property can be fair. Either (i) the decision method correctly identifies the relevant property (e.g. recidivism) more often in one subgroup (e.g. black defendants) than another (e.g. white defendants); or (ii) the decision method systematically ascribes higher probabilities to individuals who have and/or individuals who lack the property in one group compared to the probabilities ascribed to individuals who have and/or individuals who lack the property in another group. Otherwise put, these decision methods seem inherently, and unavoidably, unfair. Besides introducing this problem to the philosophical community, this paper explores different possible responses to the problem and presents three principles that should be universally applied to promote fairness: (1) *Dominance:* a decision method that is better with respect to one of the dimensions of fairness and worse with respect to none is better overall; (2) *Transparency*: decision-makers who use these decision methods should be aware of the unintended differences in impact and also be transparent to the affected community about these unintended differences; and (3) *Priority to the worse off*: a decision method that is relatively better for members of a worse-off subgroup is preferable to a method that is relatively better for members of a better-off subgroup.

Tools and heuristics that support decision-making by establishing probability predictions are indispensable in many domains of the public and also the private sector. They are used to improve medical diagnoses, to assess security risks in the judicial system, to evaluate special needs as well as potential in the educational sector, to set mortgage rates, insurance rates, and so on. Essentially, these decision-supporting tools facilitate probability predictions by pairing certain observable phenomena with a probability of something being the case. For instance, a physician will consider

A. Herlitz (✉)
Institute for Futures Studies, Stockholm, Sweden
e-mail: anders.herlitz@iffs.se

B. Lundgren, N. A. Nuñez Hernández (eds.), *Philosophy of Computing*, Philosophical Studies Series 143, https://doi.org/10.1007/978-3-030-75267-5_5

it more likely that a patient has a certain disease (e.g. malignant melanoma) if the patient has a certain set of observable properties (e.g. new, unusual changes in an existing mole). A particular kind of decision-supporting tool of this kind that has grown increasingly popular is the computerized version that often relies on machine learning that is applied to large datasets. These decision-supporting algorithms (*decision algorithms* for short) are powerful tools that can be used to help decision makers achieve better outcomes in many areas. Rapid technological development in combination with increased access to large datasets make decision algorithms that evaluate alternatives with respect to a set of given decision criteria not only attractive in theory, but also practically useful. As evidence of their potency, it suffices to contemplate the fact that so-called "quantitative hedge funds" that rely on algorithmic trading – automatized trading based on pre-programmed decision algorithms – according to a *Wall Street Journal* article from 2017 are responsible for 27% of all U.S. stock trades by investors, controlled a total of $932 billion of investments, and seem to outperform conventional hedge funds at trading by having a growth in value of 5.1%/year on average compared to 4.3%/year on average for all hedge funds (Zuckerman and Hope 2017). Yet, decision algorithms can also be used to promote moral and political goals. For instance, they can help medical professionals improve the accuracy of medical diagnoses and thereby improve health outcomes, they can help the criminal system more accurately predict the risks involved in releasing suspects on bail, they can help effective altruists better predict where donations have the highest impact, and they can be used by educational institutions to better predict which students will need special attention. There is, in principle, no area that involves predictions of the consequences of different alternatives in which decision algorithms cannot improve outcomes. Furthermore, since decision algorithms themselves are neutral with respect to what decision criteria are used, there is no conception of the good (no decision criteria) that they cannot help promote: fairness, health, security, education performance, etc.

There are well-recognized – and indeed worrisome – concerns regarding how increased use of decision algorithms will lead to unfair outcomes due to the fact that the data available might bias the results (cf. Zimmermann et al. 2020). In this paper, I address a different, and on a theoretical level much more troublesome, problem, namely a recently proven impossibility theorem that shows that under normal decision conditions, no decision algorithm will simultaneously meet two very attractive fairness conditions, which implies that under normal decision conditions, *all* decision algorithms generate results that are in some sense unfair. The first section introduces two kinds of relevant fairness conditions, the impossibility theorem and the informal version of the proof that the use of all decision algorithms under normal conditions imply some unfairness in at least one of the two relevant fairness dimensions. The second section outlines and rejects some potential responses to this problem. The third section defends three general principles that ought to guide policies that make use of systematic decision methods to predict the prevalence of relevant properties in different individuals that are affected by the policy. There is a brief conclusion.

5.1 Necessary Biases

In May 2016, the American nonprofit news organization *Pro Publica* published an article with the title "Machine Bias: There's software used across the country to predict future criminals. And it's biased against blacks" (Angwin et al. 2016). The article describes the widespread use of COMPAS (the Correctional Offender Management Profiling for Alternative Sanctions), a decision support tool used by U.S. courts to evaluate the recidivism risk of a defendant. In essence, COMPAS is a statistical method that consists of matching certain kinds of characteristics of a defendant with a dataset that contains different patterns of these kinds of characteristics in order to infer an assessment of the recidivism risk of the defendant. Importantly, the approach assumes that for each individual defendant, there is a risk (probability) that the defendant will reoffend. Granted this assumption, a statistical method is used in order to establish this risk. Very simplified, COMPAS matches the characteristics of a defendant with data on criminal offenders where some have reoffended and some have not in order to determine the risk of recidivism for different defendants. This assessment is then used by the courts when they set bail and sentences, as well as when they make decisions about early release.

The authors of the *Pro Publica* article revealed that the distribution of risk scores that COMPAS generated were asymmetric over salient social groups in a way that disadvantaged some groups and benefited others: black defendants who did not reoffend received, on average, a higher risk score than white defendants who did not reoffend, whereas white defendants who reoffended on average received a lower risk score than black defendants who reoffended. In other words, black nonrecidivists were, on average, claimed to have a higher risk of reoffending than white nonrecidivists, while white recidivists on average were claimed to have a lower risk of reoffending than black recidivists. This led the authors of the *Pro Publica* article to proclaim that COMPAS was "biased against blacks."

The for-profit company that designed COMPAS, Northpointe, rejected the contention that their product was biased against any group (Angwin et al. 2016). Northpointe pointed to how those who accused the product of being biased mistakenly focused on the *errors*. If they instead had looked at the accuracy of the predictions COMPAS made, they would have seen that there is no bias. The assessments of the probability of recidivism that COMPAS makes were equally *well calibrated* to the correct outcome for both black and white defendants, i.e. for both black and white defendants, it was true that if they received a risk score r, then they belonged to a group in which an r-fraction reoffended. Of all black defendants who received a risk score $= .1$, one in ten reoffended and of all white defendants who received a risk score $= .1$, one in ten reoffended. And so on. In other words, for both black and white defendants that stood in front of a court that used COMPAS, it was true that they would be categorized into a risk group in which the fraction of recidivists corresponded to the risk score. This led the company behind COMPAS to proclaim that their product was not biased.

Clearly, the authors of the *Pro Publica* article and the company behind COMPAS have different conceptions of what it is for a decision algorithm to be biased. Conversely, they have competing conceptions of what it is for a decision algorithm to be fair. The *Pro Publica* article and the response from Northpointe reveal that two different general determinants can be invoked to evaluate whether a decision algorithm such as COMPAS is fair or not:

Equality in accuracy: For each group of people who is ascribed a risk score and all subgroups of salient social groups who are ascribed the risk score, the fraction of individuals in the group who are positive instances reflects the risk score. For instance, if a decision algorithm that predicts recidivism ascribes a risk score .6 to a group of people, then 60% of these people are recidivists and 60% within all subgroups of salient social groups are recidivists. (This is the determinant Northpointe relied on in their defense of COMPAS).

Equality in inaccuracy: The distribution of risk scores generated by a decision algorithm is such that for all salient social groups, the average risk score ascribed to those who lack the sought-after property is the same for all groups and the average risk score ascribed to those who have the sought-after property is the same for all groups. For instance, if the average risk score ascribed to black defendants who are nonrecidivists is .3 and the average risk score ascribed to black defendants who are recidivists is .7, then the average risk score ascribed to white defendants who are nonrecidivists is .3 and the average risk score ascribed to white defendants who are recidivists is .7. (This is the determinant that the authors of the *Pro Publica* relied on when they accused COMPAS of being biased.)

For situations in which the predictions have significant implications for the individuals, these determinants of fairness are commonsensical and seem widely supported also by the most prevalent views in moral and political philosophy as at least pro tanto criteria of fairness. When a decision algorithm is used to predict the likelihood of recidivism and this prediction is used by a court to set bail, sentencing and early release, the predictions have significant implications for the individuals. Inaccurately being deemed a likely recidivist constitutes a significant harm to an individual in that it leads to higher bail, longer sentences and smaller chance for early release. Conversely, inaccurately being deemed a likely nonrecidivist constitutes a significant benefit to an individual in that it leads to lower bail, shorter sentences and greater chance for early release. Each of the determinants of fairness reflects a way in which harms and benefits can be unequally distributed across salient social groups without there being a plausible justification for the inequality.

However, in a recent paper, Joel Kleinberg, Sendil Mullainathan and Manish Raghavan prove that as long as the property that is being predicted (e.g. recidivism) is unevenly distributed across groups and the decision problem does not admit perfect prediction, there is no decision algorithm that *can* be fair in all of the senses outlined above simultaneously (Kleinberg et al. 2017).[1] Kleinberg, Mullainathan and Raghavan start their argument by presenting a more precise framework and by specifying the properties that we want a decision algorithm to have in order to be fair. First, they note that decision algorithms are designed to estimate some *probability* that some property obtains. Second, they use the term "positive instance"

[1] For discussions of similar findings, see Chouldechova 2017 and Loi et al. 2019.

to describe an individual who has the relevant property, and the term "negative instance" to describe an individual who does not have the relevant property. Using this framework, the following fairness conditions can be posed on a decision algorithm:

> *Well-calibrated*: If the algorithm identifies a set of people as having probability z of constituting positive instances, then approximately a z fraction of this set should indeed be positive instances. Moreover, this condition should hold when applied separately in each group as well. For example, if we are thinking in terms of potential differences between outcomes for men and women, this means requiring that a z fraction of men and a z fraction of women assigned probability z should possess the property in question (Kleinberg et al. 2017: 2).
>
> *Balance for the positive class*: The average score (i.e. the probability score) received by people constituting positive instances should be the same in each group. A violation of this means that those constituting positive instances in one group consistently receive lower probability estimates of possessing the property in question than those constituting positive instances in another group. For example, women who have increased risk of developing some disease might consistently receive lower probability estimates of having increased risk of developing the disease than men who have increased risk of developing the disease.
>
> *Balance for the negative class*: The average score received by people constituting negative instances should be the same in each group. A violation of this means that those constituting negative instances in one group consistently receive higher probability estimates of possessing the property in question than those constituting negative instances in another group. For example, women who do not have increased risk of developing some disease might consistently receive higher probability estimates of having increased risk of developing the disease than men who do not have increased risk of developing the disease.

Equality in accuracy is reflected by what Kleinberg, Mullainathan and Raghavan call *Well-calibrated. Equality in inaccuracy* is in Kleinberg, Mullainathan and Raghavan's framework captured in two other conditions. First, *Balance for the positive class* which states that the inaccurate probability predictions of the decision algorithms should be similar in magnitude for those who possess the sought-after property in each salient social group. Second, *Balance for the negative class* which states that the inaccurate probability predictions of the decision algorithms should be similar in magnitude for those who do not possess the sought-after property in each salient social group.

In terms of these conditions, COMPAS was well-calibrated, but it was not balanced either for the positive class or for the negative class. White recidivists were on average given lower scores than black recidivists (violation of *Balance for the positive class*), and black nonrecidivists were on average given higher scores than white nonrecidivists (violation of *Balance for the negative class*). The balance conditions reflect the idea that once one accepts that there will be inaccuracies in the predictions the impact of the mistake should be unrelated to which group the individual belongs to.

Having presented these conditions, Kleinberg, Mullainathan and Raghavan prove that in situations where the property that is assessed is unevenly distributed across groups (unequal base rate) it is impossible to design a decision algorithm that meets all three conditions unless the decision problem admits of perfect predictions, i.e.

an algorithm that only makes accurate non-probabilistic predictions is available. Furthermore, they show that the incompatibility of the conditions that they reveal applies also to approximate versions of the conditions. In their paper, they offer both a formal and an informal proof of this. Here, I will only present the informal proof and set aside their discussion of approximate versions.

To present the argument, it facilitates to have a more concrete example. Consider, thus, a population of 24 men and 24 women, and the task of predicting who has skin cancer and who does not. Since it is unknown who has skin cancer, but it is known that some do and some do not, risk scores, r, are ascribed to the individuals. To simplify, assume that only two risk scores are ascribed: $r = .25$, $r = .75$. Let "–" denote a negative instance, someone who does not have cancer, and "+" denote a positive instance, someone who has cancer. A probability assessment ascribes probabilities in the following way:

$r = .25$	$r = .75$
18–	6–
6+	18+

Eighteen negative and six positive instances are ascribed a risk score of .25, and six negative and eighteen positive instances are ascribed a risk score of .75. This seems reasonable. The fraction of positive instances reflects the risk score so that one fourth of the people who are ascribed risk score .25 are positive instances and three fourths of the people who are ascribed risk score .75 are positive instances.

Yet, assume that skin cancer is more common among the men compared to the women, that eight men and four women have skin cancer, and the distribution of risk scores above is distributed in the following way across the sexes:

Men	Women
$r = .25$	$r = .25$
6–	12–
2+	4+
$r = .75$	$r = .75$
4–	2–
12+	6+

This seems problematic. The distribution of risk scores is fair in that regardless of whether a patient is a man or a woman, if they are ascribed a risk score .25 they belong to a group of people in which one fourth of the individuals have skin cancer, and if they are ascribed a risk score .75 they belong to a group of people in which three fourths of the individuals have skin cancer. Yet, the average score ascribed to men who have cancer is higher than the average score ascribed to women who have cancer.

Assume that a healthcare provider can provide screening for cancer only to 24 patients (due to scarcity). If the healthcare provider relied on the decision algorithm in order to identify the most efficient allocation of the screenings, i.e. to identify the people with highest risk of cancer, the result would be that whereas only 60% of the women who have cancer receive the screening, 85.7% of the men who have cancer receive the screening. A man with cancer has a higher chance of receiving the screening than a woman with cancer. This seems unfair.

To mitigate this, it seems desirable to have a decision algorithm that could generate a different distribution of risk scores, for instance the following:

Men	Women
$r = .25$	$r = .25$
6−	12−
3+	3+
$r = .75$	$r = .75$
4−	2−
11+	7+

Relying on a method that ascribes risk scores in this way implies that 70% of the women who have cancer will receive the screening and 72.3% of the men who have cancer will receive the screening. This seems fairer.

However, the total risk score ascribed to women is here greater than the total sum of positive instances in this group, and the total risk score ascribed to men is lower than the sum total of negative instances in this group. The sum total of the risk scores ascribed to women is 10.5 although only 10 women have cancer, and the sum total of the risk scores ascribed to men is 23.5 although 24 men have cancer. This means that being ascribed a risk score .75 does not mean that one belongs to a group in which 75% has cancer. Furthermore, men who are ascribed risk score .75 are less likely to have cancer than women who are ascribed the risk score .75.

Consider what this implies in case the healthcare provider has three alternatives. Either (1) provide no treatment; (2) ascribe slow screening; (3) ascribe fast screening. Furthermore, assume that scarcity implies that the healthcare provider can ascribe slow screening to 23 individuals and fast screening to only one individual. If the decision maker knows that it is more likely that a woman who has risk score .75 actually has cancer than a man who has risk score .75, then she optimizes the health outcome by giving the fast screening to a woman. A woman who has been categorized as having risk $= .75$ of having cancer has a higher chance of receiving the fast screening than a man who has been categorized as having risk $= .75$. This seems unfair.

Kleinberg et al. presents a proof that either of these unfairnesses will always occur. To present this proof some more specific notation is required:

Let N_t denote the total number of people in group t.

Let P_t denote the number of people in group t who are positive instances. For men this $P = 6$, and for women $P = 4$.

Let x_t be the average score given to a member of group t who is a negative instance.

Let y_t be the average score given to a member of group t who is a positive instance.

A *score* of an individual is the probability that the individual is a positive instance that is ascribed by the decision algorithm.

The *total score* of a group is the sum of the scores ascribed to the individuals in the group.

A *bin* is a classification category associated with a specific score so that each individual in a single bin is ascribed the same score by the decision algorithm.

The first step of the informal proof is to describe that when the decision algorithm is well-calibrated the total score given to a group, t, equals the number of positive instances in the group, and this can be represented by a linear equation where the unknowns are the average scores to the negative and positive instances, x_t and y_t. The second step of the informal proof is to note that, in light of the balance conditions, the equation for each group must have the same unknowns, which means that the set of linear equations constitute a system of equations. A system of equations is a set of two or more equations with the same unknowns, and a solution to a system of equations is a set of values for the unknowns that satisfy every equation in the system. The third step of the informal proof is to note that there can only be two solutions to the system of equations, and these solutions are only attainable in extreme and radically unlikely situations: when positive instances are evenly distributed across the groups or when the decision algorithm is flawless and only makes correct predictions.

Consider first how the scores ascribed to members of a group equals the positive instances and can be represented by a linear equation with the average scores to the negative and positive instances as unknowns when the decision algorithm is well-calibrated. Members of a group t are allocated to different bins by the decision algorithm. To be well-calibrated, the decision algorithm must imply that the expected total score to people in t who get placed in a particular bin b is the same as the expected number of people from group t in the bin who are positive instances. Aggregating over all bins, the total score given to all people in group t is the same as the total number of people in t who are positive instances, i.e. P_t.

This allows us to present the total score ascribed to a group t as an equation:

$$(N - P_t)\ x_t + P_t y_t = P_t$$

$(N - P_t)\, x_t$ is the negative instances in t multiplied with the average score ascribed to the negative instances in t. $P_t y_t$ is the positive instances in t multiplied with the average score ascribed to the positive instances in t. To be well-calibrated, the sum of the scores given to individuals in t must equal P_t and it is analytically true that the sum of the scores given to a group of individuals is the sum of the average score given to the individuals, i.e. $(N - P_t)\, x_t + P_t y_t$ must equal the positive instances in t, i.e. P_t.

We can now think of the scores ascribed to different groups as linear equations like the one above. If we are interested in only two groups, *1* and *2*, we get two linear equations. The unknowns are x_1, x_2, y_1 and y_2, and the line for each group represents the pairs of values for x and y that are possible without violation of the well-calibrated condition. They are the pairs of x and y that provide a solution to the equation for a particular P_t.

To meet the balance criteria, x and y must be the same for both groups. *Balance for the negative class* stipulates that x must be the same for all groups: $x_1 = x_2$. *Balance for the positive class* stipulates that y must be the same for all groups: $y_1 = y_2$. This means that we have two equations with the same unknowns, x and y, i.e. a system of equations.

We can now see that there are only two conditions under which each of the three fairness conditions can be met. Either $P_1 = P_2$, in which case the linear equations are identical for both groups and the balance conditions are met for all possible x and y. $P_1 = P_2$ means that the property that is being assessed is evenly distributed across groups *1* and 2. The second solution to the system of equations, which allows for $P_1 \neq P_2$, is $x = 0$ and $y = 1$. If $x = 0$ and $y = 1$, then the decision algorithm is perfectly accurate. The average score ascribed to positive instances is 1, and the average score ascribed to negative instances is 0.

Under all other conditions, either the decision algorithm fails to be well-calibrated:

$$(N{-}P_1)\ x_1 + P_1 y_1 \neq P_1$$

Or

$$(N{-}P_2)\ x_2 + P_2 y_2 \neq P_2$$

This means that for either group *1* or for group *2*, it will be true that the decision algorithm implies that there is a set of people in the group who is ascribed the probability p of having some property, although it is false that a p-fraction in this set of people actually have the property. For instance, the decision algorithm might ascribe a 10% risk of developing skin cancer to a set of women although it is not true that 10% of these women will develop skin cancer, whereas 10% of the men who get ascribed a 10% risk of developing skin cancer actually do develop skin cancer.

Or, one (or both) of the balance conditions is violated:

$$x_1 \neq x_2$$

Or

$$y_1 \neq y_2$$

This means that either the average score received by positive instances or the average score received by negative instances in one group is greater than the average score received by positive or negative instances in the other group. For instance, the decision algorithm might ascribe a higher average score to men who develop skin cancer than the average score it ascribes to women who develop skin cancer (i.e. more women than men who should be diagnosed as high risk will be misdiagnosed), or it might ascribe a lower average score to men who will not develop skin cancer than it ascribes to women who will not develop skin cancer (i.e. more women than men who should be diagnosed as healthy will be misdiagnosed).

It is hard to overestimate the relevance of this result. Not only does it prove that all decision algorithms such as these that are developed by computer scientists and statisticians will violate at least one of the fairness conditions in all situations in which the property that is being predicted is unevenly distributed across different groups and the decision problem does not admit of perfect predictions. In fact, it shows that all systematic decision making that categorizes individuals into different probabilistic groups will be in some sense unfair if the relevant sought-after property is unevenly distributed across groups. The problem is not tied to computing but arises anytime some systematic and imperfect classification system is applied to a set of individuals where properties are unevenly distributed across groups. Decision algorithms that predict recidivism and assist the court system are by necessity in some sense biased and unfair, but so are all screening tests that are designed to identify special health needs across populations where the health needs are unevenly distributed across groups, all assessment tests that aim at finding children who need special educational resources early, and so on, and so on. The computerized decision algorithms are the tip of an iceberg of biases in decision-supporting tools and heuristics in society.

In their conclusion, Kleinberg, Mullainathan and Raghavan provides an illustration of their results:

> [S]uppose we want to determine the risk that a person is a carrier for a disease X, and suppose that a higher fraction of women than men are carriers. Then our results imply that in any test designed to estimate the probability that someone is a carrier of X, at least one of the following undesirable properties must hold: (a) the test's probability estimates are systematically skewed upward or downward for at least one gender; or (b) the test assigns a higher average risk estimate to healthy people (non-carriers) in one gender than the other; or (c) the test assigns a higher average risk estimate to carriers of the disease in one gender than the other. The point is that this trade-off among (a), (b), and (c) is not a fact about medicine; it is simply a fact about risk estimates when the base rates differ between two groups (Kleinberg et al. 2017: 17).

Analogous biases and unfairness will arise in all decision contexts where a decision metric is applied to a set of individuals in order to identify some relevant property and this property is unevenly distributed across groups. The only exception to this is when the decision metric is flawless and only makes perfectly accurate non-probabilistic predictions.

5.2 Possible Solutions

There are various ways of trying to respond to the incompatibility of the fairness conditions. One could dismiss one (or both) of the conditions and hold it (them) to be irrelevant to fairness. One could attempt to counteract the unfairness in some complementary stage of the decision process. And one could categorically reject decision-supporting tools that like decision algorithms facilitate probability assessment and replace decision making that relies or is informed by such tools with some other kind of decision making that is not necessarily unfair. In this section, I will discuss each of these three responses in some further length and argue that neither of them provides a satisfactory, general solution to the problem.

First, consider how one could reject one (or both) of the fairness conditions. If either what I have called *Equality in accuracy* or *Equality in inaccuracy* were irrelevant to the fairness of a decision algorithm, then the impossibility result presented in the previous section fails to show that fair decision algorithms are impossible. Without the balance conditions which reflect *Equality in inaccuracy*, x and y do not need to be the same for all groups, which means that there is no system of equations with the same unknowns and one can find innumerable solutions to both of the linear equations, thus meeting the *Equality in accuracy* condition. Without the well-calibration condition which reflects *Equality in accuracy*, one is not warranted to claim that the total scores given to different groups should be the same as the number of positive instances in the group, and again there are innumerable ways of meeting *Equality in inaccuracy* even if the decision algorithm is imperfect and the relevant property is unevenly distributed across groups.

However, those who care about distributions between groups will find it hard to justify rejecting any of the fairness conditions without any further reasons to do so. Consider, first, what rejecting *Equality in accuracy* amounts to. *Equality in accuracy* secures that no matter what group an individual belongs to, if she is ascribed a probability p of having some property, then she belongs to a set of individuals in which a p-fraction is expected to have the property in question. Thus, for instance, it is *Equality in accuracy* that secures that regardless of whether an individual is white or black, if the individual is ascribed a 50% risk of having malignant melanoma by some decision algorithm that assists medical assessments and decision making, then he or she belongs to a set of people in which 50% actually have malignant melanoma. If a decision algorithm fails to meet *Equality in accuracy* then it could be true for individuals in one groups that when the decision algorithm indicates that they have a 50% risk of having malignant melanoma they in fact belong to a set of people in which many more than half of the people have malign skin cancer. In other words, rejecting *Equality in accuracy* as a condition of fairness means that we accept it as fair that a decision algorithm systematically and predictably entails that individuals in one group are given risk scores that do not reflect the prevalence of the sought-after property. When the sought-after property is some health conditions this means that health needs in this group are either neglected so that members of the group to a greater extent suffer health problems than members of some other group, or that health needs are disproportionally misdiagnosed in this group. When the sought-after property is recidivism it means that a higher fraction of recidivists in one group will be overlooked and go free from consequences recidivists in other groups must bear. Whether one receives better healthcare or avoids some harm due to the fact that the court system is biased in one's favor, receiving these benefits in virtue of group-belonging and nothing else is unfair. *Equality in accuracy* is, prima facie, a valid condition of fairness since it secures against these unfair distributions of benefits. One needs a substantive, normative reason to reject it and to favor distributions of benefits that benefit some groups.

Consider, second, what rejecting *Equality in inaccuracy* amounts to. *Equality in inaccuracy* secures that individuals that have the sought-after property are not ascribed a higher likelihood of having the property merely in virtue of being members of some group, and likewise that individuals that lack the sought-after property are not ascribed a higher likelihood of having the property merely in virtue of being members of some group. Thus, for instance, it is *Equality in inaccuracy* that secures that a black person that lacks the sought-after property will not be systematically ascribed a higher likelihood of having the property just because she is black, and that a white person that has the property will not be systematically ascribed a lower likelihood of having the property just because she is white. If a decision algorithm fails to meet *Equality in inaccuracy* it will like COMPAS systematically ascribe a greater likelihood that members in one group who lack the property has it compared to members in another group who lack the property, and systematically ascribe a lower likelihood that members of one group who has the property has it compared to members in another group who has it. In the healthcare context, it is *Equality in inaccuracy* that ensures that a decision algorithm does not systematically ascribe greater likelihood that someone has a disease when the decision algorithm misdiagnoses patients depending on group-belonging of patients. In other words, rejecting *Equality in inaccuracy* as a condition of fairness means that we accept it as fair that a decision algorithm systematically and predictably entails that individuals that have been misdiagnosed by the decision algorithm are given risk scores that are higher in some group and lower in some group. When the sought-after property is some health condition this means that health needs are disproportionally misdiagnosed in some group. When the sought-after property is recidivism it means that a higher fraction of nonrecidivists will be treated as recidivists in one group. Whether one gets a medical misdiagnosis or receives inaccurate treatment by the court system, being exposed to these harms in virtue of group-belonging and nothing else is unfair. *Equality in inaccuracy* is, prima facie, a valid condition of fairness since it secures against these unfair distributions of harms. One needs a substantive, normative reason to reject it and to favor distributions of harms that harm some groups.

It might of course be true that there are many choice situations in which *Equality in accuracy* and *Equality in inaccuracy* do not seem important because the consequences of mistakes are so insignificant. For instance, when some social media platform uses a decision algorithm in order to determine which advertisement to show to which individuals one might not care much about the fact that this decision algorithm fails to meet *Equality in accuracy* and/or *Equality in inaccuracy*.[2] Yet, this is not because these conditions fail to reflect fairness conditions. Rather, it relates to the fact that we care more about distributions of certain goods than others. When something like "effective advertisement" is distributed, many of us simply do not care much about fair distributions. By contrast, when healthcare, punishment and education resources are distributed, we do care about fair distributions. For each of the conditions, there are situations in which a violation of them is associated with grave unfairness, and therefore neither of them can be rejected as irrelevant.

A different idea that might strike readers as plausible is to attempt to counteract the unavoidable biases with decision methods that take the biases into account and introduce some counterweights that make up for the skewed results of the decision algorithm. For instance, if a decision algorithm is known to be more likely to generate, on average, higher risk scores to black defendants who are nonrecidivists than to white defendants who are nonrecidivists when they estimate the risks of recidivism, either the programmers or the courts could take this into account and apply a discount rate to the risk scores ascribed to black defendants. This might be thought of as a kind of affirmative action, intended to correct for an existing unfairness that black defendants suffer. If accurately designed, this discount rate could cancel out the specific bias of the decision algorithm that disfavors black defendants. Similarly, of course, one could counteract biases in medical diagnostics by introducing fixed-rate likelihoods that people in some group suffer from some medical condition, and if this is accurately designed the specific bias of the medical diagnostics method that disfavors the group will be cancelled out.

As attractive as this response might seem, it is problematic and it is important to recognize that this does not mean one avoids violating the fairness criteria. The reason is that any such systematic attempt at counteracting specific biases with the introduction of fixed counterweights will implicate that the overall assessment will be biased in some other way. If the initial decision algorithm violates one of the balance conditions but meets the calibration condition and a decision maker introduces fixed-rate counterweights that successfully counteracts the initial bias the decision algorithm together with the new counterweights will violate the calibration condition. The total score ascribed to the members of the group will no longer equal the number of positive instances in the group after these weights have been introduced. This means that *Equality in accuracy* is violated. The problem does not reside in the decision algorithms as such, but in the fact that *any* systematic decision method will fail to meet the fairness conditions. Any attempt to counteract biases will lead to the violation of a fairness criterion. This might be warranted in certain contexts, but one needs a good reason to violate a prima facie valid fairness criterion. In the next section, I will suggest that concerns for the worse off might be such a reason and propose that this might in some contexts support a kind of affirmative action that justifies violating prima facie valid fairness criteria in certain ways.

[2] See also Loi et al. 2019.

An alternative response will be to dismiss decision algorithms and all systematic decision methods all together. Systematic decision making that relies on the application of principles that when matched with the pertinent characteristics of the choice situations help decision makers determine what to do is not the only kind of decision making there is. One could reject this kind of decision making and replace it with something like "the decision makers judgment". In principle, the result of judgments that cannot be systematized might not be biased in any direction. For instance, one might argue that court systems should not use decision algorithms which help them determine the likelihood of recidivism such as COMPAS, and that judges instead should use their judgment and make their decision in some way that is not systematic in the sense that it applies some general principles to determine likelihood of recidivism. The pattern of decisions that a judge who relies on her own discretion and makes decisions on a case-by-case basis forms might in principle not be skewed in favor of any socially salient group.

Although it might in principle be possible to avoid unfairness if one replaces systematic decision making with decision making that relies on the wisdom on the decision maker, rejecting systematic decision making is not an appealing response to the incompatibility of the fairness conditions (although there might be other reasons to favor human decision making, e.g. transparency, accountability). Systematic and principled decision making, whatever challenges it might face, tend to be more resistant to abuse and also in general safeguard better against influence of biases than the alternative. Furthermore, there are areas where it seems completely indispensable to rely on systematic and principled decision making in order to make anything that even resembles sound decisions.

Consider first the risk that a decision-making method is abused by decision makers if one rejects principled reasoning. To see this, consider the problems with casuistry, which was once the dominating ethical theory in the Western World. In brief, casuistry is a normative theory that commands that decision makers analyze each case by first identifying what seems to be the relevant characteristics of the case, and then, second, by comparing the case such as this has been understood with some paradigm case where the decision maker knows what the correct decision (cf. Jonsen and Toulmin 1988). The history of casuistry is not a history of successful, fair decision making. Consider, for instance, two examples from Blaise Pascal's devastating criticism of his contemporaries' casuistic analyses. First:

> It is said in the Gospel: 'Give alms of your superfluity.' Several casuists, however, have contrived to discharge the wealthiest from the obligation of alms-giving. This may appear [a] paradox, but the matter is easily put to rights by giving such an interpretation to the word *superfluity* that it will seldom or never happen that any one is troubled with such an article. This feat has been accomplished by the learned Vasquez, in his *Treatise on Alms*, c. 4: 'What men of the world lay up to improve their circumstances, or those of their relatives, cannot be termed *superfluity*; and accordingly, such a thing as superfluity is seldom to be found among men of the world, not even excepting kings (Pascal 1952: 36).

Second:

> Pope Gregory XIV decided that assassins are not worthy to enjoy the benefit of sanctuary in churches and ought to be dragged out of them; and yet our four-and-twenty elders affirm that

> ‘the penalty of this bull is not incurred by all those that kill in treachery.’ This may appear to you a contradiction; but we get over this by interpreting the word *assassin* as follows: ‘Are assassins unworthy of sanctuary in churches? Yes, by the bull of Gregory XIV they are. But by the word assassins we understand those that have received money to murder one; and, accordingly, such as kill without taking any reward for the deed, but merely to oblige their friends, do not come under the category of assassins’ (Pascal 1952: 36).

A different, and equal if not bigger, problem with the idea of replacing systematic and principled decision making with decision making that relies on the judgment of the decision maker is that this increases the risk that widespread biases and arbitrary framing influence the outcome of the decision-making process. What Pascal pointed out in relation to casuistry is that these decision methods can be easily maliciously abused to promote certain interests. A different problem with decision making that only relies on the judgment of the decision maker is that for this to have fair outcomes, the decision maker must be so-to-speak wired in a fair way so that she does not have any unintentional inclinations to favor some people over other people, or react in different ways to arbitrary framing aspects so that unfair outcomes are generated.

Empirical research over the last decades provides overwhelming reason to suspect that few if any human beings can be expected to be “wired in a fair way” in this sense. First, there is a large amount of research that show that implicit biases are widespread. Implicit biases are subconscious associations that lead to negative evaluations of an individual based on irrelevant features such as race or gender. For instance, a recent review article shows that 35 out of 42 published scientific articles found evidence of implicit biases in healthcare professionals (Fitzgerald and Hurst 2017). Decision-making processes that rely solely on the judgment of the decision maker will plausibly generate outcomes that are more influenced by implicit biases than decision-making processes that rely on rules and guidelines.

Second, the groundbreaking research by Amos Tversky and Daniel Kahneman and the large body of research that has followed reveal that how choice situations are framed tend to have significant impact on what attitude humans take to them (cf. Kahneman 2011; Tversky and Kahneman 1981). Whereas a systematic and principled decision-making process can be designed to be neutral with respect to framing, decision making methods that leave significant room for the judgment and attitude of the decision maker is much more exposed to the risk of being affected by framing effects. This significantly increases the risk for unfair outcomes.

Finally, there are several areas where pure judgment seems unable to at all ground sound decisions. Consider the healthcare sector. One of the prerequisites of modern medicine is that physicians do not merely rely only on their judgment when they diagnose patients. Rather, they apply a principled and systematic reasoning process and follow rules and guidelines to identify the health problems of their patients and to decide on what treatment is required. It might be true that the application of these rules and guidelines leads to unfair outcomes (and the impossibility theorem discussed in the previous section indicates that this is true by necessity), but this unfairness can hardly be taken as a decisive reason to categorically reject modern medicine.

The basic justification for using systematic decision methods such as decision algorithms is that these are better than non-systematic decision methods both with respect to fairness and with respect to promoting the good. The fact that they are not perfect with respect to fairness can hardly be used as decisive argument in favor of a decision method that has a record of being even worse both with respect to fairness and with respect to promoting the good.

5.3 Accepting Biases

In light of the incompatibility of the fairness conditions and also the fact that alternative decision strategies often are worse than biased systematic decision methods one wonders how to address the fact that these decision methods must be biased. This will plausibly be highly context-dependent since the severity of the different kinds of unfairness that might arise differs depending on context. For instance, inaccurately being ascribed a high probability of having a deadly but curable disease is a greater harm if it requires a lengthy and painful process to determine whether one actually has the disease or not compared to if it can be easily and quickly tested for. In the first case, an individual might worry for months that she is dying, while in the second case, it might be a matter of hours. This sort of differences in severity that are tied to context must be taken into account when one decides how to deal with the unfairness of systematic decision methods. Nevertheless, I will here present three general principles that I believe applies broadly and should guide policy that rely on systematic decision methods that generate probability predictions: *Dominance, Transparency* and *Priority to the worse off*.

First, consider:

> *Dominance:* A decision method that is better with respect to one of the dimensions of fairness and worse with respect to none of the dimensions is better overall.

Fairness is not a binary criterion. Different decision methods can be more or less fair also when neither is perfectly fair. Since each of the fairness considerations seems separable from the other in the sense that a reduction of unfairness in one dimension will not change the value of the unfairness in other dimensions, a dominance criterion can be used to reflect the desire to minimize unfairness. The disvalue of some amount of *inequality in accuracy* is unrelated to the amount of *inequality in inaccuracy*, and the disvalue of some amount of *inequality in inaccuracy* is unrelated to the amount of *inequality in accuracy*. This means that we can safely say that if one decision method is better than another in one of these dimensions and not worse in the other, then the it is better overall in terms of fairness.

Since *Equality in inaccuracy* relies on two dimensions, reflected in *Balance for the positive class* and *Balance for the negative class*, and since it is indeterminate how these dimensions relate to each other so that it would be imprudent to say anything general about this, the dominance criterion can be further specified over three dimensions:

> *Dominance, specified:* A decision method that is better with respect to one of the three dimensions *Well-calibrated*, *Balance for the positive class* and *Balance for the negative class*, and worse with respect to none of the dimensions is better overall.

Second, consider:

> *Transparency*: Decision makers who use decision methods that are unfair should themselves be aware of the unintended differences in impact, and they should be transparent about the differences in impact to the community that is affected.

That unfairness is prevalent in certain sectors due to widespread reliance on predictive decision methods with built-in biases constitutes a significant social challenge. There are several reasons to embrace transparency with regards to this (cf. Binns 2018a; Wong 2019; Zimmermann et al. 2020). First, awareness of a problem is required to mitigate the problem. As illustrated above, decision algorithms can be more and less unfair. A prerequisite for working toward the least unfair option is awareness of the problem. Decision makers must be made aware of the problems in order to be able to mitigate it, and those affected by the problem should be made aware of the problem so that they get a chance to promote mitigation of the problem. Second, transparency is a prerequisite for compensation. If some group is systematically disadvantaged by certain well-established decision practices, members of this group have a well-grounded claim on being compensated for this disadvantage in other contexts. For instance, if women are systematically disadvantaged by certain well-established healthcare practices, they have a well-grounded claim on being prioritized when new healthcare practices are developed. To make these claims, they must be aware of how they are disadvantaged in the first place. Third, transparency enables public deliberation. Biases that are built-in in various decision systems across societies constitute a social problem. Solutions to this problem should be the topic of public deliberation, and in order for the public to be able to deliberate about the problem they must be aware of its extent. Transparency helps keeping the public informed.

The fact that biases are unavoidable when decision algorithms are construed does not mean that *all types* of biases are unavoidable when decision algorithms are construed. The use of COMPAS had relatively better consequences for white defendants than for black defendants, but the fact that some bias and thus some unfairness was unavoidable does not mean that *this particular* bias and unfairness was unavoidable. COMPAS could, in principle, have had relatively better consequences for black defendants than for white defendants. This reveals that those who design decision systems are making a choice of which unfairness to accept. This is a normative choice between which bad is the lesser one. To guide this choice, I believe the following principle is reasonable:

> *Priority to the worse off*: A decision method that is relatively better for members of a worse-off subgroup is preferable to a decision method that is relatively better for a better-off subgroup.

This principle sounds prioritarian, but in most contexts it is supported by a wide range of distributive views. Egalitarians, prioritarians and many sufficientarians are in complete agreement that if one needs to choose between two unequal and to some extent unfair outcomes and the total amount of good is the same in both outcomes, one ought to choose the outcome that is best for the worse off. Egalitarians favor this outcome because it reduces inequality (Binns 2018b; Segall 2016; Temkin 2003). Prioritarians favor this outcome because they believe benefits to the worse off matter more (Adler 2012; Andric and Herlitz 2021; Eyal and Herlitz 2021; Holtug 2010). And many sufficienatarians would favor this outcome as long as the individuals are badly off enough because they agree with prioritarianism that benefits to the worse off matter more when distributions to the badly off are evaluated (Crisp 2003; Shields 2012). Yet, not only consequentialist principles agree on this point. Also a view such as Scanlonian contractualism will support the principle since it leads to the practice against which the smallest complaint can be charged (Scanlon 1998; Herlitz 2019; Frick 2015).

The *Priority to the worse off* principle can be applied in two ways. First, it can guide the choice between which of the fairness conditions one violates. Should *Equality in accuracy* or *Equality in inaccuracy* be violated? What is best for the worse off? Second, it can guide the choice between different ways in which the conditions are violated. Should *Equality in accuracy* be violated so that it harms men or so that it harms women? Should *Equality in inaccuracy* be violated so that it harms men or so that it harms women? The *Priority to the worse off* principle ought to be applied at both of these levels. It ought to guide the choice of what fairness condition to violate, and it ought to guide the choice of how to violate the fairness condition. This secures that whatever prima facie unfairness is generated by a decision system, it works to the relative benefit of the worse off group, similar to affirmative action policies.

Some might object to the *Priority to the worse off* principle in this context by invoking the view that we ought to keep different spheres of justice separated (cf. Walzer 1983). Decision systems in for example the healthcare system or in the judicial system should not be designed based on general features of affected individuals in other, unrelated, areas, or so some might argue. Why ought we, for instance, care about the fact that women on average make less money than men when we decide whether to disproportionally harm women or men in the healthcare system? Why ought we care about the fact that black people on average die at a younger age than white people when we decided whether to disproportionally harm black or white people in the judicial system?

This objection seems largely misguided. First, it is not obvious that we actually ought to keep different spheres of justice separated. When egalitarians, prioritarians, sufficientarians and concractualists speak of who is worse off and the priority this commands, they tend to speak of who is worse off generally. And from the perspective of fairness, how well off people are generally seems to be what matters. Second, it is not obvious that we are able to keep different spheres of justice separated as neatly as proponents of this idea might hope. What would be included in the health sphere and would should not be included? Income, education, living conditions and social class all seem to affect health, so should they be included in the health sphere or not? What would be included in the judicial sphere? Again, income, education, living condition and social class all seem to affect individuals in terms of whether or not they are more likely to commit crime. Should we include these aspects in the judicial sphere then? Third, and most devastating for this kind of objection, it is hard to see how an appeal to different spheres of justice can help one reject the view that priority should be given to the worse off. Even if one accepts that features in different spheres should be kept separate, the *Priority to the worse off* principle can still be applied. Who are worse off in the health sphere can guide how to design decision methods in healthcare, and who are worse off in the judicial sphere can guide how to design decision methods in the judicial sphere.

5.4 Concluding Remarks

In this paper, I have presented three general principles that I believe ought to guide policy-making that relies on predictive decision making in light of the fact that systematic decision methods such as decision making that relies on decision-supporting tools like decision algorithm is by necessity unfair. I have argued that policy makers who face this troublesome reality should be guided by three goals: unfairness should be minimized, unavoidable unfairness should be public, and unavoidable unfairness should benefit the worse off.

Besides the main philosophical argument of the paper, I hope that the paper can achieve two things. First, I hope that moral and political philosophers will take an increased interest in the unavoidable biases that arise in decision-making processes that rely on support tools that ascribe probabilities to different outcomes. These decision-making processes are ubiquitous in modern societies. They constitute a core element of modern healthcare. They are indispensable for social workers who make decisions regarding when and how to intervene to help and protect vulnerable people in society. They are used widely in the educational sector, from assessments of special needs in very young children to college admission processes. And so on, and so on. If all decision-making processes that rely on support tools that ascribe probabilities to different outcomes in all contexts are biased in one way or another so that they constitute and ground unfair decision making and policy, surely this deserves the attention of ethicists and political philosophers.

Second, I hope that the paper reaches those working in specific areas where particular decision methods are being used and that it inspires them to take an interest in the philosophical work on fairness. This includes statisticians, computer scientists and policy makers and policy researchers of different kinds. One of the aspects of the identified impossibility theorem is that it reveals that neither of these professions can avoid normative decision making. If a decision algorithm by necessity is unfair, then the designer of the decision algorithm effectively decides how her algorithm is unfair. This is something that professionals in these areas ought to be aware of, and also ascribe importance to. Whether they like it or not, they are addressing fundamentally philosophical issues on a daily basis, and they ought to take this seriously.

References

Adler, Matthew. 2012. *Wellbeing and its fair distribution*. Oxford: Oxford University Press.

Andrić, Vuko, and Anders Herlitz. 2021. Prioritarianism, timeslices, and prudential value. *Australasian Journal of Philosophy*. https://doi.org/10.1080/00048402.2021.1920043

Angwin, Julia, Jeff Larson, Surya Mattu, and Lauren Kirchner. 2016. Machine bias: There's software used across the country to predict future criminals. And it's biased against blacks. *Pro Publica*, May 23.

Binns, Reuben. 2018a. Algorithmic accountability and public reason. *Philosophy & Technology* 31: 543–556.

———. 2018b. Fairness in machine learning: Lessons from political philosophy. *Journal of Machine Learning Research* 81: 1–11.

Chouldechova, Alexandra. 2017. Fair prediction with disparate impact: A study of bias in recidivism prediction instruments. arXiv:1703.00056v1.

Crisp, Roger. 2003. Equality, priority, and compassion. *Ethics* 113: 745–763.

Eyal, Nir, and Anders Herlitz. 2021. Input and output in distributive theory. *Noûs*. https://doi.org/10.1111/nous.12392

Fitzgerald, Chloë, and Samia Hurst. 2017. Implicit bias in healthcare professionals: A systematic review. *BMC Medical Ethics* 18: 19.

Frick, Johann. 2015. Contractualism and social risk. *Philosophy and Public Affairs* 43: 175–223.

Herlitz, Anders. 2019. The indispensability of sufficientarianism. *Critical Review of International Social and Political Philosophy* 22 (7): 929–942.

Holtug, Nils. 2010. *Persons, interests, and justice*. Oxford: Oxford University Press.

Jonsen, Albert, and Stephen Toulmin. 1988. *The abuse of casuistry: A history of moral reasoning*. Berkeley: University of California Press.

Kahneman, Daniel. 2011. *Thinking, fast and slow*. London: Penguin.

Kleinberg, Joel, Sendhil Mullainathan and Manish Raghavan. 2017. Inherent trade-offs in the fair determination of risk scores. arXiv:1609.05807v2.

Loi, Michele, Anders Herlitz, and Hoda Heidari. 2019. A philosophical theory of fairness for prediction-based decisions. *SSRN*. https://doi.org/10.2139/ssrn.3450300.

Pascal, Blaise. 1952. *The provincial letters, Pensées, scientific treatises*. Trans. Thomas M'Crie, W.F. Trotter and Richard Scofield. London: Encyclopaedia Brittanica.

Scanlon, Thomas. 1998. *What we owe to each other*. Cambridge, MA: Harvard University Press.

Segall, Shlomi. 2016. *Why inequality matters: Luck egalitarianism, its meaning and value*. Cambridge: Cambridge University Press.

Shields, Liam. 2012. The prospects for Sufficientarianism. *Utilitas* 24: 101–117.

Temkin, Larry. 2003. Egalitarianism defended. *Ethics* 113: 764–782.

Tversky, Amos, and Daniel Kahneman. 1981. The framing of decisions and the psychology of choice. *Science* 211: 453–458.

Walzer, Michael. 1983. *Spheres of justice: A defence of pluralism and equality*. New York: Basic Books.

Wong, Pak-Hang. 2019. *Democratizing algorithmic fairness. Philosophy & Technology*. https://doi.org/10.1007/s13347-019-00355-w.

Zimmermann, Annette, Elena Di Rosa, and Hochan Kim. 2020. Technology can't fix algorithmic injustice. *Boston Review*, January 9.
Zuckerman, Gregory, and Bradley Hope. 2017. The quants: The quants Run Wall street now – Software-driven trading, once a novelty, is becoming dominant. *The Wall Street Journal* (May 22).

Chapter 6
Castigation by Robot: Should Robots Be Allowed to Punish Us?

Alan R. Wagner and Himavath Jois

Abstract Autonomous robots are currently being developed for tasks that may require those robots to assume a position of authority over humans. Our work examines the ethical boundaries of human-robot interaction in the context of robot-initiated punishment of humans. We observe that positions of authority often require the ability to punish in order to maintain societal norms. If autonomous robots are to assume roles of authority, they too, must be capable of punishing individuals that violate norms. This work constructs a discussion regarding permissible robot behavior, particularly from the perspective of robot-administered punishment, examining the current and future use cases of such technology and applying a consequence-based approach as a starting point for analysis.

6.1 Introduction

Antoine de Saint-Exupéry, a pioneering French aviator, proclaims in The Little Prince that, "The machine does not isolate man from the great problems of nature but plunges him more deeply into them" (Saint-Exupéry 1943). While he states this in response to the great destructive power of aircraft during World War II, perhaps a similar statement could be made about some of the autonomous systems in development today. Autonomous robots are currently being developed to work alongside police and military personnel, assist in teaching, and provide aspects of healthcare. The applications envisioned by the developers of such robots are not intended to create further issues for man, but, as these systems become more

A. R. Wagner (✉)
Department of Aerospace Engineering/Rock Ethics Institute, The Pennsylvania State University, University Park, PA, USA
e-mail: azw78@psu.edu

H. Jois
Department of Aerospace Engineering, The Pennsylvania State University, University Park, PA, USA
e-mail: hxj5142@psu.edu

B. Lundgren, N. A. Nuñez Hernández (eds.), *Philosophy of Computing*,
Philosophical Studies Series 143, https://doi.org/10.1007/978-3-030-75267-5_6

intelligent, we as a society must decide what types of behavior will be considered "out-of-bounds" for them. For some application domains, it is conceivable that the autonomous agent could hold a position of authority over a human or humans. It has also been shown that authority rests on the threat of punishment for non-compliance (Axlerod and Hamilton 1981; Fehr et al. 2002). Therefore, if we are willing to entertain the possibility of autonomous systems serving in an authoritative role, we must also consider affording these systems with the ability and power to punish. Yet, we must then examine, and ultimately decide, where to draw the line between permissible and impermissible behavior by an autonomous system serving as an authority. *Our central contention is that there are some situations that warrant allowing a robot to punish humans, but that limitations must be placed on how and when an autonomous robot can punish.* To be clear, we are not focusing on the possibility or necessity of punishing robots themselves, or their creators, for liability or related issues. Rather, we seek to understand the ethical boundaries of having robots *initiate and administer* punishments to humans.

Our objective here is to provide an introductory overview of the issues that surround this complex topic. The purpose of this paper is thus to introduce and develop some of the preliminary concepts necessary for addressing questions related to if, how, and when an autonomous system should be allowed to punish human beings. Our motivation here is to begin to organize the discussion surrounding permissible robot behavior. These topics are important because robots are being develop to assume increasingly complex social roles, such as classroom assistants (Carey and Markoff 2010) and policing roles (Metz 2014). We believe, and argue in this paper, that these types of roles will demand that the robot act as an authority. Having and acting as an authority suggests that robots may need to be able to punish people in order to maintain order and serve these roles. This work is meant to serve as an initial foray into a new, and likely controversial topic.

For the sake of common ground, it is necessary that we establish the boundaries and necessary definitions relative to this problem used throughout. Building on Arkin's definition of a robot, we use the term *robot* to mean a physically embodied and contextually situated mechanical and electronic system that acts autonomously with meaningful purpose, using cues from its environment (cf. Arkin 1998). Although a robot may have been pre-programmed with a set of behaviors and perceptual abilities, these systems autonomously determine when to enact such behaviors and may have control over parameters associated with a behavior, thus allowing a robot to mediate its reaction to a situation. The fact that the robot is embodied, autonomous, and potentially capable of creating and/or adapting punishment to individuals and situations is a distinguishing feature of robot punishment.

Further, we use the term *punishment* to represent a specific behavior or behaviors selected for the purpose of adding cost to one's action, thereby shaping their decisions away from such behavior. We do not consider here the reasons that might be used to justify a particular punishment, simply because such an investigation could be the subject of several papers. Yet, we consider punishments of both positive and negative type. Positive punishments mean that the punishment adds an aversion to the environment. A child being required to do extra chores around the house for speaking inappropriately would be an example of a positive punishment, since the extra chores are being added to the child's situation in response to their behavior. Negative punishments are the opposite; they remove a reward from the environment. For example, a student losing driving privileges after failing a class would be a negative punishment. Some punishments, such as serving time in prison, include both positive and negative punishments. Punishments can include verbal reprimands and/or physical castigations. Verbal and physical punishments can also be both positive and negative. For example, positive verbal punishment could suggest that the individual's performance was below standard. Verbal punishments tend to affect a person emotionally, whereas physical punishments can impact a person physically and emotionally. For a robot administering punishment, the difference between robot punishment and robot behavior in general is that punishment implies that the robot acts in a manner that is undesirable or intentionally unpleasant for the human.

Admittedly, the broadness of our use of the term "punishment" forces us to include contexts in which the punishment is minor and/or the punishing agent simply reacts to a stimulus. The sound emitted, for example, when one fails to use their seatbelt. Although it is common to think of this sound as a safety feature and not as a punishment, warning sounds are developed to act as an aversion, a negative reinforcement, in order to encourage compliance. Therefore, these types of minor stimuli could be considered as punishments depending on the recipient's perspective. Because we present an overview of the topics surrounding robot punishment, we use a definition of punishment that includes even mild aversions with the purpose of casting a wide net. Future work may later focus on specific types of punishment such as physical punishment or only aversive punishments.

Additionally, it is important to distinguish punishment from a robot and punishment from a *machine*. As before, a chime used as a reminder to buckle one's seatbelt or close the refrigerator door might fit the description of a machine-initiated punishment. This paper, however, is focused on robot-initiated punishment, which is distinct from machine punishment in that robots use perceptions about their dynamically changing environment and the human to alter their behavior. Hence, whereas a machine punishment may simply be a response to an environmental signal, robot-initiated punishment suggests a more dynamic, adaptive, flexible and generalizable punishment response. Hence, while the very general topic of machine harms and benefits has long been studied, we seek to expand this discussion into the domain of robotics and AI, where the robot may tailor punishments to the individual and the context and may need to evaluate right from wrong in order to determine who should be punish and how they should be punished.

To outline the arguments set forth in this paper, we will first demonstrate that there are indeed situations in which a robot must hold *authority* over a human being. Having authority over a human presumes an ability to punish on the part of the robot. Moreover, authority is often associated with physical power over another individual and, likewise, punishment is typically associated with physical punishment (Hobbes 1968). We will then examine which situations warrant the use of punishment by a robot. Next, we consider robot punishment from the perspective of an ethical framework, using it to present initial arguments for limitations on robot punishment, concluding the work with a broad discussion of these limitations in the hope of fostering further discussion in this area.

6.2 Authoritative Roles for Robots

We contend that some robotics applications may require that the robot have an authority role. We do not mean that the robot is necessarily authoritative in that it has extra or expert knowledge of a subject. By authority role, we mean that the robot holds a legitimate position of relative power over a human as part of a social hierarchy. We assume that the robot is given its authority from some governing body, but the details of how or why the robot is given authority are beyond the scope of this paper. Consider, for example, a robot tasked with leading people to safety during an emergency. This robot must have the authority to verbally (and perhaps physically) demand that people move to an exit quickly in order to ensure the safety of others (Robinette et al. 2016). Alternatively, consider a robot security guard. This robot must have the ability to demand that trespassers exit the premises, to alert human authorities if necessary, and perhaps even to apply a loud sound as an aversion if trespassers do not comply (Metz 2014). In military contexts, research has argued that it may be ethically necessary for robots to report war crimes committed by soldiers (Arkin 2009). Finally, robots are currently being developed to serve in supporting roles for classroom teachers (Park et al. 2011; Sharkey 2016). Although there are many tasks that robots might perform to support classroom teachers, some of these roles must allow the robot to admonish students that are not on task.

Admittedly, many of these applications are in their infancy and, as such, the most state-of-the-art robots today still lack the autonomy and ability to perform these roles. Even so, especially with respect to military and policing applications, we are quickly approaching the day when autonomous robots will assume the roles of warfighters and security personnel. Although human authority figures clearly imbue these robots with their authority, the immediate actions taken will often be determined by the robot. Moreover, the robot's power will allow it to adjudicate conflicts with humans. For instance, a robot physically restraining or preventing the escape of a burglar or an emergency evacuation robot needing to verbally warn evacuees of their authority when demanding that the evacuee follow the robot, just as humans currently do (Kuligowski 2009).

While society may not wish to imbue these machines with authority, not doing so might limit the machine's usefulness in critical and potentially lifesaving applications. An emergency evacuation robot, for instance, that cannot act as an authority is likely to be ignored. A patrolling security robot that cannot detain intruders is useless in its role. Hence, although we may cringe at the idea of allowing these machines to have authority over humans, doing so is necessary if we hope to reap the benefit of these applications. The next section discusses why an ability to punish is necessary for maintaining an authoritative social position.

6.3 The Necessity of Punishment

We argue that a robot that acts as an authority but does not have the power to punish has no real authority and humans will quickly recognize its lack of authority. Robotic applications that require authority, such as military and policing applications, also require the threat of punishment in order to maintain that authority. For humans a well-documented authority bias exists which tends to influence people to comply with the directions of an authority (Milgram 1974). Society-at-large empowers certain human agents with the power to punish in order to maintain social order and cooperation (Fehr et al. 2002), where that social order is fostered by implicitly agreed upon social norms (Henrich and Boyd 2001; Sethi and Somanathan 1996). Neuroscientific evidence has even shown that the reward centers of the brain are activated when people are effectively punished for transgressions (De Quervain et al. 2004). Overall, the evidence suggests that people prefer a strong authority (with a number of caveats) with the ability to punish in order to maintain social order and buttress social norms (Fehr and Fischbacher 2004). Likewise, social identity theory posits that specific intergroup behaviors, such as punishment, demonstrate status differences within a group (Tajfel and Turner 1979). Hence, suggesting that for any individual, human or robot, to be an authority figure implies that that individual is endowed with the ability to punish. Disallowing punishment by an authority will serve as a signal to the group that the individual is not, in fact, an authority.

As a second argument for robot punishment, we contend that if we assume robots will one day constitute a significant proportion of social roles in society (educational helpers, policing activities, search-and-rescue activities, etc.) then these robots need to support community social norms, which implies meting out minor punishments to norm transgressors. Social norms are standards of behavior that develop within a community and are based on widely shared beliefs describing how individuals of the community should behave in a given situation (Fehr and Fischbacher 2004b). This definition of social norms does not imply that individuals agree or disagree with the norm. Although people may disagree with a norm, the norm may still exist and govern how people should behave in a situation. In order for these rules to be effective, they must be enforced, even if only occasionally. For example, queuing at the back of the line rather than cutting directly to the front is a norm that is typically followed by most individuals. It has long been shown that in order for a norm to be effective there must be some threat of punishment for norm non-compliance (Axlerod and Hamilton 1981; Fehr et al. 2002). This threat of punishment can be critical for maintaining the norm and supporting societal interaction. For example, the concept of strong reciprocity states that members of a society should be kind to those that are kind, and punish those that are unkind (Fehr et al. 2002). Norm enforcement is also an important mechanism for preventing free riders. The free-rider problem notes that there are some things that benefit all members of a community, but require an optional contribution, i.e. parks, schools, street lights (Hardin 2013). Individuals that do not contribute but that nevertheless benefit are known as free riders. Free riding is the best strategy for an individual so long as only a small portion of a society adopt a free riding strategy. Theoretical examinations of this topic have shown that if many individuals adapt a free riding strategy, societal cooperation collapses (Pareto 1935; Sethi and Somanathan 1996).

For better and for worse, punishing free riders solves the free riding problem and maintains cooperation (Fehr and Gächter 2002; Henrich and Boyd 2001; Ostrom et al. 1992). As an example of what can happen when the threat of punishment disappears, police strikes in Liverpool in 1919 and Montreal in 1956 resulted in a massive increase in crime. In fact, neuroscientific studies have shown that individuals feel a sense of reward and happiness when they witness the punishment of a free rider (Price et al. 2002). Yet, punishment is also costly for the punisher in that it may increase social conflict, requires energy and resources to punish, and may result in retribution (Yamagishi 1986). Behavioral economists argue that altruistic punishment allows societal level cooperation to flourish in spite of these costs (Boyd et al. 2003; Fehr and Gächter 2002). Altruistic punishment describes that fact that individuals punish even if the punishment is costly and yields no material gain or advantage (Fehr and Gächter 2002).

Given that the threat of punishment underpins social norms, as robots play a larger role in everyday societal activity, it is important that these robots also support the existing system of social norms. Note that we are not presupposing that the existing system of norms is a ‘good’ set of norms, nor are we suggesting that the robot should only support ‘good’ norms. Rather, by supporting the existing system of norms we realize that this may include both good and bad norms from each individual’s perspective. Nevertheless, although others may disagree, supporting the current system is important in order to maintain consistency. Hence, it must then be possible for a robot to punish violators of social norms. It should also be noted that punishment of social norm violations can be as simple as an unhappy glance, wide-eyed staring, a simple gesture (such as a “*Shh*”), or a verbal statement. In spite of their simplicity, these are all still punishments (of positive type, specifically) in the sense that they are aversions put into a social environment to control behavior.

To summarize our arguments, in certain applications, robots need to have the ability to punish in order to demonstrate that the robot is a credible authority figure. In addition, allowing a robot to enact minor social-norm-enforcing punishments contributes to social norm maintenance and thus supports societal level cooperation. Next, we consider specific applications that may warrant the use of a robot to administer punishments.

6.4 Applications Potentially Warranting Robot Punishment

Many robots are being developed today to assist in educational, policing, search-and-rescue, and military applications. This section looks at each of these domains in turn and examines whether punishment by a robot should be permitted and, if so, what level of punishment may be acceptable.

Educational robots are being developed to take on a variety of different roles (Sharkey 2016). Although most of these educational robots are simply tools for learning, such as Lego Mindstorms, some robots are being developed as a limited capability type of teaching assistant. Japanese researchers, in particular, have investigated the possibility of teacher-robot teams in which the robot interacts with students by sharing stories or engaging in activities (Edwards et al. 2016; Edwards and Cheok 2018). Part of the motivation for this application is to give the teacher a short hiatus from teaching. Presumably, like playing a movie for the children, the teacher would press some button and the robot would begin engaging the students in some type of interactive activity. Admittedly, contemporary robots are not quite up to this task yet.

Leading children during an interactive activity requires some social management skill on the part of the leading adult and recognition by the children that the adult is the authority and, as such, may mete out punishment for those not behaving. While these punishments are minor, such as requests to be silent, or not to hit one-another, or, at worst, ending the child's participation in the activity, they are often necessary for maintaining control of the activity and for maximizing student learning. We argue that, if a robot is meant to lead such tasks, it will need to have the authority to mete out the same punishments. If it cannot mete out those punishments, then the activity may devolve into chaos and the intrinsic learning goals will not be achieved. Although the robot might simply refer potential punishments to the teacher, this would defeat the purpose of having the robot and prevent the teacher from taking a meaningful hiatus.

We can also consider educational training of adults, pilot training for example. Recent research has investigated the use of a robotic arm placed in the co-pilot seat as an autonomous trainer (Moseman 2017). Arguments for such a system include the limited variability among human trainers and greater ability of the system to recognize human mistakes, ultimately with the goal of improving pilot training and safety. Yet, operant conditioning theory suggests that training typically requires both the use of positive rewards and aversions in order to motivate pupil learning (Skinner 1953). This situation may thus require that the instructor, whether human or autonomous robot, use verbal punishments such as slight and medium castigations ("*YOU MUST complete your pre-flight checklist!*") in order generate the best possible learning. Not unlike grumpy driving instructors, a robot that does not point out mistakes and *make the student uncomfortable for making them* risks bigger mistakes in the future.

Many robots are being developed for military applications. As these robots become more capable, it is conceivable that they could be used to protect military personnel and civilians. The Loyal Wingman, for example, is a drone that autonomously serves as a wingman for a piloted vehicle in a variety of different types of missions. It has been suggested that these aircraft will contribute to A2D2 (Anti-Access/Area Denial) missions. These missions require that attacking aircraft deal with a wide variety of 'pop-up' threats. Pop-up threats are unknown at the mission planning stages and must be managed reflexively during the attack. Pop-up threats can also lead to confusion and induce aircraft to accidentally attack civilian targets by mistake. Autonomous drones can be more maneuverable, can be outfitted differently than human-piloted craft, and are able to react more quickly than human-piloted craft. Given these advantages, it is reasonable to argue that a Loyal Wingman should be given authority to identify pop-up threats and either engage them, if in a position to do so, or, alternatively, task human-piloted aircraft with engaging these threats. Giving the drone this authority could prevent coalition and civilian causalities. However, lacking the ability to reprimand human pilots that fail to obey the drone's commands, the drone's authority would likely be ignored.

Policing brings related questions of authority, dispute adjudication, and the consequences of ignoring an authority figure to the forefront. Policing actions require authority. Human police employ a wide variety of techniques to demonstrate their authority including the use of badges and uniforms and authoritative behaviors such as loud speech and the use of commands. A policing robot can similarly use body form and other perceptual cues to indicate its authority. It may also need to employ behavioral measures, such as loud verbal commands, to ensure compliance. Finally, the system may also need to take punitive measures, such as using aversions like horns to disperse a crowd in order to ensure its policing commands are obeyed, or perhaps even physical restraints.

Finally, search and rescue applications offer very compelling arguments for developing robots that assume authority and have the ability to punish. As with other application areas, robots developed for search and rescue vary from intelligent tools to autonomous robots used to guide evacuees during an emergency (Robinette et al. 2016). Certainly, for emergency evacuation applications the robot must assert its authority in order to convince evacuees to move to an exit. It may also need to verbally reprimand individuals for not following directions. Preventing the robot from using these reprimands would likely reduce compliance creating a more dangerous situation for all evacuees. It therefore behooves all evacuees if the robot is afforded the authority to reprimand individuals ignoring evacuation directions.

In summary, there are robot applications that require the robot to act as an authority. To ensure compliance with the robot's directives as an authority figure it is necessary for the robot to be able to punish transgressions and, in doing so, potentially increase safety of evacuees, civilians or the learning of students.

6.5 Ethical Justification and Limitations of Robot Punishment

Is it ethically justified for a robot to punish a human? This section provides a brief investigation of the possible ethical justifications for the use of punishment by a robot. Undoubtedly, we could examine punishment from the perspective of several different ethical frameworks. We leave such work for the future, focusing instead on a consequence-based approach in order to lay the foundation for later, more refined arguments. Additionally, we are not concerned with using this ethical framework to serve as justification for different types of punishments themselves, but rather, to examine justifications for robot-administered punishment *only*.

In the educational scenarios considered, our consequence-based approach can be used to justify the robot's authority to intervene or offer punishments. For example, if two children are arguing in the classroom, our consequence-based approach would allow a robot to intervene and administer punishments against the two children because the benefit of stopping the argument for the rest of the students in the group could outweigh the, presumably minor, physical or mental pain that would be felt by the two arguing students. For the pilot training example, robot-administered punishment for mistakes is justifiable if it improves the pilot's training and results in safer flights. Considering the use of punishment by a Loyal Wingman, the robot would be justified in meting out punishment if it increases safety for the greatest number of people.

It is also important to note that there may be situations of rule-enforcement in society that intrinsically imply the necessity for empathy. As Hartzog et al. write, "Inefficiency and indeterminacy have significant value in automated law enforcement systems and should be preserved. Humans are inefficient, yet more capable of ethical and contextualized decision-making than automated systems. Inefficiency is also an effective safeguard against perfectly enforcing laws that were created with implicit assumptions of leniency and discretion" (Hartzog et al. 2015). While these authors are speaking about contemporary automated law enforcement systems, such as traffic ticketing systems, the main idea is that "perfectly" enforcing a law may restrict freedoms result from human enforcement and decision-making. Certainly, for many of the societal norms, such as queuing schemes or protection against free riders, norm enforcement does not require constant punishments for every single infraction. In these scenarios, often simply the *threat* of punishment is enough to enforce the societal norm. In addition, transgressors of these types of norms are not always punished consistently. For example, a human enforcer may choose to verbally reprimand someone for queuing improperly, or physically move that person to the back of the line and force them to wait for their turn. Empathy towards the transgressor dictates the severity of the punishment administered. In the case of queuing, violation of the norm does not carry much cost to the transgressor; therefore, the "gray area" for what type of punishment should be administered is much smaller. For other applications, such as enforcement of rules in an elementary school classroom, the cost of punishment is high. Deciding the type of punishment and how to administer it to the transgressor requires empathy and contextual understanding, and often reshapes how a rule is enforced. Autonomous robots simply do not currently possess the emotional prowess to properly weigh these considerations.

This is not to suggest that all uses of punishment by a robot are justifiable. In fact, we argue that most uses of punishment should not be allowed. Consider, for example, the use of a robot as legal arbiter or judge. While there might be benefits for allowing the robot to select punishments for those convicted of a crime, robots lack the empathy and human experience to weigh the mitigating and aggravating circumstances that might afford more lenient or more stringent punishment. The eighteenth century philosopher David Hume noted the connection between empathy and morality (Hume 1978). Hume states that moral approval and disapproval are fostered by the psychological mechanisms related to sympathy, which enable one person to relate to and feel the emotions of another person. These psychological mechanisms are mediated by shared experience, psychological distance, and other proximal, though ultimately human, factors (Cikara et al. 2014; Liberman et al. 2007). Artificial machines, however, do not currently have the emotional processing capability to empathize with many human experiences,[1] especially as empathy is built upon shared experiences in which dependence on a caregiver was required. While humanoid robots such as "Octavia" are capable of displaying facial expressions that are commonly linked to certain emotions, this machine does not feel those displayed emotions as a living human would. Therefore, because robots do not currently feel emotions, to the best knowledge of the engineers and scientists creating them, the actions of such robots should be restricted to situations in which emotional empathy is not required.

Some might hold that a sufficiently intelligent machine with understanding of the relevant facts could act as a judge without emotions or empathy. We concede that for some simple criminal violations, knowledge of facts alone, without any consideration of the emotions, may be sufficient. Still, we contend that emotions and the emotional pleas of the victims or the accused play an important and consequential role towards judging a verdict. Emotional displays of remorse, for example, tend to generate sympathy and possibly a lighter sentence. Moreover, most legal systems are designed to delegate legal judgment to officials whose training and experience allows them to weigh the cold facts of the case against, often emotion laden, mitigating factors. The legal system is also underpinned by a variety of difficult to measure principles, such as equal protection under the law. In the absence of a precise way to measure equal protection, humans tend to rely on personal experience and intuition to evaluate fairness with regard to this and other principles. Even if a robot has the knowledge of a human's emotions in a judgement situation, it cannot be expected to empathize with the human as a human judge would, in part because of the difficulty of generating a factual understanding of emotions that would match feeling an emotion. Thus, we contend that the type of rational, deliberate reasoning currently available to robots limits their ability to arbitrate legal disputes, preventing full consideration of all factors that would be considered by a human judge, and thus implies that their use for meting out punishments should be limited to minor punishments.

Hence, we propose that robot punishment should be limited to only those situations in which the use of punishment is critical for the application, and either the application of punishment is sufficiently minor (verbal reprimands) or the overall benefit from the punishment is sufficient to warrant its use. We conjecture that, while it is difficult to quantify the costs and benefits of a particular punishment to a

[1] There is also existing literature on the topic of machine consciousness, emotions, and empathy. We do not consider these more tangential areas of research here.

human community, approaching situations where a robot would require the ability to punish, yet has the potential to cautiously overstep its bounds, could allow for the ethically sound use of robots in authoritative roles.

Our ethical justification for robot punishment is meant to be a starting point in order to cultivate debate on this topic. Clearly, a more sophisticated investigation of this topic will be warranted as robots are placed in authority roles and required to mete out punishment. A more thorough examination of topic may consider robot-mediated punishment from the perspective of different ethical frameworks.

6.6 Conclusions

This article has argued that there are applications for a robot that, in order for the robot to be successful, require giving it the power to punish humans. This argument rests on the connection between the ability to punish and authority. Having authority presumes that one has the power to punish. We have also discussed the importance of limiting the robot's ability to punish. In most cases, these limits should take the form of restricting what types of punishments the robot can enact. It will likely also be necessary to limit who can be punished, for example, ensuring that verbal admonishments are not meted out to individuals incapable of understanding why they are being punished.

Our discussion focuses on the use of medium and mild reprimands for the purpose of social facilitation. The use of such reprimands raises fewer ethical concerns, but nevertheless, offers an interesting entry point towards broader considerations of robot-initiated punishment. In addition to the numerous technical questions related to how and when a robot should initiate a reprimand, this topic may also arouse debate related to the extent that a machine should be used to shape the behavior of an autonomous person. In the extreme, even mild reprimands applied ubiquitously could reshape human behavior into a narrowly tailored norm-driven exemplar, thus hindering human creative expression and individuality.

Additionally, applying behaviors that would result in permanent physical or emotional damage to humans seems unnecessary and unethical. Although we have argued that punishment is important for both learning and maintaining social order, we *do not* advocate for the violent use of robots to promote learning or order. Using robots to violently punish individuals that violate social norms is an important ethical dilemma that we may face in the near future. For example, should a robot tasked with quelling a riot use tear gas to disperse rioters? The costs and benefits of doing so are intimately intertwined with how we value the safety and rights of the different stakeholders involved. Hence, the robot may simply become a tool for ensuring some stakeholder concerns over others, raising important ethical concerns regarding freedom and autonomy.

We want to emphasize that our work is *not* meant to serve as a justification for extreme forms of punishments by a robot. Nevertheless, we recognize that it could be appropriated for this purpose. Specifically, we acknowledge that allowing robots to punish people could lead to nightmarish applications in which robot punishment is used to motivate excessive human labor or serve as a form of tightly-controlled house arrest. The possibility of these types of applications has persuaded us to examine this topic and to develop this paper.

Yet we must also guard against the inclination to assume that all forms of robot-initiated punishment are always wrong. This paper has attempted to advocate and justify a middle ground in contrast to a more reactionary application of moral absolutism to this topic. We believe that additional work in this area will spark community discussion and deliberation towards defining acceptable boundaries for robot-initiated punishment.

Acknowledgements This material is based upon work partially supported by the National Science Foundation under Grant No. IIS-1849068. Any opinions, findings, and conclusions or recommendations expressed in this material are those of the author(s) and do not necessarily reflect the views of the National Science Foundation.

References

Arkin, R. 1998. *Behavior-based robotics*. Cambridge, MA: MIT Press.

———. 2009. *Governing lethal behavior in autonomous robots*. CRC Press.

Axlerod, R., and W. Hamilton. 1981. The evolution of cooperation. *Science* 211 (4489): 1390–1396.

Boyd, R., H. Gintis, S. Bowles, and P.J. Richerson. 2003. The evolution of altruistic punishment. *Proceedings of the National Academy of Sciences*. 100 (6): 3531–3535.

Carey, B. & J. Markoff. 2010. Students, meeting your new teacher, Mr. Robot. In *The New York Times*. https://www.nytimes.com/2010/07/11/science/11robots.html/. Accessed July 2020.

Cikara, M., E. Bruneau, J.J. Van Bavel, and R. Saxe. 2014. Their pain gives us pleasure: How intergroup dynamics shape empathic failures and counter–empathic responses. *Journal of Experimental Social Psychology* 55: 110–115.

De Quervain, D.J., U. Fischbacher, V. Treyer, and M. Schellhammer. 2004. The neural basis of altruistic punishment. *Science* 305 (5688): 1254.

Edwards, B.I., and A.D. Cheok. 2018. Why not robot teachers: Artificial intelligence for addressing teacher shortage. *Applied Artificial Intelligence* 32 (4): 345–360.

Edwards, A., C. Edwards, P.R. Spence, C. Harris, and A. Gambino. 2016. Robots in the classroom: Differences in students' perceptions of credibility and learning between "teacher as robot" and "robot as teacher". *Computers in Human Behavior* 65: 627–634.

Fehr, E., and U. Fischbacher. 2004a. Third-party punishment and social norms. *Evolution and Human Behavior* 25 (2): 63–87.

———. 2004b. Social norms and human cooperation. *Trends in Cognitive Sciences* 8 (4): 185–190.

Fehr, E., and S. Gächter. 2002. Altruistic punishment in humans. *Nature* 415: 137–140.

Fehr, E., U. Fischbacher, and S. Gächter. 2002. Strong reciprocity, human cooperation and the enforcement of social norms. *Human Nature* 13: 1–25.

Hardin, R. 2013. The free rider problem. In *The Stanford encyclopedia of philosophy*. Stanford, CA: The Metaphysics Research Lab, Stanford University.

Hartzog, W., G. Conti, J. Nelson, and L. Shay. 2015. Inefficiently automated law enforcement. *Michigan State Law Review* 2015: 1763–1796.

Henrich, J., and R. Boyd. 2001. Why people punish defectors: Weak conformist transmission can stabilize costly enforcement of norms in cooperative dilemmas. *Journal of Theoretical Biology* 208 (1): 79–89.
Hobbes, T. 1968. *Leviathan*, 1588–1679. Baltimore: Penguin Books.
Hume, D. 1978. *A treatise of human nature*, 1739–1740. Oxford: Oxford University Press.
Kuligowski, E.D. 2009. *The process of human behavior in fires*. Gaithersburg: US Department of Commerce, National Institutes of Standards and Technology.
Liberman, N., Y. Trope, and E. Stephan. 2007. Psychological distance. In *Social psychology: Handbook of basic principles*, 353–381. The Guilford Press.
Metz, R. 2014. Rise of the robot security guards. In *MIT Technology Review*.https://www.technologyreview.com/s/532431/rise-of-the-robot-security-guards/. Accessed 29 March 2020.
Milgram, S. 1974. *Obedience to authority: An experimental view*. New York: Harper & Row.
Moseman, A. 2017. A Robot Copilot Just Flew – and landed – a 737 Sim. *Popular Mechanics*. http://www.popularmechanics.com/flight/news/a26532/darpa-robot-autopilot-737-landing/ . Accessed 29 March 2020.
Ostrom, E., J. Walker, and R. Gardner. 1992. Covenants with and without a sword: Self-governance is possible. *American Political Science Review.* 86: 404–417.
Pareto, V. 1935. *The mind and society*. New York: Harcourt, Brace.
Park, E., K.J. Kim, and A.P. Del Pobil. 2011. The effects of a robot instructor's positive vs. negative feedbacks on at-traction and acceptance towards the robot in classroom. In *International Conference on Social Robotics*, 135–141. Berlin/Heidelberg: Springer.
Price, M.E., L. Cosmides, and J. Tooby. 2002. Punitive sentiment as an anti-free rider psychological device. *Evolution and Human Behavior* 23 (3): 203–231.
Robinette, P., A.R. Wagner, and A.M. Howard. 2016. Investigating human-robot trust in emergency scenarios: Methodological lessons learned. In *Robust intelligence and trust in autonomous systems*, 143–166. Boston: Springer.
Saint-Exupéry. 1943. *The little prince*. New York: Reynal & Hitchcock.
Sethi, R., and E. Somanathan. 1996. The evolution of social norms in common property resource use. *The American Economic Review*: 766–788.
Sharkey, A.J. 2016. Should we welcome robot teachers? *Ethics and Information Technology.* 18 (4): 283–297.
Skinner, B.F. 1953. *Science and human behavior*. New York: Macmillan.
Tajfel, H., and J.C. Turner. 1979. An integrative theory of intergroup conflict. In *The social psychology of intergroup relations*, 33–47. Monterey: Brooks/Cole.
Yamagishi, T. 1986. The provision of a sanctioning system as a public good. *Journal of Personality and Social Psychology* 51 (1): 110.

Chapter 7
Implementing Algorithmic and Computational Design in Philosophical Pedagogy

Rocco Gangle

Abstract This paper argues that using diagrammatically compositional forms of representation may enhance the teaching and learning of computational and algorithmic aspects of philosophy and that this potentially extends to broader ramifications for the field of philosophy as a whole. Several concrete implementations that supplement traditional textual methods with compositional diagrams are introduced, drawing from work sponsored through a Davis Foundation education grant focusing on digital liberal arts and critical thinking in the humanities. The paper concludes by suggesting how such diagrammatic representation may serve as material for higher-level and philosophically sophisticated reflection by exploiting deep connections between algebraic and logical relations on the one hand and compositional diagrams of various types on the other, in particular as mediated by the use of category theoretical tools.

7.1 Introduction

The discipline of philosophy occupies a special role in pursuing and refining the Digital Humanities research program. With respect to the rapidly evolving curricula of Digital Humanities, philosophy has unique potential both for cultivating critical understanding of new digital phenomena and also for incorporating digital tools into new modes of research and teaching.

The present paper aims to address a particular challenge and opportunity for philosophy that arises in the broader context of Digital Humanities research and teaching. Namely, this paper targets the problem of how to cultivate general habits of algorithmic and computational thinking in students across disciplines while working with the specific content and materials of philosophy. The use

R. Gangle (✉)
Endicott College, Beverly, MA, USA
e-mail: rgangle@endicott.edu

B. Lundgren, N. A. Nuñez Hernández (eds.), *Philosophy of Computing*,
Philosophical Studies Series 143, https://doi.org/10.1007/978-3-030-75267-5_7

of compositionally diagrammatic representation and reasoning is proposed as a way explicitly to express and analyze the compositionality of algorithmic and computational structures that may otherwise remain merely implicit in philosophical arguments and objects of inquiry. The employment of such diagrammatic tools in the context of the undergraduate teaching and learning of philosophy may not only provide new modes of insight into algorithmically and computationally expressible features of traditional philosophical arguments and concepts; it can also facilitate critical reflection on the relevance of those ideas to algorithmic and computational processes in contemporary digital culture.[1]

Nonetheless, this paper is not intended primarily as a contribution to the literature on the pedagogical effectiveness of diagrams and does not base its argument on experimental data from the classroom. Instead, the argument is that algorithmically and computationally articulated modes of interpreting, understanding and evaluating traditional philosophical content may be fostered and guided by way of the perspicuous representational character of certain types of pedagogically employed diagrams, namely those that exhibit relevant compositional features. The fact that such compositionality *can* be represented perspicuously is demonstrated by example. However, the concrete pedagogical utility of such representational tools must remain an open question ultimately answerable only by the improvisatory collective experience of classroom application. But to be sure, if algorithmically and computationally expressible aspects of philosophical content can indeed be made directly available for visual inspection by the right types of diagrammatic representation, it stands to reason that students' awareness and understanding of what makes such aspects susceptible to algorithmic and computational expression in the first place is likely to be facilitated by the use of such representational tools, on the basis of the principle *explicit articulation is the beginning of critical analysis.*

The paper is organized as follows. Section 7.2 below outlines some of the special characteristics that position the academic discipline of philosophy in a unique way within the more general concerns of Digital Humanities scholarship. Section 7.3 then outlines some of the pedagogical virtues of using diagrammatic methods generally as tools for teaching and learning philosophy. Section 7.4, the core of the paper, provides three concrete examples of compositionally structured diagrammatic tools that may encourage algorithmic and computational modes of critical analysis of philosophical content. Section 7.5 then offers a reflective consideration of these examples and suggests how the more general use of such methods might potentially impact the field of philosophy in the future. Some concluding remarks are then provided in Sect. 7.6.

[1] Research for this paper was supported by a 2016 interdisciplinary Davis Foundation grant in Digital Liberal Arts awarded to Endicott College "to extend critical thinking across the liberal arts curriculum by systematically integrating digital literacy into the undergraduate classroom". Information about the grant may be found at http://dla.endicott.edu. The author would like to thank the participating students from PHL100, 240 and 245 courses in connection with the grant as well as collaborators on related research: Gianluca Caterina and Fernando Tohmé.

7.2 Situating Philosophy Within the Digital Humanities

The establishment of a self-identified academic field of "Digital Humanities" is less than 20 years old.[2] Much of the attention of this still-emerging scholarly field has been directed to the use of so-called "Big Data" in humanities scholarship, for instance the application of data analysis techniques such as word frequency comparisons based on large data-sets drawn from textual archives and the employment of data visualization software that allows for perspicuous representations of data distribution and connectivity networks. The primary concern in the present paper is with digital methods that are particular to, or at least particularly relevant for, pedagogical practices in philosophy because of its particular modes of inquiry and the unique affinity (within the Humanities) of its distinctive kinds of thinking with algorithmically and computationally elaborated processes.

7.2.1 Algorithms and Algorithmic Elaboration

What is meant here by "algorithmically and computationally elaborated processes" must remain, at least in part, informal. A good guideline for the way these terms are being used in the present paper may be found in Denning and Tedre (2019), which also takes an approach grounded in pedagogical and curricular needs.[3] Most importantly, treating these topics from the point of view of pedagogy is intended to bring to light the importance of examining the ways intellectual motivations arise in the first place for the formalization of processes in terms of rigorous algorithmic structures in the sense, for instance, of expressions in the lambda calculus or programs in a functional programming language like Haskell. These structures give rise to algorithmic processes as such, that is, processes that occur expressly on the basis of an algorithmic elaboration of whatever kind. The motivation for formalization may easily be taken for granted among specialists but must be gradually and self-consciously cultivated in students alongside the awakening of their more general awareness and understanding of how algorithmic and computational processes themselves are essentially organized.

For present purposes, the kind of processes at issue are those that can be elaborated in terms of discrete, determinate and unambiguous operational steps. The

---

[2] For historical surveys of the field of Digital Humanities and some of its more recent developments, see Schreibman et al. (2004), Schreibman et al. (2016), Berry (2012), Dobson (2019), and Berry and Fagerjord (2000).

[3] The reader might also consider work such as Brandom (2008) in which the functionalist approach in artificial intelligence research is treated in terms of "the algorithmic decomposability of discursive (that is, vocabulary-deploying) practices-and-abilities" (p. 27), where such algorithmic decomposition is not further defined in a rigorous mathematical manner but rather understood primitively, as a basis for analyzing philosophically relevant higher-order compositional relations.

explicit articulation of these steps to be followed, including the conditions of their possible branchings and loopings, constitutes an algorithm as such. Following a recipe for baking a cake is, in this sense, basically an algorithmic process. Baking a cake involves a sequence of operations on given ingredients, the products of some of which (mixing the eggs, milk and flour to make batter) become operationally available for later steps (pouring the batter into a baking pan). A written recipe makes that sequence of actions practically available to a minimally proficient actor, namely, a reader with the requisite basic cooking skills. Computational thinking in turn may be roughly defined as algorithms applied to well-formatted data that generate uniquely determined outputs for any given input of the proper type.

In this sense, algorithmic and computational processes are essentially *functional* and thereby *compositional* in structure. Just as mathematical functions can be composed (a function $f : A \longrightarrow B$ may be composed with a function $g : B \longrightarrow C$ to generate a function $g \circ f : A \longrightarrow C$), algorithmically and computationally elaborated processes may be composed so long as their respective types of output and input are properly coordinated. One can, for instance, compose the algorithmic process of baking a cake (P_1) with that of packing an item into a gift box for mailing (P_2) since under proper conditions the output-type of the former (a mailable cake) is compatible with the input-type of the latter (a mailable gift). The result of the composition is a new algorithmically expressible process ($P_2 \circ P_1$): how to bake and pack a gift cake.

In ordinary life, such compositionality of algorithmically elaborated processes can be especially helpful in cases where those who undertake to perform the processes involved are unpracticed or easily confused. Some people need no more instruction to complete a given task than "Bake a cake" or "Replace the ventilator". Others, less proficient, may need some kind of flowchart consisting of well-defined "blocks" of functional accomplishment (for instance in building a chair: "Construct each of the four legs by following steps L_1 to L_7, then turn to section S to attach the legs to the seat"). Someone who knows well how to bake a cake, but has never packed a gift box might perform process $P_2 \circ P_1$ but only need to refer explicitly to the algorithmic elaboration of P_2.

How are such algorithmic elaborations related to philosophical arguments? Whatever other types of answer to this question might be possible, the focus here is strictly pedagogical: How might algorithmic elaborations (and their possible *modes* of elaboration) be used specifically for the purposes of learning about and coming to understand the functioning of philosophical arguments? Although arguments themselves are not algorithms, it is obvious that when students first encounter philosophical argumentation, it can be helpful to break arguments down into discrete, ordered steps that may be worked through and understood one at a time in a way that is similar to following the instructional steps of a recipe. After all, one does not (or should not) simply learn an argument the way one learns the name of Mongolia's capital or the definition of a vector space. To be understood, an argument must be *followed*, and it must be understood before it can be evaluated or critiqued. While an argument should not be simply identified with an empirical process of reasoning, philosophical arguments are no doubt intended to evoke or guide such processes in attentive readers or listeners. The cultivation of various patterns of these processes as reflective habits of critical thinking is one of the tasks of teaching introductory philosophy. In all but the most trivial arguments, the process of reasoning to be evoked is not well-captured by a mere list of premises together with a designated conclusion. Instead, complex arguments typically package multiple subarguments that are not necessarily linearly ordered but which nonetheless according to the overall shape of the argument's hierarchical architecture, that is, the way the various component subarguments are linked together compositionally via dependencies and supports, may be understood to "flow" from certain starting points (premises as inputs) to some conclusion (as output).

The potential pedagogical connection with general algorithmic activity should be clear. Whether or not arguments themselves can or should be elaborated as algorithmic processes, the process of *following an argument* step-by-step in order to understand and evaluate it may itself be articulated in a step-by-step way at various levels of detail and sophistication for a variety of pedagogical purposes. So much, so general. For philosophy in particular, however, unlike in other humanities disciplines, the study of patterns of reasoning as valid or invalid and the questioning of the susceptibility (or not) of arguments to formalization is also an intrinsic component of the discipline itself. These factors are part of what constitute *logic* in the broad sense as part of philosophy. How subarguments and other argument components compose with one another (or fail to do so) in leading to interesting philosophical conclusions is one key subaspect of the self-reflexive logical aspect of philosophy. In this respect, the elaboration of various aspects of philosophical arguments in algorithmically tractable ways that bring compositional features of those arguments to the fore may both help to make these aspects evident and understandable for those unfamiliar with or easily confused by their complexity and also to provide definite frameworks for examining how the associated compositionally structured features of arguments work, that is, how they constitute an essential part of what makes arguments in general valid, invalid, cogent, illuminating, misleading and so on.

7.2.2 Philosophy's Special Role in the Digital Humanities

So with respect to the broader challenge of cultivating students' general understanding of and critical engagement with algorithmic and computational processes across multiple disciplines, philosophy may be seen for several reasons to occupy a special place among the Digital Humanities. First of all, the importance of formalized arguments in philosophy entails a somewhat more natural and practically immediate coordination of computational and algorithmic methods with the indigenous methods of the discipline. Formalized arguments and proofs are closely related to the mathematics of computational processes, and philosophy's frequent recourse to the former links the discipline of philosophy quite naturally to the latter.

In the second place, philosophy is in a unique position to engage the essential intrication of form and content in the Digital Humanities. As alluded to above, digital humanities treats the new modes of digital media, information storage and dissemination, and cultural creation as basic objects for analysis and interpretation. Digital "objects" in the most general sense serve as the theoretical *content* of the discipline.[4] But at the same time, the methodological focus of Digital Humanities is on computational and algorithmic tools that provide new *forms* of analysis, representation and critical interpretation of this content. At its outer edges, this interplay of content and form touches upon the digitally reflexive dimension of research in computer science and cognitive modeling, such as the development of computational models of cognition, artificial intelligence, and machine learning that are at once products and producers of algorithmic thinking. Philosophy, particularly contemporary philosophy of mind, maintains close ties to these latter disciplines and yet is also strongly linked thematically and institutionally to the traditional humanities. It thus stands in a unique position as a potential bridge between cognitive science, computer science and the Digital Humanities with respect to the intrinsic reflexivity of the latter.

Finally, there is a broadly historical and genealogical problematic at work in the relation of philosophy to the Digital Humanities more generally. Core aspects of contemporary algorithmic and computational thinking arose historically out of work by canonical early modern and modern philosophers, logicians and mathematicians, many of whom straddled the fields of philosophy, formal logic and mathematics or saw these fields as deeply intertwined. On the one hand, it is clear that the forerunners of algorithmic and computational thinking such as Descartes, Leibniz and Pascal were squarely situated in the philosophical problematics and methodologies of their time. But it often forgotten that core figures in the modern development of computational theory such as Alan Turing and Alonzo Church were also heavily invested in philosophy. And it is hardly necessary to point out that the founding figures of modern logic, in particular Peirce and Frege, understood logic as a tool with simultaneously philosophical, mathematical and scientific relevance.

The special relationship that holds between computational and algorithmic methods on the one hand and philosophy as a Humanities discipline on the other has a variety of consequences for how philosophy in fact does (and furthermore ought to) stand with respect to the general field of Digital Humanities. In particular, philosophy may play a privileged curricular role both in providing a more detailed and rigorous understanding of how computational and algorithmic processes actually work, as well as supporting the development of the relevant skills of critical thinking that students need to engage the somewhat sudden and overwhelming prevalence of digital and algorithmic aspects of many areas of contemporary culture. The present

[4] It is true that digital humanities scholarship is not *solely* concerned with digital media (for instance, a researcher in the digital humanities might apply data analysis to a traditional text, say, Melville's *Moby Dick*) but even in such cases, the traditional materials are themselves necessarily reformatted into digital objects.

paper focuses on only one specialized aspect of the many ways that contemporary philosophy can engage and contribute to the field of Digital Humanities: the use of diagrammatic representation and reasoning as a way of integrating algorithmic and computational methods into philosophical teaching and learning. Furthermore, the present focus is not on diagrammatic reasoning in general but specifically on the compositional character of certain kinds of diagrammatic notations and the way that such compositionality may, in a variety of cases, naturally represent algorithmic and computational compositionality.

7.3 Pedagogical Virtues of Diagrammatic Methods

The study of diagrammatic representation and reasoning in both formal and informal modes has increased greatly in recent decades, in part because of the computer-supported ease with which high-quality visual diagrams may be designed and incorporated into printed texts. For more extended treatments of diagram-based thinking and the pedagogical virtues of diagram-construction and diagrammatic reasoning in the classroom, see for instance (Anderson et al. 2002; Allwein and Barwise 1996; Gangle 2016; Gerner and Pombo 2010). Here, just four relevant features are singled out for brief consideration. It is important to note that these points can be no more than indicated here, not argued for in any significant detail. Such arguments may be found, for instance, in Diezmann and English (2001), Harrell (2012) and Larkin and Simon (1987). In these works and elsewhere, a variety of experimental evidence has been marshalled in support of the pedagogical utility of diagrams. See also, for example, Li et al. (2019), which in particular emphasizes the advantages of the gestural (especially the indexical and ostensive) dimensions of diagram usage in the classroom. The present paper is not intended to add new data to this kind of empirical psychological or sociological research. The point, rather, is to analyze why certain kinds of diagrams are applicable by virtue of their compositional structure in particular to the teaching of algorithmic and computational aspects of philosophical content.

The points below apply generally to the use of diagrammatic methods in a variety of curricular contexts, but are of particular importance for philosophical teaching and learning. Of these, the fourth (diagrammatic compositionality) is the central focus of the subsequent analysis.

1. The use of diagrams encourages interactive and experimental learning. When working with diagrams, students can treat the component parts of a diagram indexically, for example pointing to an element of the diagram and asking, "What about this part over here?" The spatial organization of parts of a diagram distributes information in a more open and underdetermined manner than the one-dimensional linear organization of a written text. Frequently in this regard, different aspects or interpretations of a given diagram correspond to distinct "paths" through the space that it organizes. For instance, a teacher might explain

biological lineages using an evolutionary "tree of life" diagram by tracing different paths from "trunk" to "branches" corresponding to distinct species. Diagrams are also readily experimented with; new elements can be added to a diagram and the resulting relations with its previous parts immediately seen. They are in this respect natural ways to guide and express intuitions and to make abstract relations concrete and perceptible, as for instance every mathematician (and teacher of mathematics) knows.

2. Diagrammatic materials in the classroom can be especially useful for facilitating group engagement and critical inquiry. Multiple students can work simultaneously with a single diagram (on a shared computer terminal screen, for instance), posing questions, offering hypotheses and making suggestions to one another. Here too the capacity of diagrams to support indexical reference is important. In conversation with one another, students working on a shared diagram can pose questions to one another such as "What about this branch *here*?" or make claims like "Something seems to be missing in the path from *this* node to *this* node." Also, many forms of diagrammatic representation are modular and compositional in structure (see #4 below). This can encourage student collaboration by providing both a principle of division of labor for group projects ("You work on the supports for premise 1, and I'll work on premise 2") and also a natural way to synthesize individual work into larger collective projects.
3. For many classroom settings, especially where students may be coming from highly diverse cultural backgrounds and widely differing educational levels, it is noteworthy that diagrams tend to bracket language and writing issues and instead aid in focusing directly on conceptual and structural relations. This can be especially important in philosophy, where students who may be quite intelligent but have otherwise poor writing skills or are not native speakers of the dominant classroom language may struggle to express themselves clearly in a written essay format, even when the conceptual and rational structure of their argument is what matters most. If such students have the opportunity to express their argument diagrammatically, they can demonstrate the strengths of their thinking without having to be unnecessarily self-conscious about merely grammatical and linguistic issues.
4. The final advantage of diagram-usage in the classroom to be noted here is perhaps the most mathematically and philosophically interesting: *diagrammatic compositionality*. In a very rough and general sense, a type of object or process is said to be compositional if it can be broken down into sub-objects or sub-processes of the same type (or, correlatively, if such objects or processes can be fitted together, that is, composed, in some appropriate way so as to generate new processes or objects of the same type). Not all types of diagrams are compositional, but many are. The three pedagogical examples provided below in Sect. 7.4 are all characterized by some form of compositionality. In general, compositionality allows for modes of formal analysis and synthesis that break complex phenomena down into their more manageable component parts and then build up again from the simple to the complex. Being able to think in terms of the decomposition of tasks or procedures into component sub-tasks or

sub-procedures is obviously an essential skill for algorithmic and computational design. It is also a highly useful skill for understanding philosophical concepts and constructing philosophical arguments. The emphasis on compositionality in diagrammatic reasoning in philosophical pedagogy thus provides a natural curricular bridge between philosophical thinking and algorithmic and computational modeling and design in other fields such as engineering and computer science.

These positive pedagogical aspects of diagrammatic methods in philosophy should be kept in mind while considering the following concrete examples. The reader is encouraged to imagine the diagram types described below actually being applied and used by students in a classroom setting in ways corresponding to what has been said immediately above. The pedagogical virtues of diagrams just canvassed will in this way likely be grasped more vividly.

7.4 Three Compositional Diagrammatic Examples

The following three subsections present three different types of diagrammatic representation that were used as the basis for student assignments in three different philosophy classes. In each case, the compositional features of the type of diagram used provide philosophical insight and also allow for innovative forms of student engagement and group projects in accordance with algorithmic and computational forms of thinking.

7.4.1 Introduction to Philosophy: Argument Diagrams

The first example is taken from an Introduction to Philosophy course. Students who enroll in this course are typically not planning to pursue philosophy any further and take the course in order to meet institutional general education requirements. Such philosophy courses are quite common at many different types of colleges and universities. One challenge for teaching such a course involves introducing the requisite tools for formal reasoning without taking valuable time away from the primary curricular content (whether this is organized historically, thematically, or in some other way). The software used to design these diagrams is a free, online browser-based platform called "MindMup".[5] The user interface is intuitive, and students can easily be building and editing argument diagrams themselves with no more than 20–30 min of initial training.

[5] The software is available at mindmup.com. To use the argument diagram features outlined here, the user must select "Start Argument Visualization" under the "View" tab on the website's main screen.

Fig. 7.1 A simple argument

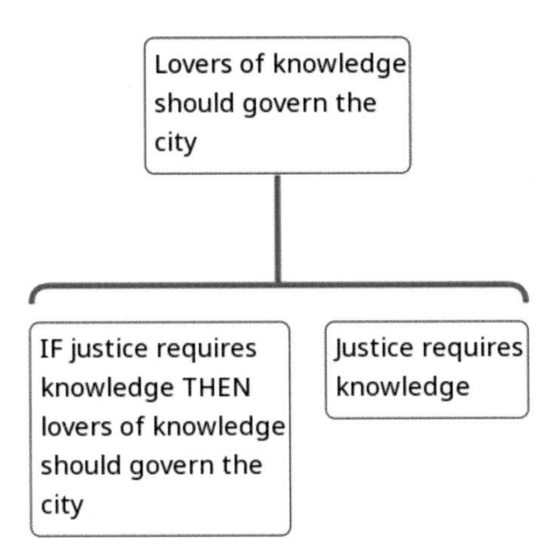

The protocols for designing and interpreting an argument diagram in this semi-formal framework are straightforward. A green bracket attached below a box is understood to mean that the conjunction of the statements in the boxes included in the bracket provide some form of reasonable support for the statement in the box to which the bracket is attached. In the example below taken from an introductory lesson in Plato's *Republic* (Fig. 7.1), the conjunction of the claims "IF justice requires knowledge THEN lovers of knowledge should govern the city" and "Justice requires knowledge" supports the conclusion "Lovers of knowledge should govern the city". The diagram is thus intended to represent the straightforward *modus ponens* deduction of the argument's conclusion as the "output" of a process of reasoning given the "inputs" of the two premises.

Part of the utility of this form of diagrammatic representation of arguments is its compositionality. The compositional principle underlying this type of diagram is essentially that of rational grounding: If A and B ground M, and C and D ground N, and M and N ground P, then A and B and C and D ground P. The diagrammatic notation here lends itself to this kind of compositional elaboration by being successively applied to the premises themselves.

An important lesson for students new to philosophy is that premises in an argument must typically themselves be justified by further reasons or arguments. It is a simple matter with this mode of diagrammatic representation to show that each premise may be attached to a bracket of its own. To build upon the example above, Fig. 7.2 shows an elaboration of the diagram in Fig. 7.1 in which each of the two premises is supported by further reasoning. Note that here the support does not consist in deductive arguments, but in more informal kinds of reasoning. A potentially effective pedagogical task, for example, could be to take the diagram in Fig. 7.2 and to ask students working in groups to reconstruct the supporting reasons for each premise such that they constitute premises for deductively valid arguments in which the relevant claim (a premise with respect to the argument in Fig. 7.1) plays here the role of a necessary conclusion. Students are thus led to see how the

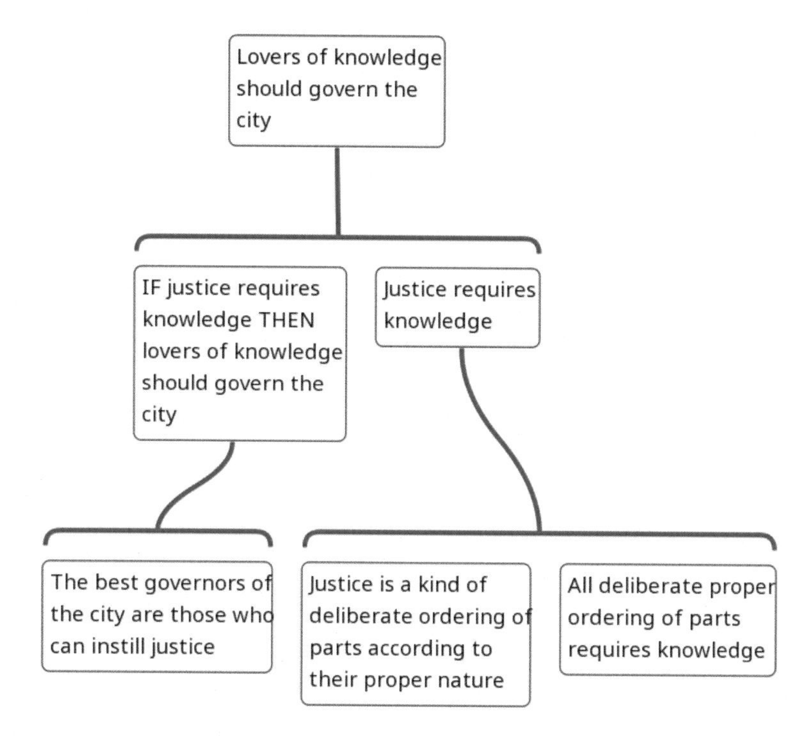

Fig. 7.2 Justifying the premises

output (conclusion) of one argument may serve as the input (premise) of another in a compositionally tractable way. Whether such an exercise turns out indeed to be pedagogically effective is of course an empirical question that must be settled in the classroom. But the fact that algorithmically elaborable features of the diagrams exist that at least potentially *could* be used pedagogically is clearly evident.

Clearly, the diagrams constructed in this manner may be extended to arbitrary depths. The software also allows for the click-and-move "grafting" of one tree into another, demonstrating to students how an independent argument may become part of a larger argumentative structure, again emphasizing the compositional character of philosophical reasoning.[6] Students may be encouraged in this respect to think of philosophical arguments in a broadly algorithmic and computational way. The form of the diagrams themselves encourages students to conceive of arguments as "flowing" upwards from premises to conclusion. Premises (or supporting reasons) appear in this respect as inputs to reasoning processes whose outputs are the relevant conclusions or claims.

[6] Other features of the software not elaborated here include the support for bracket-types that function as "objections" to claims rather than "reasons".

7.4.2 Modern Philosophy: Conceptual Algorithms

The second example is taken from a survey course in Modern Philosophy, part of a three-course elective historical sequence (Ancient and Medieval, Modern, and Contemporary Philosophy), any of which may be taken by students independently of the others. Course materials for the Modern Philosophy course are drawn from canonical writings by major figures in the Western tradition such as Descartes, Locke, Leibniz, Hume and Kant. Students often have difficulty with such texts, finding them stylistically opaque and hard to follow, before even being able to begin to understand and critique the arguments therein. In consideration of such difficulties, student assignments in this course were organized around simple and informal algorithmic flowchart diagrams. Students were asked to design flowcharts representing operational processes of reasoning in the course texts, to share and critique the flowcharts in class and then revise them. More conventional philosophical essays written by students were supplemented with flowchart diagrams expressing some aspect of the essay's own argument as an algorithmic procedure. For the purposes of this course, students used the free online diagramming software available at draw.io. Like the "MindMup" software example above, this too is a free, browser-based platform with an intuitive and easily mastered user interface. Diagrams of the same type were also used in course lectures to help students grasp the material. The example discussed here is taken from a representative introductory lecture on Descartes.

The diagram below (Fig. 7.3) expresses in an algorithmic way Descartes' basic epistemological method in Meditations 1 and 2 (Descartes and Cottingham 2017). Given some idea or claim *X* as input, the algorithmic process branches according to the answer to the question "Can *X* be doubted?" If "yes", then *X* does not count as knowledge. If "no", then it does.

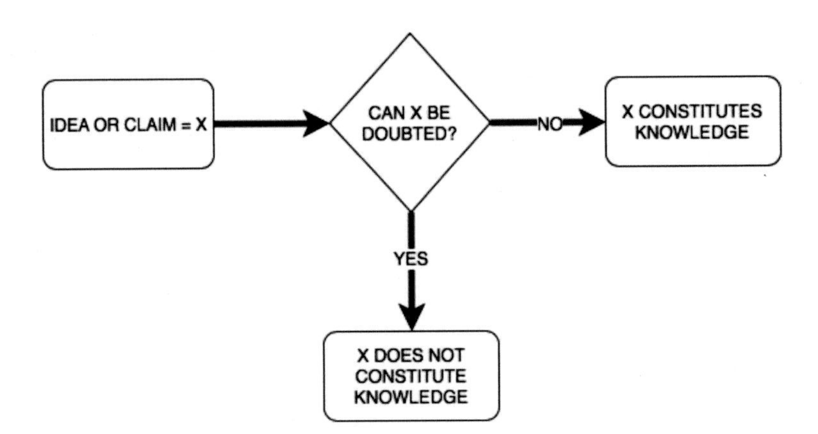

Fig. 7.3 Cartesian epistemology

To be sure, this is by no means a completely rigorously formulated algorithmic procedure. Nor does it fully capture all the nuances of Descartes' own reasoning in the *Meditations*. But in a classroom situation where students are encountering Descartes perhaps for the first time, such an informal algorithmic representation provides notable advantages over the merely verbal exposition and explanation of Descartes' method. In particular, the two spatial paths leading to the separate conclusions "X constitutes knowledge" and "X does not constitute knowledge" offer a clear visual analogy to the operational character and logical binarism of the Cartesian method of doubt.

Such algorithmic representation is intrinsically modular, and its modularity is compositional (very roughly, a linear sequence of algorithmic modules may be composed to form a single module, and a single module may be decomposed into its constituent parts). When expressed informally, operation-nodes and decision-nodes in this kind of algorithm may express processes with high degrees of internal complexity. By taking advantage of this diagrammatic feature of algorithmic design, different levels of analysis of a given argument or philosophical operation can be clearly distinguished for pedagogical purposes. For example, the central decision-node in the Cartesian example above ("Can X be doubted?") may be unpacked, as Descartes himself does at length in Meditation 1, into different ways (that is, sub-procedures or sub-processes) of calling a given claim or idea into doubt. In Fig. 7.4 below, the decision procedure for concluding whether X may or may not be doubted is specified by asking the more determinate question "Is there a conceivable world in which X is not true?". Descartes' own reasoning whereby the mere possibility of a powerful "evil genius" guarantees that *any* idea or claim thereby becomes dubitable (in Meditation 1, prior of course to the revision of the argument in Meditation 2) is indicated by the dotted oval pointing to the path "yes". The visual convention whereby this second diagram appears as "zooming in" to the content of the diamond in the previous diagram expresses in a natural and intuitive way the highly important feature of computational compositionality that becomes an essential and at times technically problematic aspect of more formal algorithmic and computational design processes (Abadi 1993).

Of course, as noted earlier, arguments are not algorithms. Yet to the extent that most non-trivial arguments may be conceived as (at the very least) partially ordered sequences of subarguments, the dependency relations of later portions of such arguments on earlier subconclusions and subresults are directly related to the ordered processes of reasoning students must struggle through as they work to understand those arguments. Indeed, it is remarkable how many canonical arguments in philosophy can be represented in an algorithmic way while remaining faithful to their internal logical structure, often with the pedagogical advantage of making the segmentations and branchings of the various component parts of the argument immediately visually apparent. Given the prevalence of algorithmic and computational processes in contemporary culture, it stands to reason that students may tend to comprehend algorithmic representations more readily than traditional textual formulations. At the very least, being able to offer different modes of representation gives students the opportunity to see traditional course material

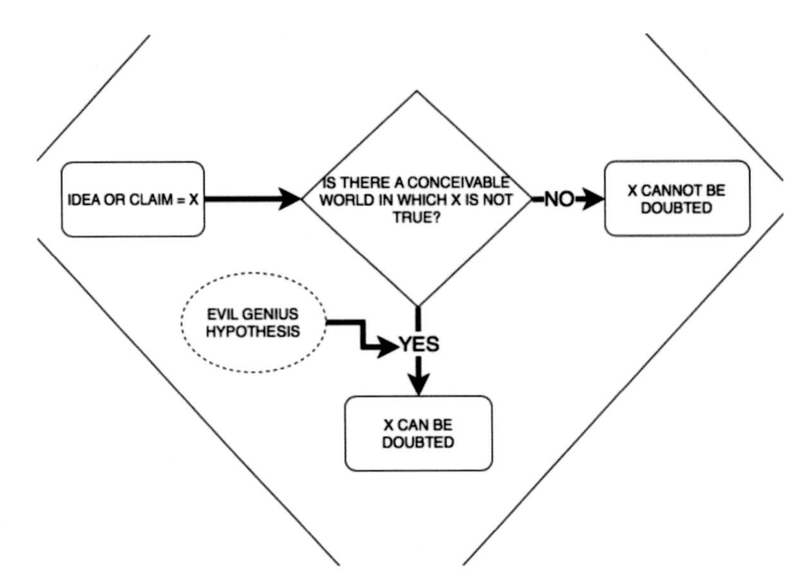

Fig. 7.4 Analyzing conditions of possible doubt

from different and non-standard points of view. In addition, the development of algorithmic thinking in philosophy curricula carries over potentially to other fields. Similar types of algorithmic representations are useful in disciplines as diverse as consumer product design, industrial engineering and molecular biology, as well as, to be sure, computer science. Utilizing these methods in philosophy pedagogy may in this way encourage interdisciplinary thinking among students.

As emphasized in the previous example, it remains an empirical question (subject to experimental verification via test data and so on) whether philosophical content is more efficiently learned or more thoroughly understood by means of the usage of such formal methods of representation in the classroom. All that is argued here is that these modes of representation at least initially serve to make algorithmic aspects of the philosophical content available visually for possible further discussion and critical analysis.

7.4.3 Contemporary Philosophy: Wiring Diagrams

The third and final example is drawn from a course in Contemporary Philosophy, the third in the historical sequence mentioned above. This course has a relatively open curriculum, and different versions of the course have included works by Husserl, Heidegger, Wittgenstein, Quine and many others. A recent full-length work of social and political philosophy is typically included to give students a feeling for the potential relevance of philosophy to contemporary collective life. The example below is based on the reading of Michael Hardt and Antonio Negri's book *Assembly* (Hardt and Negri 2017), a post-Marxist study of current social movements and political organization challenges in a global context.

Here, the diagrammatic technique used for pedagogical purposes is that of *wiring diagrams*. The formal theory of such diagrams is both quite technical and also of significant mathematical interest (see, for example, Spivak 2014, section 7.4 and Yau 2018). However, the approach used here is very informal and easily grasped at an intuitive level. The core idea of a wiring diagram representation is that "boxes" represent processes with (possibly multiple) inputs and outputs. The types of these inputs and outputs are represented by lines or "wires" attached to the boxes. Such wires may be thought of as "ports" or "channels" dedicated to whatever particular type of input (or output) is contributed (or generated) by whatever process. In the representations below, input wires are attached to the *tops* of their corresponding boxes, and output wires are attached to the *bottoms* of the boxes. The "flow" of the process as represented by the diagram is thus from top to bottom.

As with the preceding examples, here too the primary formal interest of using such diagrams to represent processes in a philosophical context is that they may be nested compositionally in an intuitive and perspicuous way. Just as a process may be composed of (or merely include) multiple sub-processes, a wiring diagram box may composed of (or include) multiple boxes, which are represented in this way by drawing them as smaller boxes inside a larger box. The "wirings" of inputs and outputs ensure the coherence of the processes' compositionality.[7] Consider the following example (Fig. 7.5):

The diagram pictures two boxes (labeled "Digital Media" and "Non-Digital Media") inside a larger box (labeled "Learning/Cognition"). The labeled inputs and outputs of the Learning-Cognition box suggest that "research questions" and "education/habits" combine in this process to generate "knowledge/comprehension" and "communication skills" as outcomes. The two interior boxes decompose and recompose this process as mediated through the sub-processes of digital and non-digital media respectively. While the diagram itself expresses very little content concerning these topics, it serves as a visual basis for analyzing their relationships in a compositional, process-oriented way in more standard student analytical and argumentative essays. Multiple students can analyze and interpret one and the same diagram in quite different ways, but they remain constrained by various features of the diagrams themselves (for instance, it is clear in the diagram above that the digital and non-digital media take the same inputs) and can critique one another's analyses on the basis of such features ("But education is also an input to the use of digital media; your argument does not take that into account").

Students focusing on different themes can construct and work with different wiring diagrams corresponding to their chosen areas of research. The following diagram (Fig. 7.6) was connected to the research project of a student interested in themes of race and racism in Hardt and Negri's text and more particularly the

[7] The examples used here are modified versions of diagrams collaboratively produced in the course.

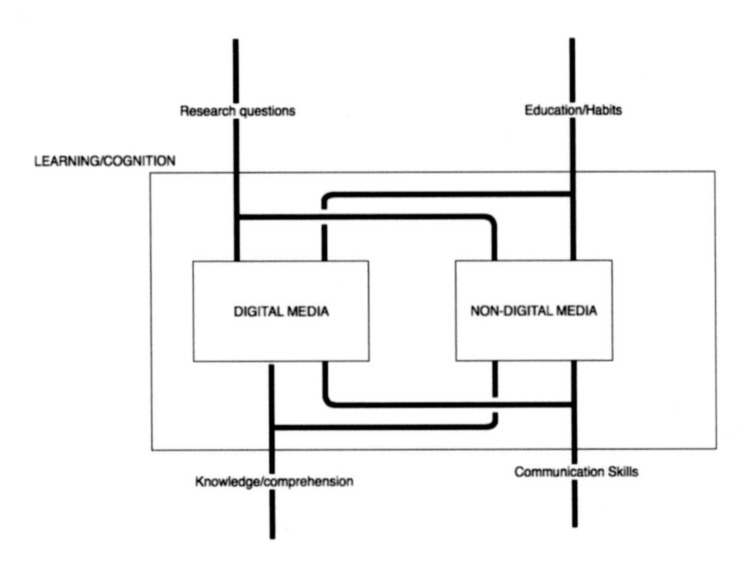

Fig. 7.5 Learning/cognition: inputs and outputs

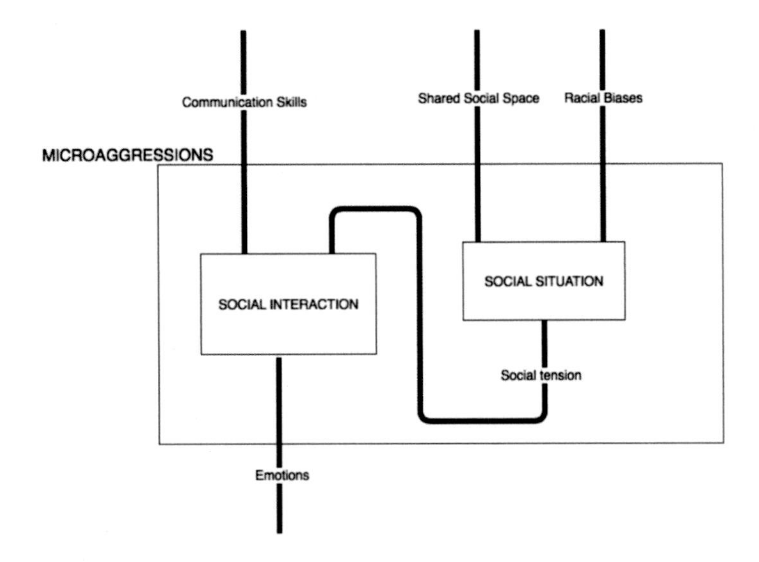

Fig. 7.6 Microaggressions: inputs and outputs

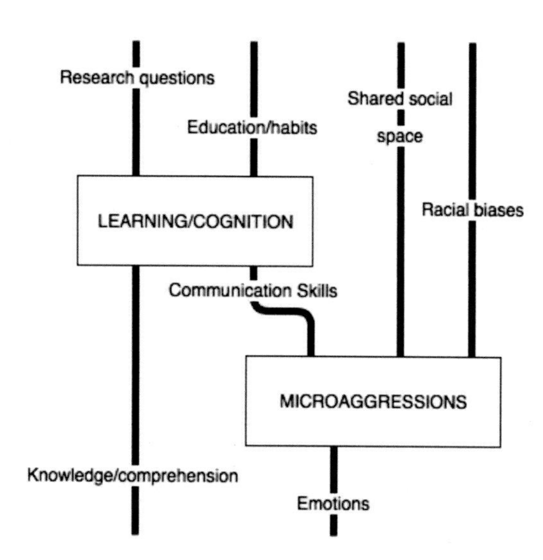

Fig. 7.7 Wiring diagrams together

notion of racially-based “microaggressions” in social interactions. The intuitive interpretation of the diagram should be more or less clear on the basis of the analysis of the diagram given above. The one special feature worth noting is the wiring together of the “Social Situation” and “Social Interaction” boxes such that the output “social tension” of “Social Situation” serves as the input of “Social Interaction”. While this is still a quite elementary structure, it should indicate somewhat how the internal linkages of the various subprocesses can be organized in potentially quite complex ways.

Finally, in accordance with the pedagogical emphasis on compositionality, the diagram in Fig. 7.7 below shows how the two diagrams above (each of which is itself the representation of a composition of processes) may be composed via the “soldering” of the output “communication skills” of the “Learning/Cognition” process with the identically typed input of the “Microaggressions” process. In this way, the two student diagrams above have been wired together to form a larger, compositional diagram. Not only does such a diagram provide a concrete linking together of separate student research projects, it at the same time suggests further questions and problems for subsequent inquiry. In this way, independent student work can be brought together explicitly into a broader interconnected context. The different levels of abstraction and detail at which analyses can be made and hypotheses conjectured become visually and structurally explicit.

In a variety of pedagogical contexts in multiple disciplines, it can be useful to have students analyze processes or operations in terms of their inputs and outputs. To cultivate a better understanding of such processes and operations, it can also be helpful pedagogically to encourage students to see how such processes may be linked with one another to form larger units as well as decomposed into their component sub-processes. This general point about analyzing processes and operations is true of philosophy as well, of course, but it may be particularly important for those philosophical contexts that are especially concerned with abstracting (and critically understanding) general forms of interactive relations that are applicable *across* disciplines. Such a tendential orientation toward formal generality is closely linked to philosophy's logical emphasis. Once the digital tools for representing wiring diagrams have been learned in one context, it is straightforward for students to apply those skills with respect to different contexts. Not only do students thereby learn tools for diagrammatic representation that are applicable across varying contexts, they also by the same token are led to recognize the common structural features that might be at work in otherwise dissimilar fields. For example, students might notice the common feature of feedback loops in certain social processes as well as chemical reactions. The recognition of this common feature could then become the basis for an examination of feedback phenomena *in general*. In this way, diagrammatic representation may help to foster passages from concrete to abstract concepts in students for whom such forms of thinking cannot necessarily be taken for granted. It cannot be known in advance whether the use of these techniques will in fact advance student learning outcomes, but at the very least such diagrammatic methods make certain compositionally tractable aspects of the course material visually explicit in a way that can then become thematized directly, that is, ostensively and indexically ("Is *this* output really the same type as *that* input?"). Having been brought to light in this way and thereby made available for analysis and discussion, these aspects of course content may be treated subsequently in their own right as subjects for more detailed study.

7.5 Analysis and Prognosis

What broader lessons concerning the use of compositional diagrammatic representations of algorithmic and computational processes in philosophical pedagogy might be learned from this sequence of examples? What future promise might the incorporation of such methods into the philosophy curriculum hold? In addition to the more general pedagogical virtues of diagrams noted in Sect. 7.3 above, there are at least two ways that the use of such diagrammatic representation systems may be especially useful for the discipline of philosophy. It was remarked above that the distinctive role of formalized arguments in philosophy places philosophy in a somewhat more direct connection with algorithmic and computational methods than other humanities disciplines. Among other things, this concern with formalization entails certain pedagogical challenges that are somewhat unique to philosophy among humanities disciplines. Outside of upper-level courses in mathematical logic, highly technical methods can usually be ignored at the undergraduate level. Yet much contemporary scholarship in philosophy requires at least some familiarity with basic logical and mathematical formalisms. Diagrammatic techniques provide a helpful mediation from informal to formal approaches to arguments and conceptual structures. Secondly, and for similar reasons, the use of diagrammatic methods in philosophy opens avenues for bridging the gap between purely pedagogical concerns in philosophy and the sophisticated topics and methods of contemporary research. These two ideas are elaborated in the following two subsections.

7.5.1 From Visualization to Formalization

The three examples given above are non-technical and in no way intended to provide rigorous formalizations of their respective subject matters. In each case, the means of diagrammatic representation itself may be taught and readily understood by untrained (and at times, alas, somewhat recalcitrant) undergraduate students in somewhat less than an hour. This process of "training" students how to use the diagrammatic medium may itself be articulated, at least to some degree, algorithmically. In most cases, this would be done informally. Nevertheless, in each case the intended meaning and use of the given informal diagrammatic mode of representation lends itself to systematic formalization in a more or less natural way, if a professor or student so desires for some particular purpose.

In this way, diagrammatic representation may be understood in its own way to anticipate formalization from the outset. Rather than asking students to generate rigorously formalized arguments directly, requiring diagrammatic representations of conceptual and argumentative structures involves a kind of "soft" formalization of the relevant material. By their very nature, diagrams are quasi-formal. The visual conventions they use necessarily impose certain constraints on the material they represent. For instance, the very form of the basic "bracket" notation in the first example above constrains the premises-to-conclusion relation to be, in general, many to one. Similar considerations with respect to the diagrammatic conventions in the other examples demonstrate how the expressive means of such diagrams already imposes certain formal or structural features that remain relatively invariant across different concrete instances. The recognition by students of such features may initially be merely visual ("All of the argument diagrams look like pyramids!"), but it then becomes a clearly-defined pedagogical problem to convert that visual recognition into more carefully formalized notations. If it is objected that the diagram-forms discussed above are not themselves formally rigorous, it should be responded that this is precisely what makes them pedagogically advantageous. The goal of philosophical pedagogy is precisely to *lead* students to grasp the inner necessity and coherence of philosophical questioning and argumentation. The use of formalization should be properly motivated before it is implemented. Informal, or better quasi-formal, diagrammatic representation is well-poised to suggest to students, through its experimental usage and critical analysis, why more formal methods might be desirable and warranted.

In addition, some of the most interesting contemporary mathematics is itself readily formalized diagrammatically. For instance, category theory uses diagrams consisting of labeled nodes (objects) and arrows between them (morphisms) as its basic representational means (see, for example, Spivak 2014). Indeed, category theory may be understood as a mathematically precise way to formalize the general notion of compositional mappings of structure that characterize mathematical and logical domains of all types (functions of sets, linear maps of vector spaces, translations of logical theories, and so on). In this respect, the compositional character of various diagram-forms that has been the particular emphasis of the present analysis readily lends itself to more precise formulation within the mathematics of category theory.

In fact, some of the more interesting and technical developments in category theory apply methods of formal mathematical representation to structures that are themselves naturally conceived as compositional diagrams. For example, wiring diagrams and string diagrams are readily represented by such mathematical gadgets as colored operads and symmetric monoidal categories (see Yau 2018 and Coecke and Kissinger 2017). While the technical details involved exceed the purposes of the present paper, it is hoped that further research will explore the explicit formalization within such enriched categories of the philosophical diagram-structures (and other similar conventional methods) discussed above.

7.5.2 Integrating Philosophical Pedagogy and Research

A second and closely linked aspect of using diagrammatic tools as a way to encourage algorithmic and computational thinking in philosophy curricula is its potential for bringing pedagogy and research into closer alignment. One of the unfortunate truths concerning academic philosophy in the current institutional and disciplinary framework is that the work of undergraduate philosophy education and the best cutting-edge research in philosophy are often almost entirely disconnected. This is especially true for some of the most exciting developments in contemporary logic and philosophy of science. Much of that work is so technical that it would be practically impossible to introduce it into the undergraduate curriculum. Yet its contemporary relevance is undeniable.

One way that research and teaching could be integrated is through the use of student diagrams as data for philosophical analysis and reflection. The program of "experimental philosophy" has been an intriguing and controversial development in recent philosophy (Alexander 2012). Whatever the merits of this approach for problems in ethics or philosophy of mind, it is clear that philosophical reflection on philosophy pedagogy can be aided by empirical data about how philosophy is taught and what methods are especially effective for bringing philosophical understanding and insight to students. If common formatting can be agreed upon, the technology exists to analyze large numbers of student diagrams according to a variety of formal criteria. Finding common structures (particularly common mistakes, such as the prevalence of "affirming the consequent" fallacies or similar errors) could provide teachers of philosophy with useful guides for what to focus on in their lessons. Such (big) data of student-level philosophical thinking might also serve as the basis for interesting philosophical reflection, just as the empirical data for experimental philosophy has led to new debates in ethics.This would be a sort of intra-philosophical experimental philosophy.

The compositional character of many diagrammatic forms of representation might in this way also have ramifications for the closer integration of teaching and constructive research. Expressing philosophical content in a compositionally diagrammatic way allows for the digital superposition of coarse-grained (elementary, schematic) and fine-grained (technical) analyses. It is rather remarkable that under the conditions of a massively technologized and digitized cultural field, the dominant form of philosophical knowledge production consists still of ordinary language-based essays and books written more or less in the same form as that of Descartes or Kant. With the technological tools already at hand (and even more so with the possibilities that might be opened up with more focused attention and initiative), it is conceivable that new ways of *doing* philosophy collectively might emerge from the development of and institutional support for structurally compositional expressive and representational methods and media.

With these possibilities in view, it is not difficult to imagine modular and highly-distributed research programs in philosophy that would facilitate collaborative work by scholars at different levels of expertise. Highly technical work in categorical logic, for example, might be integrated with informal student representations of canonical philosophical arguments. Researchers in mathematics and (especially) computer science have developed robust disciplinary infrastructures for engaging complex multi-scale research programs and also for finding ways to integrate those research programs with the environing social and economic fields. Philosophy ought to be able to pursue a similar agenda in its own disciplinary-specific manner.

7.6 Conclusion

The present paper began with a consideration of the place of philosophy within the general field of Digital Humanities and with respect to the problem of interdisciplinary integration of algorithmic and computational methods in contemporary curricula and pedagogy. It was argued that in this context the use of certain modes of diagrammatic representation and reasoning in the classroom may offer certain advantages, in particular by making compositionality, an essential feature of algorithmic and computational design, explicit as an aspect of the presentation of philosophical materials. Three concrete examples drawn from three different philosophy courses were then provided, each exhibiting a different mode of diagrammatic representation that organizes philosophical content in compositionally tractable way. Finally, the virtues of such diagrammatic representation specific to the discipline of philosophy were discussed: the possibility of a more or less "smooth" passage from informal and intuitive to formal and rigorous representation, and the opening of new possibilities for integrating teaching and research in academic philosophy.

It is hoped that these remarks and suggestions, which must remain hardly more than indicative in a paper of this length, provide incentive for increased scholarly attention to the use of diagrammatic representation and reasoning as a powerful pedagogical tool for philosophy and also to the intrinsic philosophical interest and possibilities of compositionality as an inherent feature of many of the various domains studied by philosophy as well as of the scholarly and institutional infrastructure that keeps this ancient discipline alive as a core component of the contemporary humanities and sciences.

References

Abadi, M., and L. Lamport (1993). Composing Specifications. *ACM Transactions on Programming Languages and Systems* 15(1): 73–132.

Alexander, J. (2012). *Experimental Philosophy: An Introduction*. Malden: Polity Press.

Allwein, G., and J. Barwise (eds.) (1996). *Logical Reasoning with Diagrams*. New York: Oxford University Press.

Anderson, M., B. Meyer, and P. Olivier (eds.) (2002). *Diagrammatic Representation and Reasoning*, London: Springer.

Berry, D.M. (ed.) (2012). *Understanding Digital Humanities*. New York: Palgrave Macmillan.

Berry, D.M., and A. Fagerjord (2000). *Digtal Humanities: Knowledge and Critique in a Digital Age*. Malden: Polity Press.

Brandom, R. (2008). *Between Saying and Doing: Towards an Analytic Pragmatism*. Oxford: Oxford University Press.

Coecke, B., and A. Kissinger (2017). *Picturing Quantum Processes: A First Course in Quantum Theory and Diagrammatic Reasoning*. Cambridge: Cambridge University Press.

Denning, P.J., and M. Tedre (2019). *Computational Thinking*. Cambridge: MIT Press.

Descartes, R. (2017). *Meditations on First Philosophy, with Selections from the Objections and Replies*. Translated and edited by J. Cottingham, 2nd edn. Cambridge: Cambridge University Press.

Diezmann, C., and L. English (2001). Promoting the use of Diagrams as Tools for Thinking, in *The Roles Of Representation in School Mathematics: 2001 Yearbook, National Council of Teachers of Mathematics, Virginia*, 77–89.

Dobson, J.E. (2019). *Critical Digital Humanities: The Search for a Methodology*. Urbana: University of Illinois Press.

Gangle, R. (2016). *Diagrammatic Immanence: Category Theory and Philosophy*. Edinburgh: Edinburgh University Press (2016).

Gerner, A., and O. Pombo (2010). *Studies in Diagrammatology and Diagram Praxis*. Studies in Logic: Logic and Cognitive Systems, vol. 24. Milton Keynes: College Publications.

Harrell, M. (2012). Assessing the Efficacy of Argument Diagramming to Teach Critical Thinking Skills in Introduction to Philosophy. *Inquiry* 27(2): 31–38.

Hardt, M., and A. Negri (2017). *Assembly*. Oxford: Oxford University Press.

Larkin, J.H., and H.A. Simon (1987). Why a Diagram is (Sometimes) Worth Ten Thousand Words. *Cognitive Science* 11(1): 65–100.

Li, W., F. Wang, R.E. Mayer, and H. Liu (2019). Getting the Point: Which Kinds of Gestures by Pedagogical Agents Improve Multimedia Learning? *Journal of Educational Psychology* 111(8): 1382–1395.

Schreibman, S., R. Siemens, and J. Unsworth (eds.) (2004). *A Companion to Digital Humanities*. Malden: Blackwell Press.

Schreibman, S., R. Siemens, and J. Unsworth (eds.) (2016). *A New Companion to Digital Humanities*. Malden: Blackwell Press.

Spivak, D. (2014). *Category Theory for the Sciences*. Cambridge: MIT Press.

Yau, D. (2018). *Operads of Wiring Diagrams*. New York: Springer.

Chapter 8
Our Technology Fetish

John Weckert

Abstract We have a technology fetish. That is, we have an excessive and irrational devotion or commitment to new technologies. This fetish matters. New technologies are too often developed and accepted with little critical thought. After a discussion of the fetish it will be argued that it is not inevitable and that something should be done about it because of potential harms. Finally, suggestions are made for alleviating the potential for those harms. Monitoring and surveillance technologies and artificial intelligence are used to illustrate the argument.

8.1 Introduction

The aim of this paper is to present an alternative to the predominant view that technological developments constitute progress, i.e., that new technologies are developed and accepted because they make life better. The alternative presented here is that new technologies are developed and accepted because we have a technology fetish. I cannot prove that this way of looking at things is the right way but I hope to show that it is a plausible way and that it has the advantage that it comes with less dangers because it encourages us to think about technology more carefully. In order to outline and defend this position, some issues will be raised that are too large to be adequately discussed in a paper of this length but that nevertheless need to be raised in order to make the argument. In these cases references will be given so that further work can be done. In this sense the paper is an overview of some important questions in the ethics and philosophy of technology but at the same time, it argues for a particular position.

We have a technology fetish, it matters, and we can and should do something about it, or so I will argue. We are, of course, technological creatures. We don't run, swim, fly, climb trees, dig burrows and so on very well and our senses are only

J. Weckert (✉)
Charles Sturt University, Bathurst, Australia
e-mail: JWeckert@csu.edu.au

B. Lundgren, N. A. Nuñez Hernández (eds.), *Philosophy of Computing*,
Philosophical Studies Series 143, https://doi.org/10.1007/978-3-030-75267-5_8

middling. We need technology not only to survive but also to live well. The purpose of technology as José Ortega y Gassett said, is not mere survival but to enable us to lead good lives (Ortega y Gasset 1961). (See also Alf Hornborg discussions of a technology fetish (Hornborg 1992, 2014) in a slightly different although related context.)

8.1.1 What Is Technology?

A large literature discussing the nature of technology exists and the word "technology" is used in a variety of ways. Dusek outlines three different approaches: technology as hardware (or tools), technology as rules and technology as system (Dusek 2006, 31–36). Briggle et al. mention four accounts: technology as object, as knowledge, as activity and as intention (Briggle et al. 2005). Sadjad Soltanzadeh has a slightly different account that sees technologies as being characterized by their problem-solving properties (Soltanzadeh 2016), Frederick Ferré says that it is the "practical implementation of intelligence" (Ferré 1995, 26) while for Richard Combes "technology is the deliberate practice of rearranging the world's furniture in order to maintain a decent lifestyle" (Combes 2006, 6) and technology is "the intentional exploitation of the environment for the purpose of providing needs and *perceived* wants." (Combes 2006, 11).

It is beyond the scope of this paper to discuss all of these so here "technology" will be used primarily as artefact, that is, something material manufactured by humans, for example cars, computers and spears. A related view is technology as a system. According to Ortega, the totality of human acts that manufacture tools constitute technology (94–95). In this paper "technology" is used in both senses but context should make the meaning clear.

8.1.2 What Is the Fetish?

There are various accounts of "fetish" but the important sense for the purposes here is "an excessive and irrational devotion or commitment to a particular thing". Another sense is an "object worshiped for its supposed magical powers" (Oxford 2020). Few people would admit to having a technology fetish in this sense but in practice it is not obviously very different. Karl Marx talked of a "commodity fetish" which is similar to the second sense:

> There is a definite social relation between men, that assumes, in their eyes, the fantastic form of a relation between things. In order, therefore, to find an analogy, we must have recourse to the mist-enveloped regions of the religious world. In that world the productions of the human brain appear as independent beings endowed with life, and entering into relation both with one another and the human race. So it is in the world of commodities with the products of men's hands. This I call the Fetishism which attaches itself to the products of

> labour, so soon as they are produced as commodities, and which is therefore inseparable from the production of commodities. (Marx 1867)

The claim is not that the technology fetish is the same as Marx' commodity fetish but it highlights an important feature, that is, the social aspect. Where there is a magical or religious component to a fetish, it is usually a social practice. The community worships the object because of this component. The technology fetish does have this social element and therefore arguably does have a religious aspect. Modern consumerist society as a whole invests technology with this largely unquestioned status. In summary then, two aspects of "fetish" are relevant here: there is an "an excessive and irrational devotion" to technology and it has a social element. This latter point has important consequences for how the ethics of technology should be approached.

The fetish manifests itself as an unquestioning acceptance that new technological developments are good and worth having. It also has a more personal and individual aspect. New products are embraced whether or not they have any significant benefits. Newness and novelty themselves seem to be enough. This is like working and living within a paradigm where all problems must be solved technologically and every new technological development and product is good. A technological fix is always the first option. This idea of something like a technological paradigm is not new. David Hertz talks of a "misplaced *faith* [my emphasis] in technology's ability to solve *all* problems" (Hertz 1970, 98), for Bill Joy we have a "bias toward instant familiarity and unquestioning acceptance." (Joy 2000, 4), while according to Langdon Winner there is a "technological somnambulism" (1986, 10) and he sees technologies as "forms of life" (1986, 14). Again he says:

> In the twentieth century it is usually taken for granted that the only reliable sources for improving the human condition stem from new machines, techniques and chemicals. Even the recurring environmental and social ills have rarely dented this faith. (Winner 1986, 5)

Discussing technology in terms of faith, somnambulism, forms of life and worldview (Scott 2011, see also Leo Marx 1983; Drengson 1984), suggests that this way of looking at the fetish is related to something like a Kuhnian paradigm. According to Thomas Kuhn, apart from times of scientific revolutions, scientists operate within a paradigm, working on solving problems that arise within that paradigm without questioning basic assumptions (Kuhn 1970). In this sense, a paradigm is a kind of worldview; a structure within which we work without normally questioning the structure itself. This model fits well the current view of technological development. There is very little questioning of this development; it is assumed that it is good and here to stay. Using the analogy of the Ptolemaic view of the universe illustrates the current situation. In the Ptolemaic system the earth was the centre of the universe and all the stars and planets revolved around it. This was the generally accepted and unquestioned view and is analogous to the way that technological development is viewed now. It did not work so well and many adjustments needed to be made for it to fit in with observations, commonly by adding epicycles, often nested ones, to the orbits of the planets. This way of solving problems is similar to the efforts made to regulate the use of technologies in order to mitigate various harmful effects and the efforts to find new technological ways of solving problems created by previous technologies. It is also similar the efforts made to explain away the problems. The Ptolemaic system was challenged by Copernicus, Galileo and others but their challenges were initially ridiculed in much the same way that challenging technological development is now (see Brecht 2015).

What the notion of a fetish adds to the paradigm account is that it highlights the degree of devotion or commitment to technological development and new technological products and the strength of the belief in technological progress. This is similar to, but stronger than, Kuhn' account of commitment to a scientific paradigm. This is not merely an intellectual commitment but something that is almost loved and worshiped (see Kuhn 1970, on commitment to paradigms, eg. pages 25 and 100).

8.2 Evidence of a Fetish

As stated at the beginning of this paper, the perspective of the fetish is an alternative way of looking at our approach to technology. In many cases of new technologies the fetish does not matter because the technology is clearly beneficial with few or no particularly harmful consequences. The case to be considered here, monitoring and surveillance technology, is one where the consequences can be dire but despite expressed concerns and warnings the relevant research, development and use of the technologies are embraced enthusiastically anyway. The fetish gives a plausible explanation for this.

8.2.1 Monitoring and Surveillance Technology

The idea of keeping populations under surveillance is nothing new. Totalitarian governments were doing it long before modern computing and communication technology was around. George Orwell's novel *1984* was published in 1949 and in it Big Brother managed to keep an eye on everyone with technology that was primitive by the standards of today. Ben Elton's *Blind Faith* published in 2007, has technologies much like current ones so gives a more realistic picture of what is possible now. These novels show that the awareness of the dangers of intrusive surveillance is not new, so there should be no surprise at the use of new technologies for this purpose.

Most of the technologies required for monitoring and surveillance are familiar (O'Brien 2020). The Internet, smartphones, surveillance cameras, drones, credit cards and so on are commonplace. When buying groceries and paying with a card, information about us is, or can be, logged and matched with other information about us. Medical records, banking details, smartphone communications, web searches, social media activity, online shopping, music downloads and so on are all stored electronically and so are available to relevant authorities and businesses or unscrupulous hackers. Much of this data is being actively collected and mined purely for business in what is called by Shoshana Zuboff surveillance capitalism (Zuboff 2019). The justification given for all of this data collection by corporations is that it improves our shopping or other business experiences or that it makes life better in some other way.

Governments and government agencies, including those in law enforcement, also collect our data. The justification for this is generally that it helps in keeping us safe by assisting in fighting terrorism and crime, and making government services more efficient and useful to the general public. During the current COVID-19 pandemic, smartphones are being used to check on close personal contacts (Australian Government 2020) and drones are being developed for sensing those in a crowd might have high temperatures and heart rates (Daly 2020). The justification for this is that authorities are better able to see where the virus is most prevalent, who is in contact with whom, how people are moving around, and so on, and therefore have more information on how to control movements and thereby try to restrict the virus' spread.

Perhaps the most ambitious data collection plan reported to date is the Chinese Social Credit System. In 2014, China's State Council released a road map for a comprehensive "social credit system" which would assign citizens, firms, and organizations a credit score based on multiple economic and social categories. Pete Hunt describes it like this:

> The data-driven system would help meet market objectives by effectively extending financing options to the country's large unbanked population, and ideological objectives by addressing rampant corruption, profiteering, and mistrust in the country—or as early documents promised, to "allow the trustworthy to roam everywhere under heaven while making it hard for the discredited to take a single step." (Hunt 2018)

This rather dystopian picture of the system that has been widely reported in the Western press, has been challenged as being inaccurate. It is primarily aimed at regulating corporate behaviour in China and is not as all pervasive or widespread as often reported, although it is not without its dangers (Trivium China 2019; Kobie 2019; Zuboff 2019, 308ff). Even if it is currently relatively benign, it does point in directions that state monitoring and surveillance could go.

As can be seen from the discussion above, we are now living in a world where widespread electronic monitoring and surveillance is commonplace, by governments, in workplaces, by businesses and in public places. This has become possible because new technologies have been developed and accepted without very much examination, at least by the general public and by industry.

Concerns of course have been raised about privacy and data protection for many years, despite the statement of the chief executive officer of Sun Microsystems, Scott McNealy in 1999 "You have zero privacy anyway, get over it." (Sprenger 1999). While these expressed worries have had some effect on regulations regarding the use of the technologies, the development and acceptance of the technologies has not diminished. Deborah Johnson discussed privacy problems in computer use in 1985 (Johnson 1985) and Tom Forester and Perry Morrison in 1990 (Forester and Morrison 1990). Since then there have been many others including Jeroen van den Hoven (1999), Helen Nissenbaum (2001), Roger Clarke (2019) and Shoshana Zuboff (2019). This is only a small selection of the vast literature on this topic but enough to show that it has been discussed for many years and these discussions continue. All point to serious real or potential dangers in other people, including governments and corporations, gathering large amounts of information about individuals. But the lure of new technologies regularly overrides concerns about the dangers. They are explained away or just ignored. These dangers are not so much with privacy itself but rather with the use that could be made of the information gathered.

Artificial Intelligence (AI) is one of these new technologies that is becoming an increasing part of monitoring and surveillance (Evans and Webb 2020; Tighe and Andre 2020). AI has seen a resurgence in recent years with applications discussed in a wide variety of areas (Tegmark 2018). This has been facilitated by large increases in computer power, vast amounts of data being collected from the online world and developments in learning algorithms that has allowed machines to learn decision-making rather than needing to be programmed in detail as in early AI systems. Driverless cars have caught the imagination of many and their benefits and problems are frequently raised. They will, we are told, make life easier and driving safer given that most road accidents are caused by human error. One worry concerns how decisions will be made when an accident cannot be avoided, with the well-known trolley problem being used as an example (Wu 2020). Medical applications are seen as important ways of making treatments and diagnoses more efficient and currently research in AI is under way for triaging in the COVID-19 pandemic as well as AI software "that is capable of analysing patient interviews . . . and detecting whether there is a high risk of COVID-19" (Corti 2020). Other areas include finance with decisions on buying and selling shares amongst other things, in law and as an important component in monitoring and surveillance technology with data mining, profiling, facial recognition and so on.

AI has caught the imagination of many and can be seen as one of the most exciting developments in technology, spawning numerous movies and novels. Nevertheless, over the last fifty or so years numerous people have worried about its impact on humans. One of the early critics who was also one of the computing pioneers, Joseph Weizenbaum, first said in 1976:

> The very asking of the question, "What does a judge (or a psychiatrist) know that we cannot tell a computer?" is a monstrous obscenity. That this has to be put into print at all, even for the purposes of exposing its morbidity, is a sign of the madness of our times. .... Computers can make judicial decisions, computers can make psychiatric judgments. . . . The point is that they ought not be given such tasks.
>
> . . . all projects that propose to substitute a computer system for a human function that involves interpersonal respect, understanding, and love . . . are obscene.' And their '. . . very contemplation ought to give rise to feelings of disgust in every civilised person' (Weizenbaum 1984: 268–9)

Weizenbaum's worries may seem rather remote from today's concerns with monitoring and surveillance technologies but they are not so far removed from data mining and machine learning and facial recognition applications. The judgments made by these technologies regarding who might be a terrorist or financial loan risk, or if the face seen looks like that of some wanted person, should involve some interpersonal respect and understanding. Using computers for these kinds of purposes (including those mentioned by Weizenbaum) suggests that humans are like machines and perhaps that machines are better than humans. The need for human contact is reduced as is the scope for caring. Overall it is dehumanising. Similar concerns were expressed by others, including Arthur Kuflik (1999) and James Lenman (2001). While this may also be dehumanising in monitoring and surveillance, more worrying is that it enables these activities to become increasingly intrusive and widespread because many tasks can be automated.

More recently Stephen Hawking (Healy 2014), Elon Musk (Wootson 2017) and others have raised concerns about dangers with the increasing development and use of robots and development and use of autonomous weapons (Sharkey 2019). More dramatically, Nick Bostrom has warned that "superintelligence represents an existential risk to humanity," which he defined as "a risk that threatens the premature extinction of Earth-originating intelligent life or the permanent and drastic destruction of its potential for desirable future development" (Dvorsky 2015; Bostrom 2014).

Despite these and other concerns expressed in a vast array of literature, much larger than the examples mentioned here, and suggestions for future research priorities (Russell et al. 2015; Cath et al. 2018), there seems to be no appetite to change the direction of or to reduce the research, development and use of AI, with many arguing that the dangers are overstated. Luciano Floridi, while accepting that dangers exist in current applications of AI, doubts that the greater harms of superintelligence that Bostrom fears will eventuate (Floridi 2018). The benefits of continuing research and development are assumed to outweigh the dangers, or perhaps our fetish blinds us to the dangers.

8.3 Can and Should Anything Be Done?

A number of separate issues arise here; can we do anything about the fetish and can we do anything about controlling or influencing new technologies? If nothing can be done about controlling the technology then the fetish is not a problem. It is only a problem if it affects the direction of technological development. Furthermore, if something can be done about the direction of the development and about the fetish, should anything be done? In this section the focus will be on the technology, briefly looking at two supposed obstacles to controlling or directing it: technology determinism and the Collingridge Dilemma.

First, technological determinism. Technological determinism historically has meant that technology determines society although it will be used in a slightly different sense here, to be explained shortly. The historical view is exemplified in the statement by Karl Marx:

> The hand-mill gives you society with the feudal lord; the steam-mill, society with the industrial capitalist. (Marx 1935, 92)

In other words, the society that develops depends on the technology. For example, modern cities have developed because of motor cars. Jacques Ellul uses the concept of autonomous technology in a similar way:

> … neither economic nor political evolution conditions technical progress. Its progress is likewise independent of the social situation. … Technique elicits and conditions social, political and economic change. (Ellul et al. 1964, 133)

If Marx and Ellul are correct then the common catchcry of those who resist controls being applied to research or development, "You cannot stop progress" is also correct. This could mean that progress *ought* not be stopped or that it *cannot* be. This latter interpretation assumes that progress, for our purposes new technological developments, will happen regardless of what we do, so it is futile to try to stop it. Given the current state of the world and the laws of nature, the development of technology proceeds in a fixed way that is completely determined by those laws. This hard determinist view will be rejected without argument here in favour of a softer technological determinism, *normative* technological determinism where people play a role (Bimber 1994, 79. See Webster 2017). On the normative account, technology develops in what seems to be a deterministic manner simply because the values of efficiency and productivity are allowed to take precedence over all other ethical and social values. While it is true that on this account there is nothing inevitable about the direction of technological development so it is not determinism at all in any strict sense, it is not completely wrong either. Technology is designed, developed, and used in a context, and the current context, in the Western world where much of the technological development is taking place, is one where capitalist values hold sway. The fetish plays an important part in driving the apparent determinism. If the fetish could be controlled, technology could be controlled. One reason for the continuous development of ever more powerful technologies that underlie monitoring and surveillance capabilities, including AI technologies, is the consumer demand for better smart phones, broader bandwidth for faster downloading of movies and generally faster communication, and so on. This demand, even if it is at least partly created by advertising, is an important driver of new technological developments.

Second, the Collingridge Dilemma. Suggesting that more thought should be given to the introduction of new technologies is easy but actually giving it more thought is difficult, as has been pointed out by David Collingridge in what has become known as the Collingridge Dilemma (Collingridge 1980). According to Marvin Croy's formulation of this supposed dilemma, either a technology is in a relatively early stage of development when it is unknown what changes should be made, or a technology is in a relatively late stage of development when change is expensive, difficult and time-consuming:

> If the former, then control is not possible
> If the latter, then control is not feasible
> Therefore, either controlling technology is not possible, or controlling technology is not feasible. (Croy 1996)

It is not possible because in the early stages of development the uses of a technology are unclear. Prediction is possible but is unreliable and therefore what changes should be made to minimise its harm are unknown. On the other hand, once the technology is well developed changing, it is not feasible because it is difficult and expensive. A recent example of the difficulty of predicting in the early stages is the research that led to the molecular motor that won the 2016 Nobel Prize for Chemistry:

> In terms of development, the molecular motor is at about the same stage as the electric motor was in the 1830s, when researchers proudly displayed various spinning cranks and wheels in their laboratories without having food processors," said the Nobel committee. In other words, what their development means for sectors like healthcare, energy and industry will be for future generations to discover. (Samuelson 2016)

This is not a real dilemma however, because some plausible predictions can be made early in development and changes are possible throughout development not just at the end, particularly when these new technologies are based on previous ones. It may be reasonable to say that not much can be predicted yet about the uses of molecular motors but the case of the 5G wireless network, for example, is different. We know already what technologies are important for surveillance so it is more than an educated guess that new technologies that enable faster data transmission, more sensitive input devices and greater interconnectivity between people and things will lead to more intrusive surveillance.

Something then can be done about the direction of technological development, even if this is not easy. But should it be?

Even if the technological development can be controlled, to some extent it does not follow that the fetish can be, so we still have two questions; should anything be done about the fetish and should we try to control the technological development?

It was suggested in the previous paragraph that a danger with the fetish is that it leads to uncritical acceptance of new technologies. Benefits are assumed and dangers overlooked or downplayed so there is reason to try to control or modify it. We will return to this in the following section and here look at some issues with the technologies themselves.

At one level it seems obvious that if a new technology causes or is likely to cause harm or not improve life that it should not be developed or at least not used.

The argument in this section will not be so much that we should actively intervene to control technological development because of the dangers but rather to criticise arguments that interventions are always to be avoided because these interventions might lead to even more harm. In other words, the focus is on questioning arguments proposing that technology and the science that supports it should be given free reign.

It is necessary here to consider more carefully the relationship between science and technology because at least some technological developments depend on scientific knowledge. If technological developments depend on or grow out of basic scientific research, then in a sense that research is the first step in technological development. Philip Kitcher (2001) supports the view of a close link and argues that it is a myth that some science is so pure that it is not contaminated by applied science or technology. Advocates of technoscience also see no sharp distinction (Hottois 2015). This close relationship does seem to be the case with molecular motors. Any technological innovations using those motors will be based on the basic research that led to the Nobel Prize.

If this is so in some cases even if not in all, then consideration must be given to placing limitations on scientific research as well as to the development and use of technologies. The former as well as the latter perhaps should be subject to oversight. One who disagrees with this is Australian scientist Sir Gustav Nossal who, speaking of scientific research, writes:

> To free the human spirit from ignorance and superstition must be good, no boundaries should be placed around such a search. (Nossal 2007, 6)

Michael Polanyi wrote in 1962:

> Any attempt at guiding scientific research towards a purpose other than its own is an attempt to deflect it from the advancement of science. ... You can kill or mutilate the advance of science, you cannot shape it. (Polanyi 1962)

Another enthusiastic supporter of this view was Vannevar Bush who had a large influence on science policy in the USA. He wrote:

> Scientific progress on a broad front results from the free play of free intellects, working on subjects of their own choice, in the manner dictated by their curiosity for exploration of the unknown. Freedom of inquiry must be preserved under any plan for Government support of science ... (Bush 1945)

This view still has currency, at least amongst many scientists and there are some good reasons for it. Placing restrictions on or trying to control scientific research certainly carries risk. One notable example is the Lysenko Affair in Russia in the first part of the twentieth century where Soviet ideology and scientific research became entwined with disastrous results for Soviet agriculture. (Kean 2017). A less dramatic case was the decision to stop funding AI research in Britain as a result of the Lighthill Report of 1973, which stifled research for quite a number of years (Crevier 1993). Another example of government interference was the Japanese 5th Generation project which was an attempt to develop computers and machine reasoning based largely on logic programming (Pollack 1992). These last two might be thought to be examples of technological development and not science but at these theoretical levels no sharp distinction can be drawn between the two.

However, despite the problems, given the relationship between science and technology, restrictions being placed on scientific research is something that must be addressed in this context (see Kitcher 2000). The link is too strong as can be seen the close relation between research in nanoscience and developments in nanotechnology.

If we look at technological innovation as including both the scientific research and the technological development then we can see that this development is already guided to some extent. Government funded projects are determined by government funding priorities. These priorities are commonly based on what it is thought that the nation needs most. Not all projects can be funded, given limits to the funding available, so the projects considered to be more important for national prosperity or health are more likely to be funded. Given that many projects are expensive, these priorities and this funding determine this research and development. Industry funded research and development is even more guided. Industry focuses on certain kinds of projects which help the industry to be more profitable and also controls are already placed on development of some technologies, eg. chemical weapons. The use of technologies is controlled, at least to some extent, so it can be, even if it is with some difficulty and some dangers like mistakes in predictions and some loss of scientific freedom.

While no hard logical link exists between research and products, if research is successful in generating new knowledge and if this knowledge can be used for the development of new technologies for which there appears to be a market, those technologies will probably be developed into new products or enhance products. And given the technology fetish, these products will most likely be adopted, whether or not they are particularly beneficial or even if they are harmful.

This discussion of research has some similarities to discussions of the precautionary principle. One standard definition is:

> Where an activity raises threats of harm to the environment or human health, precautionary measures should be taken even if some cause and effect relationships are not full established scientifically. (Wingspread 1998)

In other words, if some activity is likely to lead to serious harm then some precautionary measures should be taken even if there is no conclusive evidence that the harm will occur.

One common criticism of the precautionary principle concerns the dangers of making predictions about the results of research or development, another that the principle contains a paradox. Predicting is hazardous but as we saw earlier, some predictions can be reasonable. The paradox is supposedly that precautionary action can lead to more harms than allowing the research or development to proceed (Clarke 2009). These criticisms show that care must be taken when applying the precautionary principle but not that it is of no use. The paradox can be avoided however, at least in many instances, by considering whether it is better to take precautionary action or not to, that is, by considering what harms might result from the precautionary action. It is argued by Clarke and others that anything useful that the can be achieved by the precautionary can be achieved by cost-benefit analysis without invoking any of the problems of the former. Cost-benefit analysis is seen as the alternative to the precautionary principle. It is not altogether clear however, that these two are the real alternatives in practice. The actual alternative to the precautionary principle seems to be rather something like the Bravado Principle, which seems to be the default position:

> Where an activity appears to have benefits, do it and worry about any harmful consequences if and when they arise, (Weckert 2012, 12)

Technological development then can be controlled to some extent, even if doing this is difficult, but what about the fetish? Can anything be done about that?

8.4 Where to from Here?

Hopefully future research will help to answer this question. This section will merely point in some directions which might suggest answers or at least provide background for further research. It will attempt to show a possible way forward.

We need technology if we are to survive and live well. Many of our problems can best be fixed with technology. It has been argued however, that we do have a technology fetish but also that something can be done about the direction of technological developments and innovation and that something should be done. Something can be done because is not determined in any hard way but is driven by particular values in a capitalist society, for example efficiency, profitability and consumerism and that the best way to fix problems is through technology. So the direction that technology takes can be altered. It has also been argued that something should be done about both the fetish and the direction of technology because of dangers in current directions. The fetish helps direct where the technology goes because the fetish and the values just mentioned are closely related.

It was suggested earlier that the technology fetish is similar to a paradigm in which technology is seen as progress, the first choice in solving problems and something to be desired. In order then to remove the fetish a new paradigm is required, which will involve something like a gestalt switch (Kuhn, 85). Achieving this will not be simple. According to Kuhn, a paradigm change only occurs when the old one no longer works satisfactorily, it contains too many anomalies, and when a new and better one is available. That will involve looking at things in a different way, a gestalt switch, or a change in our worldview.

The difficult question now is, what needs to be done to bring about this change? As we have seen, for many decades the dangers of various technologies have been raised by philosophers, scientists, sociologists, technologists and others, with little noticeable effect. Much has been written on the dangers of monitoring and surveillance but technologies for even more invasive monitoring and surveillance have continued to be developed and used. For over fifty years concerns have been expressed about AI technologies but that has had little effect on its acceptance. The excitement of the new appears to override warnings or worries about any dangers. Perhaps the reason that they are not believed or taken more seriously is because, as Mercier and Sperber say, "we have a boundless ability to produce reasons for just about anything we believe" (Mercier and Sperber 2018, 252). This is particularly true if some favourite beliefs are challenged (see Quine 1970, 7; Lakatos 1974, p. 133 on this). I will find reasons why my belief that I need a new smart phone are valid even when the evidence suggests otherwise; it is expensive, it has few features that my old one does not, my old one works well and so on.

Two recent discussions suggest possible ways forward. Freya Mathews raises the question why, after many years of environmental ethics, the natural environmental is suffering more rather than improving. She also talks in terms of a paradigm shift or Copernican Revolution:

> The call for a specifically environmental ethic, for a new Copernican Revolution in ethics that would overturn anthropocentrism the veritable cornerstone of Western civilisation and reconfigure human identity as essentially ecological, requiring epochal reform of economies and polities, seemed by the early twenty-first century to have pretty much fizzled out. (Mathews 2019, 2)

In a similar vein, Rebecca Huntley discusses this with respect to human induced climate change. The science has been around for decades but many people still do not accept it. She writes:

> ...now that the climate science has been proven to be true to the highest degree possible, we have to stop being so reasonable and find more and more 'irrational' ways to talk about climate change. (Huntley 2020, 28)

In other words, reasoning with people has failed in many cases regardless of the evidence being overwhelming (see also Sabine Roeser 2006 on risk acceptability).

Both Mathews and Huntley (and Roeser) suggest alternatives to just using reason. Mathews talks of creating:

> A new formation, introduced to take the place of traditional religions but serving Earth-friendly values (6)

This new formation could be

> An ecological cosmology, socially organised into place-based congregations of commitment, richly informed with natural history and the relevant sciences, and anchored in hands-on practices of conservation (8).

Huntley's suggestion is rather less dramatic. In order to convince people of the importance of climate change, she thinks, "we need to stop being reasonable and start being emotional"(43). Her argument is not that reason has no place but simply that in order to change minds, people's values, fears, hopes and so on must be taken into account.

Can these suggestions be of any help in knowing what to do about the technology fetish? Both Mathews and Huntley see underlying values as important and a greater focus on these may help in getting people to see and question this fetish.

In typical ethical discussions of, for example, surveillance or AI technologies, actual or potential beneficial and harmful consequences and their ethical implications are explored. Many of these attempt to warn the public of the dangers but as we have seen, this has had little effect. It might be better to first look at some big questions and then see what comes out of that. It should be remembered that the fetish is societal as well as individual. Society has the fetish by virtue of a large percentage of its members having it. The fetish is perpetuated because the structures and institutions of the society promote it.

Asking big questions with respect to technology is nothing new. If technology is for the improvement of life then of course questions about what constitutes a good life must be asked. This will involve asking questions about how societies should be structured. According to Langdon Winner, the important question in technology is this, "As we "make things work", what kind of *world* are we making?" (Winner 1986, 17). Answering this involves asking big questions which explore underlying values and it is these values that influence the direction of technological development.

What are some of these big questions? An important question concerning surveillance is "Do we want to live in a surveillance society?" The answer seems to be "yes and no". It clearly depends on a variety of factors such as how intrusive it is, the extent to which it increases safety, the amount of trust we have in those doing the surveillance and whether it has some specific purpose and is for a limited time only. This leads us to ask in what sort of society we want to live. Presumably a safe one but not necessarily where the safety comes at the cost of too much reduction in privacy and autonomy or where we are so safe that life becomes boring. There is satisfaction in overcoming danger.

AI is making life easier and more efficient in various ways but bigger questions here include what is important in being human, are we being dehumanised, how do we get enjoyment and satisfaction and how easy do we want life to be. It seems better if life is not too hard but we need challenges.

Paradigm change can only occur when an alternative is available and a brief look at Indigenous Australian society prior to European settlement can help our thinking on alternatives as well as on big issues.

Around 1770, Captain James Cook wrote that these Indigenous people

> are far happier than we Europeans. They live in Tranquillity which is not disturb'd by the Inequality of Condition: the Earth and sea of their own accord furnishes them with all things necessary for life, they covet not Magnificent Houses, Household-stuff &c' ... they seem'd to set no Value on any thing we gave them, ... this in my opinion argues that they think themselves provided with all the necessities of Life and that they have no superfluities. (quoted in Flood and Wylie 2006, 15)

We develop new technologies in order to improve life but if Cook was correct, the Indigenous Australians thought that they already had "all the necessities of Life" so looking for new technologies was not a high priority. According to anthropologists Ronald and Catherine Berndt:

> We place so much emphasis on material goods, material wealth, that it is difficult for us to understand people who do not share this interest in accumulating goods almost for their own sake. (Berndt and Berndt 1999, 107ff)

Indigenous writer, Mary Graham, outlines two basic Indigenous precepts; "The land is the Law" and "You are not alone in the world". The first is described as:

> The land is a sacred entity, not property or real estate; it is the great mother of all humanity.

And the second

> Aboriginal people have a kinship system which extends into land.

This ties groups to the land and people to the group. Individuals get their identity by being part of the group, therefore they are not alone (Graham 1999, 106). All is connected and nothing should be exploited, especially not the land.

A good case can be made that this is also the best way to view and understand their technology, an observation made by John Charles Ryan:

> Indigenous technological practices also involve environmental and cultural understandings, particularly of country. (Ryan 2015)

These Indigenous technologies have been called gentle technologies (Pascoe 2012) and lo-technologies (Ryan 2015, 6). They had minimal impact on the earth, something that follows both from seeing the earth as sacred and as having kinship relations with humans.

Two related aspects of indigenous thought or cultural understandings are central, the relationship with the land and nature, and cultural values, in particular here, the lack of materialism and consumerism.

The purpose here is not to suggest that Indigenous Australians had a perfect life or that we should move to their lifestyle. That would be impossible in any case. Its purpose is to provide a sharp contrast to modern life in a capitalist society in order to aid in thinking about what values are important. Is our materialism and consumerism which drives the technology fetish and is in turn driven by it, really conducive to living well? Do we need to place so much strain on our environment in order to live well? We evolved along with all other mammals but now most of us live in an environment and have a lifestyle very different from other mammals and in many cases quite remote from the natural world. Have we moved too far away from the natural world for our own good? There is certainly evidence that getting back into nature is good for both our physical and mental health (Twohig-Bennett and Jones 2018). Perhaps our answers to these questions will lead us to something like Mathews' proposal for protecting the environment (Mathews 2019).

The suggested way forward here in switching from the technology fetish paradigm to something which considers new technologies more carefully, is to think more about the big issues and use very different perspectives, such as Indigenous ones, as aids to more creative thinking about technology. We might still have a love of technology but not "an excessive and irrational devotion or commitment" to it. This will likely lead to a greater emphasis on the natural world, perhaps a "Terrestrial Turn in Philosophy of Technology" (Lemmens et al. 2017).

8.5 Conclusion

Perhaps the ethics of technology could be studied more fruitfully if it included examination of some Indigenous cultures that were very different from our current one and where no technological fetish exists. Demonstrating such alternatives may help in understanding the importance of asking the big questions and this in turn may help reveal the importance of ethical issues in technology. We are technological creatures but also mammals and it is not clear that our fetish with new technologies is always leading us in ways that allow human mammals to flourish. We may be moving too far too quickly from our evolutionary roots. As Indigenous writer Bruce Pascoe has said:

> The incredible advances in science and engineering need to be analysed against the direction they take us before we applaud every new toy. (Pascoe 2017, 237)

Acknowledgements I would like to thank the editors of this volume and the anonymous reviews of this chapter for their helpful comments.

References

Australian Government. 2020. *COVIDSafe help*. https://www.health.gov.au/resources/apps-and-tools/covidsafe-app/covidsafe-help.

Berndt, R.M., & C.H. Berndt. 1999. The world of the first Australians: Aboriginal traditional life. *Past and Present,* 5. Canberra: Aboriginal Studies Press.

Bimber, B. 1994. Three faces of technological determinism, in Smith, Merritt Roe, and Marx, Leo (eds), *Does Technology Drive History: The Dilemma of Technological Determinism*, Cambridge, MA, MIT Press, 79–100.

Bostrom, Nick. 2014. *Superintelligence: Paths, dangers, strategies*. Oxford: Oxford University Press.

Brecht, B. 2015. *Life of Galileo*. London: Bloomsbury Publishing.

Briggle, Adam, Carl Mitcham, and Martin Ryder. 2005. Technology: overview. In *Encyclopedia of science, technology and ethics*, ed. Carl Mitcham, 1908–1912. Detroit: Thomson, Gale.

Bush, V. 1945. *Science, the endless frontier*. North Stratford: Ayer Company Publishers.

Cath, C., S. Wachter, B. Mittelstadt, M. Taddeo, and L. Floridi. 2018. Artificial intelligence and the 'good society': The US, EU, and UK approach. *Science and Engineering Ethics* 24 (2): 505–528.

Clarke, S. 2009. New technologies, common sense and the paradoxical precautionary principle. In *Evaluating new technologies*, 159–173. Dordrecht: Springer.

Clarke, R. 2019. Risks inherent in the digital surveillance economy: A research agenda. *Journal of Information Technology* 34 (1): 59–80.

Collingridge, David. 1980. *The social control of technology*. New York: St. Martin's Press.

Combes, R. 2006. A taxonomy of technics. *International Philosophical Quarterly* 46 (1): 5–24.

Corti. 2020. *Fighting COVID-19 with artificial intelligence*. https://www.corti.ai/.

Crevier, D. 1993. *AI: The tumultuous history of the search for artificial intelligence*. New York: Basic Books.

Croy, M.J. 1996. Collingridge and the control of educational computer technology. *Society for Philosophy and Technology Quarterly Electronic Journal* 1 (3/4): 107–115.

Daly, Nadia. 2020. A 'pandemic drone' and other technology could help limit the spread of coronavirus and ease restrictions sooner, but at what cost? *ABC News*. https://www.abc.net.au/news/2020-05-01/new-surveillance-technology-could-beat-coronavirus-but-at-a-cost/12201552.

Drengson, A.R. 1984. The sacred and the limits of the technological fix. *Zygon* 19 (3): 259–275.

Dusek, V. 2006. *Philosophy of technology: An introduction*. Vol. 90. Malden: Blackwell.

Dvorsky, George. 2015. *Experts warn UN panel about the dangers of artificial superintelligence*. https://www.gizmodo.com.au/2015/10/experts-warn-un-panel-about-the-dangers-of-artificial-superintelligence/.

Ellul, J., J. Wilkinson, and R.K. Merton. 1964. *The technological society*. Vol. 303. New York: Vintage books.

Evans, Michael, and Carolyn Webb. 2020. Australian police using face recognition software as privacy experts issue warning. *The Age.* January 19. https://www.theage.com.au/national/australian-police-using-face-recognition-software-as-privacy-experts-issue-warning-20200119-p53ssj.html.

Ferré, F. 1995. *Philosophy of technology*. Athens: University of Georgia Press.

Flood, J., and D. Wylie. 2006. *The original Australians: Story of the aboriginal people crows nest.* Crows Nest: Allen & Unwin.

Floridi, Luciano. 2018. True AI is both logically possible and utterly implausible. *Aeon Essays. Aeon*, 15 May. https://aeon.co/essays/true-ai-is-both-logically-possible-and-utterly-implausible.

Forester, T., and P. Morrison. 1990. *Computer ethics: Cautionary tales and ethical dilemmas in computing*. Cambridge, MA: Mit Press.

Graham, M. 1999. Some thoughts about the philosophical underpinnings of aboriginal worldviews. *Worldviews: Global Religions, Culture, and Ecology* 3 (2): 105–118.

Healy, Blathnaid. 2014. *Stephen Hawking warns artificial intelligence could end humankind.* https://mashable.com/2014/12/02/stephen-hawking-artificial-intelligence-bbc/.

Hertz, D.B. 1970. The technological imperative—Social implications of professional technology. *The Annals of the American Academy of Political and Social Science* 389 (1): 95–106.

Hornborg, A. 1992. Machine fetishism, value, and the image of unlimited good: Towards a thermodynamics of imperialism. *Man* 27: 1–18.

———. 2014. Technology as fetish: Marx, Latour, and the cultural foundations of capitalism. *Theory, Culture & Society* 31 (4): 119–140.

Hottois, G. 2015. Technoscience (trans: Lynch JA). In: Holbrook, Britt, J (ed. in Chief) *Ethics, science, technology engineering: A global resource* (pp. 334–337). Gale, Cengage Learning, Farmington Hills, MI.

Hunt, Pete. 2018. China's great social credit leap forward, *The Diplomat.*https://thediplomat.com/2018/12/chinas-great-social-credit-leap-forward/.

Huntley, Rebecca. 2020. *How to talk about climate change in a way that makes a difference.* Sydney: Murdoch Books.

Johnson, Deborah. 1985. *Computer ethics*. Upper Saddle River: Prentice Hall.

Joy, B. 2000. Why the future doesn't need us. In *Nanoethics. The ethical and social implications of nanotechnology* (17–30).

Kean, Sam. 2017. The Soviet Era's deadliest scientist is regaining popularity in Russia. *The Atlantic.*, 19 December. https://www.theatlantic.com/science/archive/2017/12/trofim-lysenko-soviet-union-russia/548786/.

Kitcher, P. 2001. *Science, truth, and democracy*. Oxford University Press.

———. 2007. Scientific research–who should govern? *NanoEthics* 1 (3): 177–184.

Kobie, N. 2019. The complicated truth about China's social credit system. *Wired.* https://www.wired.co.uk/article/china-social-credit-system-explained.

Kuflik, A. 1999. Computers in control: Rational transfer of authority or irresponsible abdication of autonomy? *Ethics and Information Technology* 1 (3): 173–184.

Kuhn, T.S. 1970. *The structure of scientific revolutions*. 2nd ed. University of Chicago press.

Lakatos, Imre, 1974, "Falsification and the methodology of scientific research programmes" in Lakatos, Imre and Musgrave, Alan, *Criticism and the growth of knowledge*, Cambridge, Cambridge University Press, 91-196.

Lemmens, P., V. Blok, and J. Zwier. 2017. Toward a terrestrial turn in philosophy of technology. *Techné: Research in Philosophy and Technology* 21 (2/3): 114–126.

Lenman, J. 2001. On becoming redundant or what computers shouldn't do. *Journal of Applied Philosophy* 18 (1): 1–11.

Marx, Karl. 1867. Das Capital vol. 1 part one section 4. Online https://web.stanford.edu/~davies/Symbsys100-Spring0708/Marx-Commodity-Fetishism.pdf.

Marx, L. 1983. Are science and society going in the same direction? *Science, Technology, & Human Values* 8 (4): 6–9.

Mathews, Freya. 2019. Why has environmental ethics failed to achieve a moral reorientation of the West? *ABC Religion & Ethics*.https://www.abc.net.au/religion/why-has-environmental-ethics-failed-to-achieve-a-moral-reorient/11216540.
Mercier, Hugo, and Dan Sperber. 2018. *The enigma of reason: A new theory of human understanding*, UK, Penguin Random House.
Nissenbaum, H. 2001. How computer systems embody values. *Computer* 34 (3): 120–119.
Nossal Sir. G. 2007. *Introduction, ethically challenged: Big questions for science,* Mills J, series editor, the Alfred Deakin Debate, The Miegunyah Press, Melbourne.
O'Brien, Matt 2020. Surveillance is everywhere at the CES gadget show, *The Age*, Jan 7. https://www.theage.com.au/technology/surveillance-is-everywhere-at-the-ces-gadget-show-20200107-p53pep.html.
Ortega y Gasset, José. 1961. Man the technician, *History as a system: And other essays toward a philosophy of history*, translated by Helene Weyl, W. W. Norton & Company, New York, 1961, 87–161.
Oxford. 2020. *Oxford Dictionary* on Lexico.com, https://www.lexico.com/definition/fetish.
Pascoe, Bruce. 2012. *Aboriginal agriculture, technology and ingenuity*, Chapter 2. http://education.abc.net.au/home#!/digibook/3122184/bruce-pascoe-aboriginal-agriculture-technology-and-ingenuity.
———. 2017. *Convincing ground: Learning to fall in love with your country*. Canberra: Aboriginal Studies Press.
Polanyi, M. 1962. The republic of science. *Minerva* 1 (1): 54–73.
Pollack, Andrew. 1992. Fifth generation' became Japan's lost generation. *The New York Times* (June 5). https://www.nytimes.com/1992/06/05/business/fifth-generation-became-japan-s-lost-generation.html.
Quine, W.V. 1970. *Philosophy of logic*. Englewood Cliffs: Prentice-Hall Inc.
Roeser, S. 2006. The role of emotions in judging the moral acceptability of risks. *Safety Science* 44 (8): 689–700.
Russell, S., D. Dewey, and M. Tegmark. 2015. Research priorities for robust and beneficial artificial intelligence. *AI Magazine* 36 (4): 105–114.
Ryan, J.C. 2015. "No more boomerang": Environment and technology in contemporary aboriginal Australian poetry. *Humanities* 4 (4): 938–957.
Samuelson, Kate. 2016. 5 things to know about 'Molecular Machines', *Time*, https://time.com/4519282/molecular-machines-nobel-prize/.
Scott, D. 2011. The technological fix criticisms and the agricultural biotechnology debate. *Journal of Agricultural and Environmental Ethics* 24 (3): 207–226.
Sharkey, A. 2019. Autonomous weapons systems, killer robots and human dignity. *Ethics and Information Technology* 21 (2): 75–87.
Soltanzadeh, S. 2016. Questioning two assumptions in the metaphysics of technological objects. *Philosophy & Technology* 29 (2): 127–135.
Sprenger, P. 1999. Sun on privacy: 'Get over it'. *Wired News* 26: 1–4. http://archive.wired.com/politics/law/news/1999/01/17538.
Tegmark, Max. 2018. Benefits & Risks of Artificial Intelligence. *Future of Life Institute,*https://declara.com/content/N5vPEl71.
Tighe, Alex, and Andre, Julia. 2020. The Australian behind Clearview AI, a facial recognition software, says it is being used here. *Australian Broadcasting Corporation*, Tuesday 17 March, https://www.abc.net.au/news/2020-01-23/australian-founder-of-clearview-facial-recognition-interview/11887112.
Trivium China. 2019. *Understanding China's social credit system*. socialcredit.triviumchina.com.
Twohig-Bennett, C., and A. Jones. 2018. The health benefits of the great outdoors: A systematic review and meta-analysis of greenspace exposure and health outcomes. *Environmental Research* 166: 628–637.
Van den Hoven, Jeroen. 1999. Privacy and the varieties of informational wrongdoing. *Australian Journal of Professional and Applied Ethics* 1: 30–43.

Webster, M.D. 2017. Questioning technological determinism through empirical research. *Symposion* 4 (1): 107–125.

Weckert, J. 2012, In defence of the precautionary principle, *IEEE Technology & Society Winter*, 12–17.

Weizenbaum, Joseph. 1984. *Computer power and human reason: From judgement to calculation*. Harmondsworth: Penguin Books.

Wingspread. 1998. *Wingspread statement on the precautionary principle*. www.gdrc.org/u-gov/precaution-3.html.

Winner, L. 1986. *The whale and the reactor: A search for limits in an age of high technology*. Chicago: University of Chicago Press.

Wootson, Cleve R. 2017. Elon Musk doesn't think we're prepared to face humanity's biggest threat: Artificial intelligence, *Washington Post*, July 17, https://www.washingtonpost.com/news/innovations/wp/2017/07/16/elon-musk-doesnt-think-were-prepared-to-face-humanitys-biggest-threat-artificial-intelligence/.

Wu, S.S. 2020. Autonomous vehicles, trolley problems, and the law. *Ethics and Information Technology* 22 (1): 1–13.

Zuboff, S. 2019. *The age of surveillance capitalism: The fight for a human future at the new frontier of power: Barack Obama's books of 2019*. London: Profile books.

Chapter 9
Models, Explanation, Representation, and the Philosophy of Computer Simulations

Juan Manuel Durán

Abstract The philosophical study of computer simulations has been largely subordinated to the analysis of sets of equations and their implementation on the computer. What has received less attention, however, is whether simulation models can be taken as units of analysis in their own right. Here I present my own experimental work investigating this issue. This article explores the capacity of programming languages to represent target systems and submits that, in a number of cases, the representation of simulation models differs in non-trivial ways from sets of equations. If my claim is correct, then a few important methodological and epistemological concerns emerge that need our attention. This article finishes by briefly addressing some implications for the philosophy of computer simulation.

9.1 Introduction

Philsosophical studies in computer simulations have advanced in two predominant directions. One takes computer simulations to be instrumental to finding the solutions to a set of equations (e.g., Humphreys 1990; Hartmann 1996; Parker 2009); the other claims that computer simulations are a *special kind* of mathematical model[1] (e.g., Humphreys 2004; Weisberg 2013; Lenhard 2019). At their cores, both interpretations assume that simulations can be analyzed as sets of equations, with the latter interpretation being a bit more lenient towards changes in the simulation models that do not necessarily have mathematical representation. This article is my

[1] Mathematical models consist of a set of equations that describe certain aspects of physical reality, broadly conceived. Mathematical modeling is the specific practices attached to such models.

J. M. Durán (✉)
TU Delft, Faculty of Technology, Policy and Management, Delft, The Netherlands
e-mail: j.m.duran@tudelft.nl

B. Lundgren, N. A. Nuñez Hernández (eds.), *Philosophy of Computing*, Philosophical Studies Series 143, https://doi.org/10.1007/978-3-030-75267-5_9

own experimental work on a third position, one that takes that simulation models—that is, the models at the basis of computer simulations—must be approached as units of analysis in their own right, irrespective of the equations implemented.

The exploration of this third position begins with a simple intuition: if simulation models are instrumental to finding the solutions to a set of equations, then the expected epistemic value of representing, explaining, and predicting target systems should be the same or, at least, very similar to the equations implemented. But this does not seem to be the case for many simulation models, including very simple ones. Simulations aggregate and disaggregate information about target systems, add and subtract components in the model, and involve methodologies that are utterly at odds with interpreting them as sets of equations. For instance, simulations represent high-level dynamics of systems that, in practice, are not—or cannot be—represented in mathematical formalism. Simulations also represent target systems using numerous variables, each of which can take a wide range of values and interact with each other in non-linear ways. Even if we could reconstruct such a simulation as a set of equations, we would be able to confirm neither that the equations behave in the way intended nor that the reconstruction captures the same patterns of behavior of the target system as the simulation. Finally, simulation models sometimes introduce non-physical quantities and representations that might not make physical sense but which fulfill different purposes. An example of this is the so-called 'Arakawa operator,' which was introduced to compensate for instabilities in simulations of the atmosphere at the expense of involving a non-physical assumption of conserved kinetic energy (Lenhard 2019).

As a result, the ways in which a simulation describes a target system, explains it, and predicts future states diverges in non-trivial ways from sets of equations. If this intuition is correct, then we might be analyzing simulations under the wrong magnifying glass. To show why this might be the case, I propose a twofold strategy. First, I will show that, even in cases where a given simulation only requires the implementation of a set of equations, the changes in syntax (i.e., from mathematical formalism to programming languages) entails non-trivial changes in the representation of the target system. This first discussion draws significantly from technical elements of computer science as well as studies on philosophy of computer science. Second, I will discuss the extent to which changes in the syntax give us good reasons to revisit studies in scientific representation and scientific explanation that try to accommodate computer simulations.

This article is influenced by at least two important traditions in philosophy. The first one springs from the philosophy of science. In the late twentieth century, philosophers were debating the divorce of models and theory (e.g., Cartwright 1983; Morrison and Morgan 1999), and more recently the debate took a turn to the limits of representation and the introduction of non-epistemic values in scientific modeling (Boon and Knuuttila 2009; Knuuttila 2011, 2017). My claims are motivated by similar reasons, but with an appropriate recast of actors. The second influence stems from debates within the philosophy of computer science, particularly those surrounding the definition of computer algorithms and computer software. Within this branch of philosophy, three chief paradigms emerge. The *mathematical paradigm*, which conceives algorithms as mathematical entities and thus mathematics as the natural discipline for their analysis; the *software engineering paradigm*, which defines computer science as an engineering discipline, and software engineering as the right discipline for understanding software applications; and the *scientific paradigm*, which takes computer science to be a branch of (empirical) science (Colburn 2000; Eden 2007; Turner 2014). Against the backdrop of these paradigms, the aforementioned interpretations of computer simulations can be placed within the mathematical paradigm. I call them *the received view of computer simulations* as a way to reflect their value as the standard interpretation of the specialized literature. Opposing this view is the claim that computer simulations are better understood from the software engineering and scientific paradigm perspectives. Conceivably, the latter is more suitable than the former for addressing computer simulations, as it conveniently combines formal methods, software engineering practices, and scientific experimentation. But unfortunately this is not the place for splitting hairs. The goal of this article is to explore whether the philosophical study of computer simulations could be moved away from the grips of mathematics, and place it within the scope of computer science and engineering. For the purposes of this paper, then, computer simulations are more a product of engineering than of mathematics.

The discussion proceeds as follows. Section 9.2 presents and discusses *the received view of computer simulations*, the most accepted viewpoint of computer simulations in the literature. At its core is the assumption that simulation models are built up from and grounded in sets of equations. This first section comes with a short discussion on the reception of the received view by philosophers of science and how it influenced their philosophical analysis of computer simulations. It also offers a few reasons as to why I believe this approach falls short of fully grasping the philosophical value of computer simulations.

Section 9.3 makes an effort to show how a new syntax drives the representational value of simulations and how this differs, in non-trivial ways, from mathematical formalism. Admittedly, this is the most exploratory part of the paper and many things can go wrong. The strategy will be to build up my case from simple examples of simulations to more complex ones. A response to an imaginary advocate of the received view and fierce opposer to my claims is included.

Section 9.4 offers examples of three different computer simulations. It starts with a simple example of a simulation of a satellite orbiting around a planet. This is the sort of simulation that is preferred by the received view and deserves our full attention, despite the fact that it is possibly the least interesting case for scientific practice. I will argue that, even for such simple simulations, the syntax used for simulation models drives representation in such a way that it differs in non-trivial ways from sets of equations. The two remaining cases are a vivid example of the complexities that pervade the actual scientific practice with computer simulations. To my mind, these two examples make a final case against the received view. Now, if my intuitions are effectively sound, then I must answer the question ‘what difference does my third position make for the philosophical study of computer simulations?’ Back in 2009, philosophers addressed the question about the philosophical novelty of computer simulations. Positions were divided. Some philosophers argued that issues in connection with simulations could be addressed by standard philosophical treatment, such as standard studies on models and experimentation (Frigg and Reiss 2009). Some others argued simulations brought about specific issues of genuine philosophical interest that have no precedent in the general philosophy of science (Humphreys 2009; Durán 2020a). My response to this question, which assumes the novelty of computer simulations in the philosophy of science, is that issues about explanation, prediction, and representation take a new form. The core claim is that under the received view of computer simulations, philosophers are required to (erroneously) pick the set of equations as the unit with explanatory and predictive force, whereas it should be the simulation model itself. These discussions can be found in Sect. 9.5. Finally, since the article also intends to be a vindication of the philosophy of computer simulation, Sect. 9.6 presents a short proposal for such a philosophy.

9.2 The *Received View of Simulation Models*

Philosophers have made numerous efforts to provide a proper characterization—or a straightforward definition—of computer simulations. Early attempts can be traced back to the work of Holstein and Soukup who, in 1961, stated that a computer simulation "necessarily involves the use of mathematical expressions and equations [...] and which are so complex as to be impossible of solution without the aid of massive electronic computers" (Holstein and Soukup 1961). A few years later, in 1966, Teichroew and Lubin defined simulation problems as "characterized by being mathematically intractable and having resisted solution by analytic methods" (Teichroew and Lubin 1966). A chief element of both interpretations, and which pervaded much of the subsequent literature, is the idea that computer simulations are the result of directly implementing a set of equations on the physical computer. That is to say, simulation models are, *mutatis mutandis*, sets of mathematical equations.[2]

We are still hearing the echoes of these definitions today. In the early 1990s, Humphreys presented a *working definition* that reads in the following way: "[a] computer simulation is any computer-implemented method for exploring the properties of mathematical models where analytic methods are unavailable" (Humphreys 1990, 501). As we can appreciate, Humphreys joins in considerations about the mathematical origin of computer simulations and its capacity for inference when solutions cannot be found through standard analytic methods.

A few years later, Hartmann suggested some amendments to Humphrey's working definition while maintaining the core claim. To Hartmann's mind, there were two chief problems with Humphrey's working definition. On the one hand,

[2] For an historical analysis of the definitions of computer simulations and their philosophical significance, see Durán (2019).

any definition of computer simulations must stress the dynamic character of the equations in the mathematical model. According to Hartmann, then, "scientists reserve the term 'simulation' exclusively for the exploration of *dynamic* models" (Hartmann 1996, 84).[3] On the other hand, Hartmann pointed out that computer simulations are also successfully used even when analytic methods are available. Based on these amendments, the author offers his own definition that reads: "[s]imulations are closely related to dynamic models. More concretely, a simulation results when the equations of the underlying dynamic model are solved. This model is designed to imitate the time-evolution of a real system. To put it another way, *a simulation imitates one process by another process*. [...] If the simulation is run on a computer, it is called a *computer simulation*" (Hartmann 1996, 83). Thus understood, Hartmann's definition stands on three main claims,

(a) a *simulation* is the result of solving the equations of a dynamic model;
(b) a *computer simulation* is the result of having a *simulation* running on a physical computer;
(c) a *simulation* imitates another process

To marshal the evidence, claim (a) takes that the equations of a dynamic model are solved by means of mathematical inference—and thus the model itself is mathematical. As Hartmann puts it, "[t]he corresponding dynamic model is conveniently formulated in the language of differential equations" (Hartmann 1996, 84). When (a) is complemented with (b), we are in the presence of a *computer simulation*. Taken together, (a) and (b) entail that a computer simulation results when a set of dynamic equations is implemented on a physical computer.

To illustrate these points, consider Hartmann's example of the Boltzmann-Uehling-Uhlenbeck (BUU) model of the dynamic behavior for low energy nuclear collisions. The model is representative of a typical mathematical model in the sense that it is highly idealized, it abstracts from collisions as well as from relativity theory, and ignores certain quantum interactions, among other characteristics (Hartmann 1996, 93). Since the equations in the model are analytically unsolvable and cognitively intractable, it is a good idea to opt for a computer to do the numerical work. These intuitions are well captured by Hartmann's as well as Humphreys' definitions.

Despite these differences, Humphreys and Hartmann share a common core view on computer simulations. First, both agree that *a* sets of equations are implemented on the computer *simpliciter*. This view translates into the following: a) the model is conceived as a mathematical unity, that is, as a theoretical description of a target system constituted by a set of equations. To contrast this, consider a simulation that incorporates different kinds of models, from phenomenological to data models, including theoretical models (Durán 2020a) (I shall return to this

[3] *Exploration* here should be interpreted as mathematically finding the space of solutions to the dynamic model. This explains why all five functions of simulations described by Hartmann heavily resemble finding solutions to the underlying mathematical model (Hartmann 1996, 84–85).

point later). And (b), both authors agree that the implementation of the equations are carried out directly on the physical computer and only mediated by a minimal methodology. By 'minimal methodology' I mean that they accept that some degree of discretization and a handful of ad hoc modeling are involved in the construction of the simulation model. But these are introduced to the extent that they make the equations computationally tractable.

The second shared claim is that the representation, purposes, fitness to the target, and other philosophical features selected and used for epistemic and practical purposes of the simulation are inherited from the set of mathematical equations (Giere 2006; Boon and Knuuttila 2009; Parker 2020b). This means that the mathematical equations—but not the simulation model—is the unit that demarcates what is represented, the purposes of the simulation, etc. as well as to what extent these are the case. These claims stem from taking that the specification and coding of simulation models must be highly coupled with, and dependent upon, theories and theory-articulation strategies. The reason for this, I believe, is a profound mistrust of the epistemic merits of simulation models (tagged along, possibly, with an aggrandized confidence that mathematical equations are epistemically superior for representing target systems). Indeed, the need for computer simulations comes solely from the instrumental impossibility of finding analytic treatment of the equations. A crude realization of this can be found in Teichroew and Lubin, who state that "simulation is a technique of last resort" (Teichroew and Lubin 1966, 724).

Let us call *the received view of simulation models* the view that takes simulation models as built up from and fully grounded in sets of mathematical equations, as described above. I claim that analyses of computer simulations that have the received view at their bases will not be in a position to fully capture the philosophical value of computer simulations. This is largely the case because, as I will argue throughout this article, simulations are approached as a branch of mathematics instead of a branch of computer science and software engineering.

9.2.1 The Reception of the Received View of Simulation Models

The received view of computer simulations has produced a number of followers. Let me arbitrarily begin with Parker, who adopted Hartmann's definition for her analysis of the experimental value of simulations. In her 2009 article, she makes explicit reference to it by characterizing computer simulations as a time-ordered sequence of states that represent another time-ordered sequence of states. Interestingly, in her latest publication, she reaffirms her viewpoint that "a *computer simulation model* is a computer program that is designed to iteratively solve a set of dynamical modelling equations, either exactly or approximately, following a particular algorithm" (Parker 2020a, Section 2). Guala has also made explicit reference to Hartmann in distinguishing between static and dynamic models, time-evolution of a system, and the use of simulations for mathematically solving a set of mathematical equations (Guala 2002). Krohs, on the other hand, also adopts Hartmann's definition but uses it to account for the explanation of real-world phenomena (Krohs 2008). According to Krohs, computer simulations are only instrumental for solving a set of equations, whereas the equations themselves explain real-world phenomena. A somehow more complex picture is given by Winsberg, who advances a hierarchy of models that stretches from sets of equations to the 'computational model' (Winsberg 1999, 280). Admittedly, Winsberg is solely concerned with the application of "the practice of modeling [to] very complex physical phenomena for which there already exist good, well-understood theories of the processes underlying the phenomena in question" (Winsberg 1999, 277). His discussion, as he admits, is not necessarily relevant to simulations that do not draw on a base of accepted theory.[4] For his part, Weisberg considers computational models as "a subset of mathematical models" (Weisberg 2013, 30), albeit an especially important subset. Finally, Morrison has urged that more attention must be given to these methodologies, which are the "result of applying a particular kind of discretisation to the theoretical/mathematical model" (Morrison 2015, 219).

The common theme that, to my mind, ties all these interpretations together is that simulations are understood as implementing a set of equations on the computer *simpliciter*. Surely, such an implementation comes with a methodology, but this is a 'minimal methodology' (Durán 2019) and, ultimately, the simulation is treated as the instrumental substitute for the equations. In fact, the character of the simulation model has little bearings in its philosophical treatment and goals. Rather, the equations attract most—if not all—of the philosophical attention. An interesting case is the recent work of Parker, who advances in important ways the *adequacy-for-purpose view* of computer simulations (Parker 2020b). To this author, the evaluation of models—including simulation models—is grounded in "a representational target but with a target, user, methodology, and background circumstances [standing in a suitable relationship]" (Parker 2020b, 460). The latter depends on the design, coding, and execution of the model, which go well beyond the mere solving of mathematical equations. Parker is absolutely right in her analysis. What is troubling, however, is that she also accepts simulation models *qua* the implementation of sets of equations *simpliciter* (recall her definition mentioned earlier). Adopting her definition of simulations, to my mind, brings tensions to the adequacy-for-purpose view. For instance, it is hard to see how background circumstances relevant to the target system can always be captured by mathematical equations. The example of a simulation of spatio-temporal patterns of respiratory anthrax infection discussed on

[4] Winsberg has written extensively about computer simulations, and among his essays we can find more comprehensive analyses of simulations that include those that do not depend on well-defined set of equations.

page 232 is a case in point. Instead, taking simulation models as units of analysis in their own right square better with Parker's claims.[5]

Returning to the diverse accounts of simulations models, it is possible to find a handful of quite elaborate ones. This is the case for authors like Humphreys Humphreys (2004) and Lenhard Lenhard (2019), for instance. By 2004, Humphreys replaced his working definition of computer simulations with the notion of *computer model* consisting of the sextuple <Template, Construction Assumptions, Correction Set, Interpretation, Initial Justification, Output Representation> (Humphreys 2004, 103). Each member of this sextuple carries its own features and raises its own set of technical and philosophical difficulties. The *construction assumptions* and the *correction set* are responsible for adjusting the computational model to a desired fidelity in the representation and computation of the template—they are also responsible for the degree of accuracy in the outputs. The *interpretation* is originally employed in the construction and acceptance of the template. The *output representation* is essentially the visualization stage of the simulation (Humphreys 2004, 72ff). Finally, the *template* which, to my mind, reveals Humphreys' (partial) adoption of the received view of simulation models, is presented and discussed as a computationally tractable set of partial differential equations with the appropriate initial and boundary conditions (i.e., the theoretical template) (Humphreys 2004, 103). In other words, the template is a computationally tractable set of equations. At any rate, what is important to us is that the *computational model* as a whole intends to reverse the idea that a set of equations are directly implemented on the computer *simpliciter*—an idea strongly present in his earlier work (Humphreys 1990).

Lenhard draws a similar picture for his interpretation. He explicitly says that simulation models are not "a mere derivation from the theoretical blueprint" (Lenhard 2019, 44) that finds home in the idea that simulations offer numerical solutions to an (intractable) mathematical equations. Instead, computationally-inspired practices such as *artificiality* and *plasticity* are capable of significantly modifying the representational value of simulation models. Following Morgan and Morrison, he takes stock of the mediating role of models and claims that simulation models are "autonomous from [mathematical] models" (Lenhard 2019, 42).

Interestingly, when it comes to defining simulation models, Lenhard call them a *special* kind of mathematical modeling, suggesting in this way what seems to be a return to simulation *qua* set of equations. Indeed, he claims that simulation models have a dual-nature: they must be "counted into the established classical and modern class of mathematical modeling," but we must also take stock on how simulations models "contribute to a novel explorative and iterative mode of modeling characterized by the ways in which simulation models are constructed and fitted" (Lenhard 2019, 7).[6] It is at this point where understanding the roles

[5] In Parker's defense, an exact characterization of a simulation model might not be relevant for her purposes. It is still striking to see, nonetheless, that philosophers still pass equations for simulations.

[6] For a short review of Lenhard's book, see Durán (2020b).

of *plasticity* and *artificiality* comes in handy. Plasticity means that a simulation model can instantiate several different target systems by means of parametrization (Lenhard 2019, 70). Lenhard is absolutely right on this point. Parametrization in computer simulations does come cheap, and so does multiple representation of target systems—or at least cheaper than with mathematical equations. Artificiality, on the other hand, refers to the addition of components "that are formed for instrumental reasons" (Lenhard 2019, 10). When "a theoretical model is formulated in the language of continuous mathematics and has to be discretized," artificiality plays several important roles, particularly regarding adjusting the model to the representation of the target system (Lenhard 2019, 133). The Arakawa operator is a case in point, since it "made it possible to produce a more 'realistic' picture of atmospheric dynamics by using what were initially contraintuitive and artificial assumptions that were justified instrumentally rather than theoretically" (Lenhard 2019, 38).

Thus understood, plasticity and artificiality seem to pave the way to understanding simulation models as units of analysis in their own right. Perhaps this was Lenhard's intention all along, perhaps it wasn't. But if we were interested in understanding simulation models as something else than the mere implementation of sets of equations, then further argumentation needs to be provided. Indeed, even after considering plasticity and artificiality as core constituents of simulation models, the underlying equation-based structure in the simulation does not need to change, nor does the interpretation of simulation models as sets of mathematical equations. Thus, whereas artificiality and plasticity account for certain practices attached to computer simulations, they do not necessarily entail a rejection of the received view.

Although there are voices that root for more autonomy of simulation models from sets of equations, such as the later work of Humphreys and the recent work of Lenhard, their interpretations still appear to be contingent on sets of mathematical equations. My proposal is admittedly more radical. I want to stretch the autonomy of simulation models beyond the idea of being a special kind of mathematical model containing sets of equations and into being units of analysis in their own right. To this end, I will assume as fundamentally correct Lenhard's plasticity and artificiality of simulation models, and Parker's and Winsberg's study of non-epistemic values in the design and programming of simulation models,[7] and move the debate into analyzing first the implications of changes of syntax for the representation of a target system (i.e., from mathematics to computer languages), and later how this should alter our analysis of model representation and scientific explanation.

[7] As a reviewer correctly pointed out, running a computer simulation regarding the possible path of a hurricane, for instance, requires *inter alia* information about the weather and its patterns.

9.3 The Syntax of Simulation Models

Models are very complicated entities. They are not the simple representation of a target system, derived from some theory, and confirmed through validation procedures. They involve diverse methodologies, idealizations and abstractions, individual and social decision-makings strategies, epistemic and non-epistemic values, purposes and goals, and they are historically situated, perspectival, and limited (Morgan and Morrison 1999; Giere 2006; Massimi and McCoy 2020). Simulation models add to this mix a new syntax.

In what follows, I explore the ways in which the output of a simulation model represents a target system, and argue that it differs in non-trivial ways from the set of equations that might be at the basis of the simulation. To this end, I turn to programming languages. I believe that the quantity, quality, and accuracy of information conveyed about a target system by the programming language gives simulation models their uniqueness of character that I wish to explore.

To illustrate these ideas, take for instance any *control flow statement*, such as 'do-while loops' and 'if-then statements.' The received view takes that these statements have only *operational* or *instrumental* capacity. That is, they enable the computation of some mathematical entity. Consider for instance the following mathematical function:

$$skf(x) = \begin{cases} \exp x & \text{if } 0 < x \leq 2,147,483,647 \\ 1 & \text{if } x < 0 \\ 0 & \textit{if } x > 2,147,483,647 \end{cases} \tag{9.1}$$

Equation 9.1 can be implemented on the computer rather straightforwardly in the following way:

```
Result: if positive integer, then exp x; 1 otherwise
if x > 0 then
 | i = exp(x) ← x;
else
 | i ← 1;
end
return i;
```

Algorithm 1: Implementing Eqs. 9.1

Algorithm 1 is instrumental to the mathematical equation 9.1 in the sense that it finds a solution to the equations, it calculates over the same range of natural numbers, and it maintains similar mathematical properties such as a binary relation over the domain and co-domain.

Let us further note two interesting facts. First, the mathematical equations are set to the highest computable number for a Pentium V with a 32-bits word size processor. Any larger number would require either a larger word size processor or it will not be computable. Second, Algorithm 1 does not require an indication of the upper limit for x since it is, indeed, the highest computable number for this particular computer. If the mathematician were to range x over the entire domain of all natural numbers, the computer scientist would have to include a further conditional clause as well as an error function capable of dealing with numbers out-of-range. In Sect. 9.4, I'll be discussing a similar situation for simulation models.

Now, computer science and software engineering take that statements such as if-then conditionals can also have *representational value*. This means that certain statements both facilitate the implementation of the sets of equations and work as representational bits in the simulation model itself. This can be seen, for instance, when the conditional in the if-then statement depends on one or more procedures. Indeed, the return of a procedure or function could determine how the conditional is further evaluated (i.e., it retrofits back to the conditional and controls its evaluation). Such an if-then statement might represent the decisions made by actors in a simulation that determines the fluxes of individuals via transportation (for a full example, see Ajelli et al. 2010). For example,

```
Result: Behavior of individual based on their preferences of transportation
if decision.individuals() == ['bus', 'train', 'car'] then
|  return ← highway.individuals();
else
|  return ← nonhighway.individuals();
end

Function decision.individuals() is
|  #compute the decision of the individuals d = 'bus' or 'train' or 'car' or
|    'airplane' or 'boat'
|  return ← d;
end
```

Algorithm 2: Implementing representational if-then statements

Furthermore, in some computer languages, the data type of a variable might change according to the input for that variable. For instance, in Python, one can have the following code:

```
Data: s ← "Today's date is 09/07/2020"
Result: The algorithm only adds numbers found in s. It returns '20'.
Function addition.numbers() is
|  d ← sum(int(x) for x in s if x.isdigit())
|  return ← d;
end
```

Algorithm 3: Variables in an algorithm changing data types. Note that the value of x in the algorithm changes types (i.e., Char and Int)

Finally, nested conditionals require a rather simple form of computational implementation and are representative of a broad spectrum of options (e.g., conditionals on behavior and behavioral decisions).

```
case a_1 do
 | case a_1 block
...
case a_n do
 |  case b_1 do
 |   | case b_1 block
 |  ...
 |  case b_m do
 |   |  case c_1 do
 |   |   | case c_1 block
 |   |  ...
case a_p do
 | case a_p block
end
```

Algorithm 4: Generic representation of nested conditionals

The examples above show different ways of accounting for preferences, behaviors, patterns, and arrangements of decisions utilizing conditional statements. Some programming languages (e.g., Python, Perl, PHP, Objective-C) allow programmers to use different data structures as conditionals in the if-then statements (e.g., Algorithm 3). Others enable complicated conditional structures, such as switch-select statements and pattern matching, thus extending the ways in which the arrangement of preferences and behaviors are implemented and represented in the algorithm (e.g., Algorithm 4).

To my mind, these computational practices are not mere syntactic sugar coating of if-then statements for making the code more readable, but rather genuine ways of accounting for conditional forking in decision-making and behavior networking of target systems.[8] Take for instance the case of nested conditionals shown in Algorithm 4. These conditionals are well suited for a simulation of spatio-temporal patterns of respiratory anthrax infection in a given population (Cooper et al. 2004). The simulation can be described as having two components: a Bayesian network, which is treated with strict mathematical formalism, and a large partition of different types of nodes strategically distributed within a network. This latter network consists of a set of global nodes, *G*, a set of interface nodes, *I*, and a set of person subnetworks $P = \{P_1, P_2, \ldots, P_n\}$ (Cooper et al. 2004, 95). A fair simplification of the workings of the simulation is that it computes the actual probability of a person being exposed to anthrax. This is done by means of Bayesian networks. Now, the spatial distribution of a person exposed to anthrax, however, is determined by the series of nodes of interests. Particularly fascinating is the person node, which

[8] Fetzer has offered further arguments as to how computer models can be singled out (Fetzer 1999).

contains a network of interconnected sub-nodes such as *age decile*, *gender*, and *anthrax infection*, among others (see Figure 3, Cooper et al. 2004, 98). The *anthrax infection* node, we should note, takes up to four different states: no infection, 24 h infection, 48 h infection, and 72 h infection (for details, see Cooper et al. 2004, 98–99).

As mentioned, nested conditionals are quite suitable for this kind of simulation. In particular, the network of nodes and sub-nodes can be directly coded into the simulation through such nested conditionals (i.e., no mathematical formalism is required), and thus constitute genuine forms of representing, say, different valid paths in the proliferation and spread of the infection, and suitable states of the infection at a given time, etc.[9]

Consider now the following counter-factual situations. If the nested conditionals were removed, the simulation would simply be a handful of Bayesian networks disconnected at the high-level (i.e., at the level of relation among G, I, and the person nodes P_i). If a connection breaks, the simulation provides virtually no information. If the nodes in the network are navigated differently (e.g., due to different conditions in the if-then statement), different probabilities of contagion will be measured. Again, the probabilities are mathematical, but the high-level structure of the network is not.

In fact, neither the selection of nodes, their spatial distribution, or the fact that a given node can take up to four different states is accounted for by mathematical formalism. The practice of computer simulations involves a myriad of methodologies and design decisions that seldomly stem from or make it back to a set of equations. Aggregating details such as a high-level network and four different states of a node into a set of equations would unnecessarily complicate the interpretation of the system with no visible added value to the overall understanding of it. Programming languages, however, can easily and adequately aggregate these details into the simulation model, and therefore account for the target system in ways that a set of equations would not. For this reason, insisting that computer simulations can be reduced to and thus be philosophically analyzed as sets of equations tends to obscure the real—philosophical, technical, and conceptual—value of computer simulations.

We could further praise the expressiveness of programming languages by mentioning that new forms of representation do not come cheaply—if at all—for mathematical equations. In the case of Cooper and colleagues, the simulation monitors only a single variable at a time, such as the rate of patients visiting emergency departments. But by the grace of adding further if-then statements—or by other computational means—Cooper and colleagues can realistically consider the possibility of extending the behavior of the simulation to multivariate methods that take as input spatial data (e.g., patient zip codes) and temporal data (e.g., the

---

[9] Admittedly, nested conditionals are not the only option for representing the relations between nodes. In fact, they might not even be the best option. For instance, the *pyramid of doom* is a common problem that arises when a program uses too many levels of nested conditionals—other syntactic structures also apply, like nested indentation to control access to a function. See Accessed November 2021.

time at which patients visit the emergency room), as well as patient features, such as age, gender, and symptoms (Cooper et al. 2004, 102). Scaling up the Bayesian network to a million nodes for real-time surveillance applications is only realistic within simulation models (Cooper et al. 2004, 95).

These examples show that programming languages and the general practice of computer simulations offer a wider range of options for the instrumentation and representation of a target system. These options, in many non-trivial cases, differ from those offered by systems of equations. This is partially the reason why computers are so successful in scientific practice. They can account for complexities of the target system that, at least in practice, would be extremely difficult to reproduce with mathematical formalism.

Researchers working with computer simulations do make use of these options extended by computer language programs, as these present powerful and robust procedures for enhancing what simulation models have to offer. In the same vein, we must also call attention to trends in current scientific practice where researchers omit the mathematical and logical formalism altogether in favor of a ready-made algorithmic structure. This means that, in an increasing number of instances, researchers prefer to code the simulation model directly rather than to write a series of equations first and then code their implementation for the computer. Authors like Peck (2012), MacLeod and Nersessian (2013), and DeAngelis and Grimm (2014) show how a simulation model—or parts of it—may be nothing more than an algorithm that frames the agents' behavior. In this sense, the model's representation takes place directly at the level of the algorithmic structures and without mediating any set of equations. The representation, then, is built up from suspected relational structures abstracted from the target system and directly coded into the simulation model (Durán 2020a).

9.3.1 A Response by the Received View

Partisans of the received view can still insist on interpreting simulation models as sets of equations. One promising way to do so is by further distinguishing the *core* simulation from the *instrumentation* of it. The core simulation corresponds to the set of equations, and its instrumentation are the technical additions for making the simulation computable (e.g., using a wide range of data types, control procedures for handling errors, rounding-up routines, etc.). In this way, an if-then statement used for error control is part of the simulation model solely for the purpose of handling numerical misrepresentations, like *division by zero*. This is a weaker version of the received view, since it only commits to mathematics to the extent that it accounts for the target system. I believe that this view could be ascribed to Humphreys (2004) and, perhaps even also to Lenhard (2019).

To my mind, this way of looking at simulation models has three problems. First, it imposes a representational 'corset' to the simulation model, one that covers the same extent as the equations allow. This is a rehash of an idea presented earlier, namely, that the simulation model represents, explains, predicts, etc. to the same extent as the set of equations in the mathematical model. But scientific practice is in conflict with this idea. As I will discuss in Sect. 9.4, even for simple examples it is possible to detect non-trivial changes that escape the information encoded in the equations.

The second problem is more pressing and stems from a core assumption ingrained in the received view. To many of its advocates, the set of equations that will be computed constitute, on balance, *one* mathematical model. But this seems to be an unwarranted assumption on three accounts. First, because it is difficult to make the case that one mathematical model is implemented on the computer. Surely, this might be true for simple simulations, but it is hardly convincing that cases such as the Influenza-Like-Illness simulation (see Sect. 9.4.2) and climate simulations can possibly be accounted for by a single mathematical model. It therefore remains to be explained to what extent a mathematical model equals the simulation model. This brings me to my second point, that is, that the simulation models recast a multiplicity of models. Theoretical models, data models, causal models, phenomenological models, etc. are all coded together into a super-class that is the simulation model (Durán 2020a). In such a case, the received view must explain how the mathematical formalism captures the interaction among these sub-models. Surely, the received view will assert that each individual sub-model is mathematical in the sense here suggested. But they still have to account for the overall interaction of the simulation. Durán (2020a), and Katzav and Parker (2018) have argued against these possibilities. Finally, by taking that simulation models are the implementation of sets of equations *simpliciter*, the received view accepts that any given simulation model can be reconstructed in mathematical formalism. Thus understood, simulations are a two-way street: mathematical equations are implemented on the computer, and simulation models are reconstructed in mathematical formalism. But this strikes me as a gross simplification of computational practice. It is virtually impossible to reconstruct in detail the simulation model and what it represents back into some mathematical formalism. This assumption by the received view ignores debates in the philosophy of computer science regarding the verification debate, the different paradigms of computer science, and the different levels of abstraction (see Colburn 1999, 2000; Eden 2007; Eden and Turner 2007; Turner 2007).

The third and last problem that I see is that the received view neglects the impact that the instrumentation of a core simulation has for the overall representation of the target system.[10] One cannot represent the sequence $S(a_k)_{k=0}^{\infty} = a_i$ as an algorithm (call it $\bar{S}(\ldots))$ without a procedure for handling errors $\Xi(\bar{\mathrm{S}})$. Depending on how such a procedure is designed and coded, the system could halt, return the smallest real-number, have an assigned 'error number,' or take any other course of action. None of these options lead to the same behavior of the simulation as a whole, and

[10] Note that not only is the representation of the target system affected by instrumenting a core simulation, but also the feasibility of the computation as a whole and the accuracy of the output obtained thereafter.

thus the representation of the target system is also affected. Similarly, one cannot represent real numbers without a control procedure that checks the limit of the word-size being represented. Otherwise, the system might fail or return random values.

9.4 Three Examples of Computer Simulations

Let us now illustrate the above discussion with three concrete examples of computer simulations. The first one is a simple implementation of a classical Newtonian model of an orbiting satellite under tidal stress. This kind of model ticks all the boxes insofar as the received view of computer simulations is concerned: the simulation is largely the implementation of a set of equations, the target represented by the equations in the mathematical model is expected to be the same for the simulation model, and finally the model is essentially *a* collection of equations of the same kind. The second example is two simulations of the dynamics of an Influenza-Like-Illness. These simulations are considerably more complex than the orbiting satellite and nicely illustrate how simulations escape many of the considerations made by the received view.

9.4.1 *Example I: Simulating an Orbiting Satellite*

The first example of a simulation model corresponds to the implementation of a classic Newtonian model of an orbiting satellite around a planet which shows tidal stress. This example will help me illustrate that, even in cases of a simple set of equations, the representational output of the simulation model diverges in non-trivial ways from the original equations.

Following Woolfson and Pert (1999a, 18), consider an orbiting satellite under tidal stress which stretches along the direction of the radius vector. Some core equations in this simulation are the total energy (Eq. 9.2), and the angular momentum (Eq. 9.3)

$$E = -\frac{GMm}{2a} \tag{9.2}$$

$$H = \{GMa(1 - e^2)\}m \tag{9.3}$$

Here M is the mass of the planet and m ($\ll M$) is the mass of the satellite. The orbit is a semi-major axis represented by a, and the eccentricity is represented by e (see Fig. 9.1.).

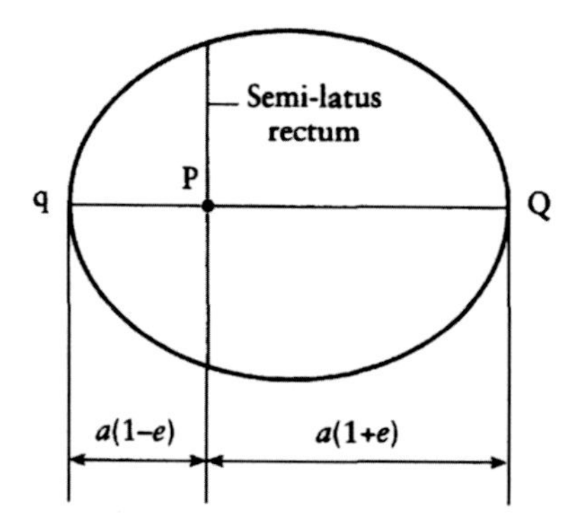

Fig. 9.1 The elliptical orbit of a satellite relative to the planet at one focus. Points q and Q are the nearest and furthest points from the planet, respectively (Woolfson and Pert 1999a, 19)

The model can be reconstructed and implemented as a simple simulation model. A few lines taken from the original algorithm used by Woolfson and Pert (1999b) are the following:

```
PROGRAM NBODY
DIMENSION CM(20), X(20,3), V(20,3), DX(20,3,0:4), DV(20,3,0:4),
  WT(4), +XTEMP(2,20,3), VTEMP(2,20,3), XT(20,3), VT(20,3),
  DELV(20,3)
COMMON/A/X, V, TOL, H, TOTIME, DELV, XT, VT, NB, IST, TIME, IG,
  XTEMP, +VTEMP, CM
DATA WT/0.0, 0.5, 0.5, 1.0/
IST = 0
OPEN(UNIT = 9, FILE = 'LPT1')
[...]
```

Algorithm 5: Partial code in FORTRAN for the simulation of the orbiting satellite under tidal stress (Woolfson and Pert 1999b)

Now, recall that my claim has two sides. First, I contest that a set of equations are directly implemented on the physical computer via some sugar coating instrumentalization (e.g., some discretization techniques and a handful of necessary ad hoc changes in the model for computational purposes). Even for simple cases, it is possible to show that the original set of equations in the mathematical model are modified in non-trivial ways. This claim touches upon representation and implementation practices, as discussed so far. Second is the claim that there are important epistemological challenges that come in connection with these representational and methodological concerns. That is, computer simulations pose challenges for the epistemology of scientific models in their own right that cannot be reduced to the epistemology of mathematical models.

Consider the first claim. The mass of the planet and of the satellite are represented in the equations by mass points M and m, respectively. Given the postulation of the presence of tidal stress, in the simulation the satellite must be represented as a "distribution of three masses, each $m/3$, at positions S_1, S_2 and S_3, forming an equilateral triangle when free of stress" (Woolfson and Pert 1999a, 19). Figure 9.2

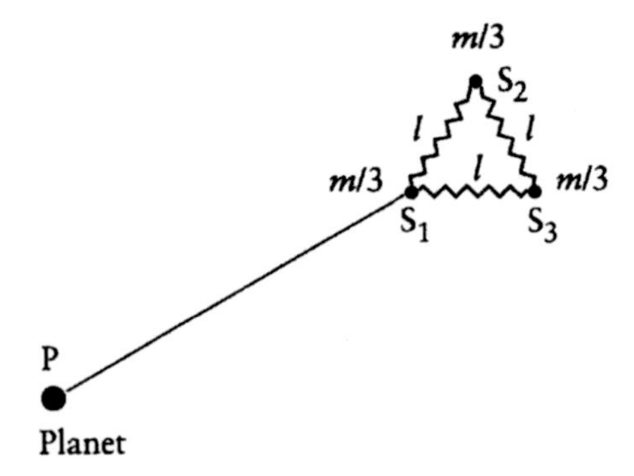

Fig. 9.2 Representation of the satellite and the planet (Woolfson and Pert 1999a, 19)

shows how the masses are distributed and connected by springs, each of unstressed length l and the same spring constant, k. Such a simple design decision that initially aimed at making the implementation of the equations of tidal stress possible, now imposes restrictions on what can be accounted for about the target system. For instance, the unstressed length l remains constant upon nearest distance to the planet (i.e., at q) and furthest distance from the planet (i.e., at Q), when one should expect real-world tidal stress to vary differently at points q and Q.

In addition, Woolfson and Pert need to complement their algorithm with subroutines for error control. In particular, if the maximum absolute error that can be tolerated in any positional coordinate (X, Y, Z) is set too low, then running the simulation can become very slow. To avoid this, the authors recommend a tolerance of 100m (Woolfson and Pert 1999a). Yet, another subroutine added to the simulation model is the control of rounding-off errors produced by the Runge-Kutta algorithm, partly responsible for the orbital eccentricity trending steadily downwards (see Fig. 9.3 on page 241). Researchers know, for instance, that with a 32-bits word size processor (e.g., a Pentium V), you can represent 2,147,483,647 as the maximum positive value. Errors in the simulation are, therefore, bound to occur. An interesting implication that follows is that we would be unable to explain the orbital eccentricity effect by just looking at the equations. It is the NBODY subroutine (see Woolfson and Pert 1999b), and not a set of equations, that is responsible for the outcome in the simulation. More on this in Sect. 9.5.

Of course, one could claim that many of these concerns can be solved by simply adding specific subroutines, such as one that controls the tidal stress upon closer approach to the planet q and furthest distance Q, or with a more powerful computer that extends the number of positive numbers. But it seems to me that these quick fixes overlook important practices ingrained in modeling with simulations, and which are irrelevant in modeling with sets of equations. In particular, no mathematician would (easily) accept that her model splits the satellite into three masses connected by an unstressed length. The reason for this is strikingly simple: she can have a more accurate representation and simpler formulation if she uses the mass point m. The simulationist, however, has no choice but to divide the satellite into three masses, for otherwise the simulation will fail to represent an important feature in the two-body interaction (i.e., the tidal stress). But even if a function $\mathscr{F}$ is added to control the variation of stress at points q and Q, *ex hypothesis* $\mathscr{F}$ constitutes an aggregate to the mathematical equations. It follows that one can only ask if $\mathscr{F}$ will introduce further representational issues, such as a rounding-off error similar to the one experienced in Fig. 9.3. Likewise, adding new subroutines for controlling the tidal stress of the satellite at different positions is alien to the methodology of most mathematical models—and to this one in particular. These considerations are especially true when the simulation models become increasingly more complex, such as the one discussed next.

One more point before moving on. The partisan of the received view can—again—object that additions to the model, such as routines of control and functions like $\mathscr{F}$ will not introduce alterations in the representation of the target system that could significantly divert from the original set of equations. Rather, they are part and parcel of operationalizing the equations. I will circle back to this issue in Sect. 9.6, but I hope that I have provided a sound response already in Sect. 9.3. My interest in returning to this point is because, this time around, I will put into question my own findings. The strategy will be to accept my premise and then ask: "if the equations and the simulation do not represent the same target system, then what does the simulation represent?" This is not an easy question to answer, and I will probably leave many readers unsatisfied with my answer, but the question needs to be asked and an initial answer needs to be provided.

9.4.2 *Example II and III: Simulating Highly Infectious Outbreak Dynamics*

The case of the satellite under tidal stress is a simple yet paradigmatic example of the kind of simulations preferred by the received view. But to fully appreciate the value of computer simulations, we must also discuss more complex cases. Here I present a brief overview of two such simulations accompanied by some remarks on the side.

Consider two simulations of an epidemic outbreak.[11] Ajelli and colleagues provide a side-by-side comparison of a stochastic agent-based model and a structured meta-population stochastic model (GLobal Epidemic and Mobility—GLEaM). The agent-based model includes a description of the Italian population through highly detailed socio-demographic data. In addition, and for determining the probability of commuting from municipality to municipality, Ajelli and colleagues use a general gravity model used in transportation theory. The epidemic transmission dynamics is based on an ILI (Influenza-Like-Illness) compartmentalization based on stochastic models that integrate susceptible, latent, and asymptomatic and symptomatic infections (Ajelli et al. 2010, 5). The authors define their agent-based

[11] This simulation is also presented and discussed in more depth in [hidden].

model as "a stochastic, spatially-explicit, discrete-time, simulation model where the agents represent human individuals [...] One of the key features of the model is the characterization of the network of contacts among individuals based on a realistic model of the socio-demographic structure of the Italian population" (Ajelli et al. 2010, 4).

On the other hand is the GLEaM simulation, a multi-scale mobility network based on high-resolution population data given by cells of 15 $\times$ 15 min of arc. GLEaM integrates highly detailed population databases worldwide with the air transportation infrastructure and short-range mobility patterns. Balcan and colleagues explain that a typical GLEaM simulation consists of three data layers. The first layer combines population and mobility networks, allowing for the partitioning of the world into geographic regions. This partition defines a second layer, the sub-population network, which accounts for the fluxes of individuals via transportation infrastructures and general mobility patterns. This sub-population network uses geographic census data and a demographic model—which in turn includes a short-range commuting model and a long-range air travel model. The demographic model and its sub-models obtain data from different databases, including the International Air Transport Association database consisting of a list of airports worldwide connected by direct flights. Finally, and superimposed onto this second layer, is the epidemic model which determines the dynamics of the disease inside each sub-population (Balcan et al. 2009). For validation purposes, the GLEaM simulation and the agent-based model are dynamically calibrated sharing exactly the same initial and boundary conditions. Such calibration is carried out by yet another module integrating both simulations (Ajelli et al. 2010, 6).

These two examples are quite eloquently highlight the misinterpretations of the received view of computer simulations. First, because we get to appreciate the myriad of models that converge into a complex computer simulation. As mentioned before, theoretical models, phenomenological models, and data models all need to be recast into a super-model, namely, the simulation model (Durán 2020a). Second, traditional mathematical modeling practice is disrupted, since no single set of equations could account for the complexities of any of these simulations.

If the above considerations are persuasive evidence that simulation models are units of analysis in their own right, then it follows that philosophical problems that come up in connection with simulations do not find treatment in more familiar contexts (Frigg and Reiss 2009). Let us now see how simulations challenge established philosophical ideas about scientific explanation and representation.

9.5 Why Should We Care? Explanation and Representation

9.5.1 A Logic of Scientific Explanation that Accounts for Computer Simulations

The above discussion is an effort to show that a reductionist view dilutes the epistemological and methodological value of computer simulations into mathematics. To illustrate this, consider the Runge-Kutta algorithm again (Woolfson and Pert 1999b). This algorithm is responsible for the orbital eccentricity trending steadily downwards as shown in Fig. 9.3. How can we explain these outputs? How can we explain the behavior of the real satellite? The received view is compelled to claim that the real-world phenomenon is explained with the set of equations. This claim follows organically from their assumptions: (a) a set of equations is implemented on the computer for instrumental purposes; (b) the output of the simulation corresponds, *mutatis mutandis*, to those of the equations, had they been solved; and (c) the output of the simulation is an accurate representation of the real-world phenomenon. It follows that the representation of the equations is confirmed, *mutatis mutandis*, by the simulation (i.e., a representational link between the equations and the target system via the simulation has been established). Since the mathematical equations have epistemic priority over the simulation (recall that "simulation[s are] a technique of last resort" Teichroew and Lubin 1966, 724) then, according to the received view, they must be better suited for explaining the target system.

These are the ideas that authors like Krohs (2008), Weirich (2011), and Woolfson and Pert (1999a, 21) advance. Krohs, for instance, takes that computer simulations

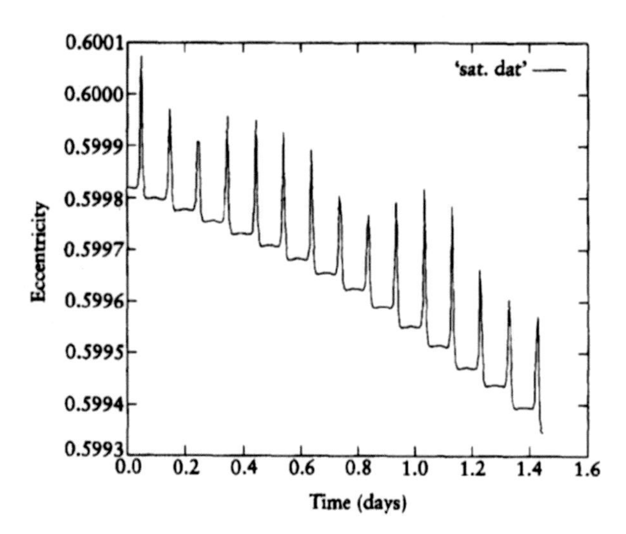

Fig. 9.3 Visualization of the orbital eccentricity. Figure 1.9 in (Woolfson and Pert 1999a, 20)

are sets of equations (classified as theoretical models) implemented on the computer which represent internal mechanisms of a real-world phenomena. In this regard he says, "[s]imulations provide numerical solutions to models. They are run primarily when models cannot be integrated analytically [...] but may, of course, be helpful also in cases where analytical methods are in fact available" (Krohs 2008, 278). He then continues, "[s]uch models may be regarded as not only describing, but also as explaining, the process under consideration" (Krohs 2008, 278). The explanatory input, then, comes from a set of equations exogenous to the simulation, instead of the simulation itself. Krohs leaves no doubt of this arrangement when he says "[t]he explanatory relation holding between simulation and real world is [...] an indirect one. In the triangle of real-world process, theoretical model [i.e., sets of equations], and simulation, explanation of the real-world process by simulation involves a detour via the theoretical model" (Krohs 2008, 284).

Opposing this view is my approach, which fosters a logic of explanation for computer simulations that includes simulations in the explanatory relation (Durán 2017). This is possible if we can show two things: first, that the received view is mistaken about sets of equations explaining real-world phenomena or, alternatively, that simulation models are units of philosophical analysis in their own right, as I am arguing here. Second, that simulation models are in a better position to take up the explanatory role of the target system.

The example used in that article is, incidentally, the one described above of a satellite orbiting under tidal stress. Consider now the visualization of the output of the simulation, which shows a series of spikes that represent the tidal stress that the satellite undergoes while orbiting (see Fig. 9.3). Researchers are naturally inclined to first explain the behavior of the simulated satellite, such as the spikes (i.e., the output of the simulation), before designing, building, and operating the real satellite. In the context of this example, I notice that the visualization shows the spikes having a steadily downwards trend, an effect that is the result of a truncation error during the computation of the simulation model. I then conclude that, if a set of equations exogenous to the simulation were to be used for explanatory purposes, as Krohs and other advocates of the received view propose, the researcher would either be unable to account for the trend downwards or, having explained the spikes with the equations, she would have wrongly ascribed the downwards trend to the real world (i.e., she would have mistakenly claimed that the satellite will eventually have a round orbit). I then go on to show how, by reconstructing the simulation model as the *explanans*, researchers are in a better position to account for the shown trend and thus be at an advantageous position for understanding the overall behavior of the satellite. Motivated by these results, I offer an account of explanation for computer simulations that, as a matter of fact, includes simulation models in the explanatory relation.

If these claims are correct, then we find ourselves with a handful of genuine philosophical issues attached to explanation for computer simulations. On the one hand, and taking my approach seriously, further arguments need to be provided on how and to what extent researchers are able to access and reconstruct the simulation model for the purpose of explanation. Although I offer a fairly detailed reconstruction, I also point out that cognitively accessing every function, procedure, and data-type in the simulation model is challenging. In this respect, the *right levels of description* for the *explanans* is a novel issue that needs to be discussed in detail if this account is to be successful. On the other hand, a central philosophical question that seeks an answer is how, by explaining the outputs of a computer simulation, researchers are able to understand the target system. These issues draw on problems related to representation, realism, and the notion of understanding in the context of computer simulations. Unfortunately, these topics have also received little attention in the specialized literature, possibly because of the influence of the received view which looms large in the philosophical literature on computer simulations.

9.5.2 A Theory of Representation for Computer Simulations

Bueno (2014) is among the few that have effectively elaborated on a theory of representation for computer simulations. He proposes to expand his—and Colyvan's (2011)—*inferential conception* for mathematics into the context of computer simulations. Bueno's strategy consists in introducing a set of minimal changes into the inferential conception such that it can still accommodate computer simulations. To this end, the focus is on the *immersion*, *derivation*, and *interpretation* steps which can also be applied, *mutatis mutandis*, to computer simulations.

Now, adapting the inferential conception makes plain its most fundamental requisite, namely, that the simulation model remains conceptually close to mathematical equations. Bueno does not hide his preference for this ontological similarity. He claims, for instance, that the simulation model is obtained by the same immersing process that gave rise to each individual mathematical equation (Bueno 2014, 379). There is, admittedly, one fundamental distinction that singles out the simulation model, that is, "theoretical assumptions are also incorporated in order to give structure and dynamism to the simulation" (Bueno 2014, 385). In other words, the simulation model incorporates theoretical structures that make the equations computationally tractable. It is important to note that, according to Bueno, such theoretical structures do not play any relevant role in the representational relation, but are rather there for computational purposes. In our terminology, Bueno's 'theoretical assumptions' are syntactic sugar coating for instrumentation purposes. In fact, Bueno is explicit in his reductionist approach. To his mind, the success of the inferential conception depends on keeping "a close interconnection between, on the one hand, the underlying model, which provides the basic structure for the simulation, and the mathematical framework, which yields the expressive and inferential resources of the model, and, indirectly, of the simulation" (Bueno 2014, 385).

In its current state, the inferential conception seems to hold onto a few assumptions that keep a tight link to the received view of computer simulations. First is the acceptance that the simulation model and the mathematical equations are ontologically on a par. This assumption comes with several implications, most prominently for us is that the inferential conception presupposes that the simulation model inherits its representation from a set of mathematical equations. This can be concluded from the immersion step, where certain features of the empirical set-up are selected to be represented in the mathematical equations first, and then by a further identical immersion step, to be represented into the simulation model (Bueno 2014, 739). This conclusion is also supported by the fact that the theoretical structures added into the simulation model are for instrumental purposes and play no representational role whatsoever.

Second, and related to the previous point, it is unclear—at least to me—that the inferential conception can accommodate a multiplicity of models by embedding them into one another. Suppose that the target system is so complex that it needs to be represented by different sets of equations (e.g., differential equations, stochastic equations, etc.). Whereas in principle this complies with Bueno's inferential conception, two problems transpire. First, a fully functional computer simulation depends on the integration of a myriad of models (e.g., phenomenological models, data models, theoretical models), modules (e.g., protocols, performance checks), and large amounts of data (e.g., databases), all packed into a fully functional simulation model (Durán 2020a). It is now up to the inferential conception to convincingly show that the integration of such heterogeneous components can effectively be immersed in the simulation model. Would it be by means of a similar immersion step as that for a set of mathematical equations? Or would it be treated as a theoretical assumption for computational purposes? If my considerations about the instrumental and representational value of programming languages are correct, then these theoretical assumptions cannot be explained away as mere syntax sugar coating. Instead, they play a central role in the representation of the target system. It then follows that the inferential conception may not be capable of accounting for representation in computer simulations. Now, if that were indeed the case, then we might need to consider coming up with a new theory of representation exclusively for computer simulation.

9.6 Final Remarks: A Very Short Proposal for a Philosophy of Computer Simulations

The main claim of this article is that simulation models are complex units of analysis that can neither be reduced to sets of equations nor, under specific conditions, to *special kinds* of mathematical models. To show this, I argued in favor of the instrumental and representative value of the new syntax brought about by programming languages.

The article also presents and briefly discusses some epistemological concerns emerging in this context. Most prominent is the discussion on a logic of scientific explanation interested in accommodating computer simulations. For the received view, explanation can only follow the literature's standards: a set of mathematical equations holds explanatory power over a phenomenon in the world. But for me, explanation is a more complex endeavor, one that effectively includes the simulation model in the explanatory relation and explains, first and foremost, the output of the computer simulation.

Admittedly, this article only scratches the surface on many of the issues here addressed, and certainly more needs to be said. Consider for instance a direct implication stemming from the central claim of the article, namely, that the representation of a simulation model differs in non-trivial ways from a set of equations. But then, what is really being represented by the simulation? Aren't equations 9.2 and 9.3 ultimately resembling the same (conceptual) target system as Algorithm 5? In a way, they are. Researchers can correct for the rounding-up of the satellite as well as the size of the rounding-off error, if they were to accept this resemblance. Similarly, researchers can also factor the three masses composing the satellite into their assessment of the target system. In a way, the equations and the simulation are on a par, representing two sufficiently similar target systems. But this rumination presupposes that researchers are capable of correcting and accounting for the differences in the representation. More complex simulations, such as those in Sect. 9.4.2, shun cognitive accessibility to the algorithm and therefore claims about resemblance can be cast as unsupported. This becomes particularly evident when researchers are unable to cross-explain with the equations the output of the simulation (i.e., either the equations are unable to account for the trend downwards in the simulation or, having explained the spikes, the researchers would have wrongly ascribed the downwards trend to the real world).

If these ideas are on the right path, then there is a tension among representational, methodological, and epistemological practices that set apart mathematical equations from simulation models. This tension needs to be acknowledged and discussed in more depth if we are to fully grasp the philosophical value of computer simulations. But where can we start? To my mind, the problem of representation is frankly overdue. Unfortunately, it is not the only one. For a more nuanced philosophical study of computer simulations, we need to look at what computer science and its philosophical study bring to the table. Philosophers of science could then approach familiar issues by incorporating the perspective of computer models and computer-based scientific practice. In this respect, it will certainly help to see what philosophers of computer science—or philosophers of science that look deeply into computational practices—have to say. The list is fortunately very long, and includes (Sloman 1978; Thagard 1988; Colburn 2000; Turner and Eden 2007; Turner 2018; Durán 2018; Primiero 2019; Curtis-Trudel Forthcoming). With this change of frame in place, simulations invite us to rethink methodological problems of modeling and representation, idealizations and fictionalizations, and answer epistemological questions about scientific explanation, epistemic opacity, inference, and prediction. These studies, then, will bring new debates on how scientific research using simulations is effectively carried out.

In fact, once simulation models have been reasonably decoupled from sets of equations, we should be able to see more sharply the relationship between computer simulations and scientific disciplines. The partisans of the received view adhere to the idea that computer simulations serve mathematics, and just like the latter, they must also serve disciplines such as physics and biology.[12] A philosophy of computer simulations could help simulations divorce mathematics by showing that the former entails ontologies, methodologies, and epistemologies that are not captured—nor could be captured—by standard mathematical equations. By accomplishing this, a computer simulation in biology, for instance, is no longer conceived as merely instrumental to a set of equations but rather as a unit with representational value and explanatory input.

I have also mentioned, albeit only in passing, that computer simulations are shaping scientific research in several new ways: they account for target systems in ways that were not possible before, they enable researchers to *see*, as it were, their objects of study from radically new perspectives, they afford inferences about empirical phenomena that enhance our scientific understanding of the world, and much more. These features of simulations have been discussed by several philosophers before me, and I have nothing new to add to the debate. What has received less attention, I believe, is the way researchers have meaningfully changed their scientific practices in light of simulations, and the extent to which computer simulations served those changes.[13] Galileo introduced the empirical method that governed scientific research for many generations—with the proper technological and methodological updates, of course. Today, this empirical method is playing a rather secondary role in many disciplines, largely reserved as a backdrop against which simulation outputs can be validated. The sheer volume of reliable data rendered by simulations and other research methods suffice, in many cases, to make traditional experimentation less prominent—if not fully outdated—for achieving specific scientific goals (Leonelli 2016; Leonelli and Tempini 2020). In such technologically-driven scenario, we are arguably in a better position to understand scientific discoveries, predictions, evidence, and explanations once we understand computer simulations in their own right.

Acknowledgments I would like to thank the editors of the proceeding, Nancy Abigail Nuñez and Björn Lundgren, for their patience and encouragement. Also, thanks go to Giuseppe Primiero for our discussions on the nature of algorithms and computer programs. Some of the ideas discussed with him ended up in this article. All mistakes are of course of my authorship. I would also like to thank Fondo para la Investigación Científica y Tecnológica (FONCYT - Argentina) - PICT 2016-1524, for their financial support. Finally, I would like to thank the section Values, Technology and Innovation, Faculty of Technology, Policy and Management at the Delft University of Technology for their unrivaled support.

[12] Of course, mathematics is a discipline in its own right, in the same sense that computer simulations is also a field in its own right within computer science and engineering.

[13] Historians of science and technology have, indeed, discussed these issues (e.g., De Mol and Primiero 2014; Haigh et al. 2016; De Mol and Bullynck 2018; De Mol 2019). Here, I am rather thinking of the philosopher interested in the social aspects of scientific practice.

References

Ajelli, M., B. Gonçalves, D. Balcan, V. Colizza, H. Hu, J.J. Ramasco, S. Merler, and A. Vespignani. 2010. Comparing Large-Scale Computational Approaches to Epidemic Modeling: Agent-Based versus Structured Metapopulation Models. *BMC Infectious Diseases* 10(190): 1–13.

Balcan, D., V. Colizza, B. Gonçalves, H. Hu, J.J. Ramasco, and A. Vespignani. 2009. Multiscale Mobility Networks and the Spatial Spreading of Infectious Diseases. *Proceedings of the National Academy of Sciences* 106(51): 21484–21489, .

Boon, M., and T. Knuuttila. 2009. *Handbook of the Philosophy of Science, vol. 9, Philosophy of Technology and Engineering Sciences*. Chapter Models as Epistemic Tools in Engineering Sciences: A Pragmatic Approach, 687–720. Amsterdam: Elsevier.

Bueno, O. 2014. Computer Simulation: An Inferential Conception. *The Monist* 97(3): 378–398.

Bueno, O., and M. Colyvan. 2011. An Inferential Conception of the Application of Mathematics. *Noûs* 45(2): 345–374.

Cartwright, N. 1983. *How the Laws of Physics Lie*. Oxford: Oxford University Press. https://doi.org/10.1093/0198247044.001.0001

Colburn, T.R. 1999. Software, Abstraction, and Ontology. *The Monist* 82(1): 3–19.

Colburn, T.R. 2000. *Philosophy and Computer Science*. Armonk: M. E. Sharpe.

Cooper, G.F., D.H. Dash, J.D. Levander, W.-K. Wong, W.R. Hogan, and M.M. Wagner. 2004. Bayesian Biosurveillance of Disease Outbreaks. In *Proceedings of the 20th Conference on Uncertainty in Artificial Intelligence*, UAI '04, 94–103, Arlington, Virginia 2004. AUAI Press. ISBN 0-974-90390-6.

Curtis-Trudel, A. Implementation as resemblance. *Philosophy of Science*, Forthcoming. https://doi.org/10.1086/714872.

De Mol, L. 2019. 'A Pretence of What Is Not'? A Study of Simulation(s) from the Eniac Perspective. *NTM Zeitschrift für Geschichte der Wissenschaften, Technik und Medizin* 27(4): 433–478. https://hal.univ-lille.fr/hal-01807956

De Mol, L., and M. Bullynck. 2018. Making the history of computing. the history of computing in the history of technology and the history of mathematics. *Revue de Synthèse* 139(3–4): 361–380.

De Mol, L., and G. Primiero. 2014. Facing Computing as Technique: Towards a History and Philosophy of Computing. *Philosophy & Technology* 27(3): 321–326.

DeAngelis, D.L., and V. Grimm. 2014. Individual-Based Models in Ecology After Four Decades. *F1000prime Reports* 6(39): 1–6.

Durán, J.M. 2017. Varying the Explanatory Span: Scientific Explanation for Computer Simulations. *International Studies in the Philosophy of Science* 31(1): 27–45.

Durán, J.M. 2018. *Computer Simulations in Science and Engineering. Concepts - Practices - Perspectives*. Berlin: Springer.

Durán, J.M. 2019. A Formal Framework for Computer Simulations: Surveying the Historical Record and Finding Their Philosophical Roots. *Philosophy & Technology*. https://doi.org/10.1007/s13347-019-00388-1

Durán, J.M. (2020a). What Is a Simulation Model? *Minds and Machines*. https://doi.org/10.1007/s11023-020-09520-z

Durán, J.M. (2020b). Calculating Surprises: A Review for a Philosophy of Computer Simulations. *Metascience*. https://doi.org/10.1007/s11016-020-00527-x

Eberhardt, C. 2014. Tearing Down Swift's Optional Pyramid of Doom. *Scott Logic*. https://blog.scottlogic.com/2014/12/08/swift-optional-pyramids-of-doom.html. Accessed November 2021

Eden, A.H. 2007. Three Paradigms of Computer Science. *Minds and Machines* 17(2): 135–167. https://doi.org/10.1007/s11023-007-9060-8

Eden, A.H., and R. Turner. 2007. Problems in the Ontology of Computer Programs. *Applied Ontology* 2(1): 13–36.

Fetzer, J. 1999. The Role of Models in Computer Science. *The Monist* 82(1): 20–36.

Frigg, R., and J. Reiss. 2009. The Philosophy of Simulation: Hot New Issues or Same Old Stew? *Synthese* 169(3): 593–613.
Giere, R.N. 2006. *Scientific Perspectivism.* Chicago: The University of Chicago Press.
Guala, F. 2002. *Models, Simulations, and Experiments*, 59–74. Dordrecht: Kluwer Academic.
Haigh, T., P.M. Priestley, M. Priestley, and C. Rope. 2016. *ENIAC in Action.* Cambridge: MIT Press. ISBN 978-0-26-203398-5.
Hartmann, S. 1996a. Modelling and Simulation in the Social Sciences from the Philosophy of Science Point of View. In *Modelling and Simulation in the Social Sciences from the Philosophy of Science Point of View*, eds. R. Hegselmann, U. Mueller, and K.G. Troitzsch, 77–100. Berlin: Springer.
Holstein, W.K., and W.R. Soukup. 1961. Monte Carlo Simulation. *Institute Paper No. 23 [Lafayette, Ind.: Institute for Quantitative Research and Economics and Management, Graduate School of Industrial Administration, Purdue University]*, 1.
Humphreys, P.W. 1990. Computer Simulations. *ProceedingPSA: Proceedings of the Biennial Meeting of the Philosophy of Science Associations of the Biennial Meeting of the Philosophy of Science Association* 2: 497–506.
Humphreys, P.W. 2004. *Extending Ourselves: Computational Science, Empiricism, and Scientific Method.* Oxford: Oxford University Press.
Humphreys, P.W. 2009. The Philosophical Novelty of Computer Simulation Methods. Synthese 169(3): 615–626.
Katzav, J., and W.S. Parker. 2018. Issues in the Theoretical Foundations of Climate Science. *Studies in History and Philosophy of Science Part B: Studies in History and Philosophy of Modern Physics* 63: 141–149. ISSN 1355-2198. https://doi.org/10.1016/j.shpsb.2018.02.001 http://www.sciencedirect.com/science/article/pii/S1355219817301648
Knuuttila, T. 2011. Modelling and Representing: An Artefactual Approach to Model-Based Representation. *Studies in History and Philosophy of Science Part A* 42(2): 262–271.
Knuuttila, T. 2017. Imagination Extended and Embedded: Artifactual Versus Fictional Accounts of Models. *Synthese* 99(3): 56–21.
Krohs, U. 2008. How Digital Computer Simulations Explain Real-World Processes. *International Studies in the Philosophy of Science* 22(3): 277–292.
Lenhard, J. 2019. *Calculated Surprises. A philosophy of Computer Simulations.* Oxford: Oxford University Press.
Leonelli, S. 2016. *Data-Centric Biology: A Philosophical Study.* Chicago: University of Chicago Press.
Leonelli, S., and N. Tempini, eds. 2020. *Data Journeys in the Sciences.* Berlin: Springer.
MacLeod, M., and N.J. Nersessian. 2013. Building Simulations from the Ground-Up: Modeling and Theory in Systems Biology. *Philosophy of Science* 4(80): 533–556.
Massimi, M., and C.D. McCoy, eds. 2020. *Understanding Perspectivism.* Milton Park: Routledge.
Morgan, M.S., and M. Morrison, eds. 1999. *Models as Mediators: Perspectives on Natural and Social Sciences.* Cambridge: Cambridge University Press.
Morrison, M. 2015. *Reconstructing Reality. Models, Mathematics, and Simulations.* Oxford: Oxford University Press.
Morrison, M., and M.S. Morgan. 1999. Models as Mediating Instruments. In *Models as Mediators: Perspectives on Natural and Social Sciences*, eds. M.S. Morgan and M. Morrison, 10–37. Cambridge: Cambridge University Press.
Parker, W.S. 2009. Does Matter Really Matters? Computer Simulations, Experiments, and Materiality. *Synthese* 169(3): 483–496.
Parker, W.S. 2020a. Evidence and Knowledge from Computer Simulation. *Erkenn.* https://doi.org/10.1007/s10670-020-00260-1.
Parker, W.S. 2020b. Model Evaluation: An Adequacy-for-Purpose View. *Philosophy of Science* 87: 457–477.
Peck, S.L. 2012. Agent-Based Models as Fictive Instantiations of Ecological Processes. *Philosophy and Theory in Biology* 4:e303. http://dx.doi.org/10.3998/ptb.6959004.0004.003
Primiero, G. 2019. *On the Foundations of Computing.* Oxford: Oxford University Press.

Sloman, A. 1978. *The Computer Revolution in Philosophy*. The Harvester Press. Sussex, England.

Teichroew, D., and J.F. Lubin 1966. Computer Simulation—Discussion of the Technique and Comparison of Languages. *Communications of the ACM* 9(10): 723–741.

Thagard, P. 1988. *Computational Philosophy of Science*. Cambridge: MIT Press.

Turner R. 2007. Computable Models. *Journal of Logic and Computation* 18(2): 283–318.

Turner, R. 2014. Programming Languages as Technical Artifacts. *Philosophy & Technology* 27(3): 377–397.

Turner, R. 2018. *Computational Artifacts. Towards a Philosophy of Computer Science*. Berlin: Springer.

Turner, R., and A.H. Eden 2007. The Philosophy of Computer Science: Introduction to the Special Issue. *Minds and Machines* 17(2): 129–133.

Weirich, P. 2011. The Explanatory Power of Models and Simulations: A Philosophical Exploration. *Simulation & Gaming* 42(2): 155–176.

Weisberg, M. 2013. *Simulation and Similarity*. Using Models to Understand the World. Oxford: Oxford University Press.

Winsberg, E. 1999. Sanctioning Models: The Epistemology of Simulation. *Science in Context* 12: 275–292.

Woolfson, M.M., and G.J. Pert. 1999a. *An Introduction to Computer Simulations*. Oxford: Oxford University Press.

Woolfson, M.M., and G.J. Pert. 1999b. SATELLIT.FOR.

Index

B. Lundgren, N. A. Nuñez Hernández (eds.), *Philosophy of Computing*,
Philosophical Studies Series 143, https://doi.org/10.1007/978-3-030-75267-5